PROTEIN-PROTEIN COMPLEXES

Analysis, Modeling and Drug Design

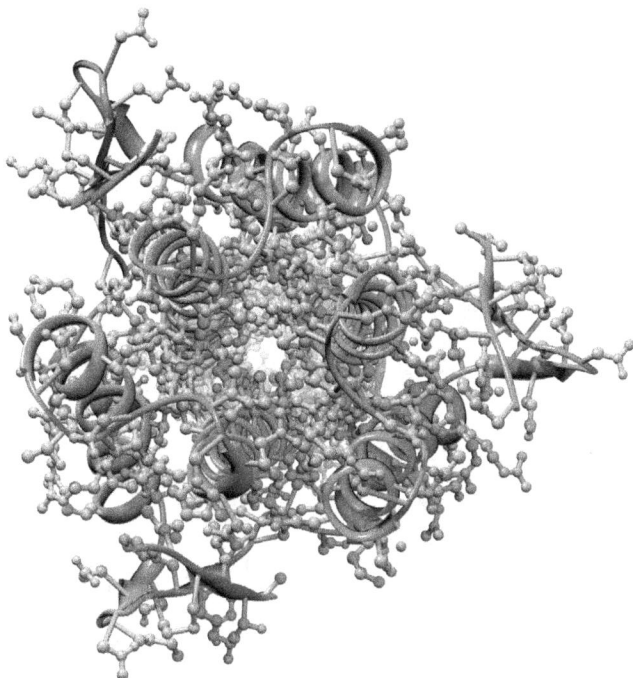

PROTEIN-PROTEIN COMPLEXES

COMPLEXES

Analysis, Modeling and Drug Design

Edited by

Martin Zacharias

Technische Universität München, Germany

Imperial College Press

ICP

Published by

Imperial College Press
57 Shelton Street
Covent Garden
London WC2H 9HE

Distributed by

World Scientific Publishing Co. Pte. Ltd.
5 Toh Tuck Link, Singapore 596224
USA office: 27 Warren Street, Suite 401-402, Hackensack, NJ 07601
UK office: 57 Shelton Street, Covent Garden, London WC2H 9HE

British Library Cataloguing-in-Publication Data
A catalogue record for this book is available from the British Library.

PROTEIN-PROTEIN COMPLEXES
Analysis, Modeling and Drug Design
Copyright © 2010 by Imperial College Press

ISBN-13 978-1-84816-338-6
ISBN-10 1-84816-338-X
ISBN-13 978-1-84816-339-3 (pbk)
ISBN-10 1-84816-339-8 (pbk)

Printed in Singapore.

CONTENTS

PREFACE

Basically all processes in a cell involve proteins and the great majority of biological functions are mediated not only by isolated proteins but also by the interaction of proteins. Powerful experimental techniques are available to systematically investigate the network of protein–protein interactions in cellular systems. However, for a full understanding of protein–protein interactions, knowledge of the three-dimensional structure of complexes formed between interacting proteins is essential. Immense progress has been achieved in recent years to elucidate protein–protein complex structures and to better understand the physical principles of complex formation. What are the driving forces for protein–protein association? What can we learn about specific recognition from studying protein–protein interfaces? How can this knowledge be used to predict protein–protein interactions and is it possible to influence protein–protein interactions by small drug molecules? These and many other questions will be tackled in the 13 chapter contributions in this volume.

Although the book covers the state-of-the-art research in the area of protein–protein complex analysis and modelling, it is not primarily directed at specialists in the field. The book is also meant to be a useful guide for students and researchers in the area of Chemistry, Biochemistry and Biophysics with an interest in proteins and protein–protein interactions. Most chapters contain significant introductory information in addition to the most recent progress in the field. Readers will gain insight into the recognition principles of proteins; how to determine, analyse and predict protein–protein interactions and complex structures, as well as learn about possibilities of interference with protein–protein interactions.

Leading researchers in the field have been selected to contribute chapters to the book. Authors were free to select the exact scope of their contribution and express their own view on the field. Possible overlapping between chapters can be profitable for the reader since key information is provided from different perspectives by leading scientists.

The first part of the volume introduces the analysis of experimentally determined structures of protein–protein complexes. Experimental protein structures contain rich information on the principles of interaction. The systematic analysis of the interface region of protein–protein complexes and the comparison with other surface regions of a protein reveal the physical characteristics of protein binding sites. A deeper understanding of the driving forces of protein–protein complex formation also requires an analysis of the thermodynamics of protein–protein association. The first part of the book includes an overview of experimental methods to investigate the thermodynamics of protein–protein binding, and also discusses theoretical methods to calculate energetic and entropic contributions. The study of the kinetics of association and dissociation of protein–protein interactions is of central importance to understanding the mechanism of protein complex formation. How the kinetics of protein–protein binding can be studied experimentally and theoretically is at the focus of a separate chapter. Proteins bind to specific sites on the surface of proteins with high affinity. The physico-chemical character of binding sites can differ from the properties of other surface regions. In addition, often the amino acids at protein binding sites are evolutionarily more conserved then the rest of the protein surface. The properties and conservation of protein functional sites and how they can be used to identify relevant amino acid residues for protein–protein recognition are discussed in the fifth chapter.

Due to the large number of putative protein–protein interactions and the transient nature of many protein–protein complexes, only a fraction of possible protein–protein complex structures can be determined experimentally. A variety of computational docking prediction methods have been developed in recent years to tackle the problem of providing at least structural models of important protein–protein complexes. A general overview of docking methods is provided, followed by chapters on how to best include experimental data or information from bioinformatics resources to high-resolution docking methodologies. Typically, modelling protein–protein complex structures is not a one-step procedure but instead distinguishes an initial exhaustive search followed by a refinement and rescoring phase. The options of refining and identifying the most realistic predicted complex structure are also introduced.

The last five chapters of the volume shift the focus from three-dimensional modelling of protein–protein interactions towards approaches that influence or interfere with protein–protein interactions. A significant fraction of protein–protein interactions – particularly in higher organisms – are mediated by reoccurring motifs or interaction patterns. Chapter 10 gives an overview of several examples of biological and medical importance. The chapter also includes a discussion of the involvement of motif-mediated interactions in diseases. Mutations in proteins may perturb interactions with other partners. However, site-directed mutagenesis can also be used to redesign protein binding regions to create new or altered protein–protein interactions. Methods to estimate changes in protein–protein affinity, due to residue substitutions at the interface, are described and the possibility to directly and specifically interfere with protein–protein interactions is at the focus of two separate chapters. The concepts are introduced and discussed on examples that are of relevance to several human diseases. Proteins can undergo conformational changes upon association. In addition, the binding process can also influence the flexibility of binding partners which may even mediate long-range allosteric communication. The analysis of such dynamical recognition processes and the possibility to influence them by drug-like molecules is the subject of the last chapter.

It is my great pleasure to thank all authors for the time and efforts they devoted to the demanding work of contributing book chapters to this volume. I am grateful to the editors of Imperial College Press for their cooperation and also to my co-workers and family for their patience and support.

Munich, July 2009

Martin Zacharias

CHAPTER 1

X-ray Study of Protein–Protein Complexes and Analysis of Interfaces

Joel Janin

Yeast Structural Genomics, IBBMC UMR 8612 CNRS,
Université Paris-Sud, 91405 Orsay, France
E-mail: joel.janin@u-psud.fr

Highly efficient procedures to express genes and prepare individual proteins for structural analysis, developed during the first round of the Structural Genomics initiatives world wide, are now being extended to protein complexes and multi-subunit assemblies. These structures are still few in the Protein Data Bank, but one can exploit the abundant information on binary protein–protein complexes and oligomeric proteins to set up appropriate methods of analysis, and derive rules on protein–protein interaction, which will be applicable to larger assemblies when their structures become available.

1.1 Introduction

Following the completion of the first complete genome sequences at the turn of the century, the question was put to structural biologists: can crystallography and NMR provide three-dimensional structures for the products of all these genes? At that time, it was estimated that a set of 10,000 experimental structures, carefully chosen, would cover the space of existing folds; the remainder could be built by homology.[1] Structural Genomics (SG) initiatives were launched in the USA and Japan in the years 2000–2001, with that goal. With the end of 2009, they will have deposited more than 8,000 new structures in the Protein Data Bank (PDB, *http://www.rcsb.org/pdb/statistics/*), and the target of 10,000 will almost certainly be reached before 2010. But meanwhile, the landscape around has changed greatly. We now realise that the diversity of DNA sequences may be orders of magnitude greater than what was thought when only a few model genomes were known. Many of the new sequences are unrelated to what we have in the databases, and therefore, many protein folds have yet to be discovered. Moreover, it has become clear that most gene products do not exist and function as single entities. Genome-wide studies of protein–protein interaction have demonstrated that cells contain thousands of macromolecular assemblies of all sizes, from simple dimers to objects that comprise tens or hundreds of polypeptide and/or nucleic acid chains.[2,3] The examples of the ribosome and the nuclear pore show that the whole assembly, not the individual chains, carries the biological function. The structural analysis should, therefore, not be limited to the isolated components.

The number of solved macromolecular assembly structures is still small compared to that of isolated proteins.[4] In this review, attempts will be described to characterise macromolecular assemblies similar to the systematic studies that SG initiatives performed on single proteins. While these studies are ongoing, we may look at simpler systems for which the PDB offers more examples: protein–protein complexes and homodimeric proteins. Their atomic structures contain a wealth of information on the chemistry and physical chemistry of the non-covalent interactions that allow polypeptide chains recognising each other and self-assembling into a functional macromolecular entity.[5–9] The methods developed to extract this information, the observations and rules derived from its analysis, will undoubtedly help us to understand the more complex systems when their structure becomes available.

1.2 Preparing Proteins for Structural Studies

The first genome-wide studies of protein–protein interactions were completed at about the same time as the SG initiatives of the first generation. As a result of that coincidence, the second generation of SG initiatives that started in 2005–2006, included several programmes that are concerned with macromolecular assemblies.[10–11] Thus, the Yeast Structural Genomics, a small-scale pilot-project that we carried out in

Orsay in 2001–2004, is now part of two programmes funded by the European Union, SPINE2-Complexes and 3D-Repertoire (*http://www.spine2.eu, http://www.3drepertoire.org*). Both combine high-resolution X-ray/NMR and medium/low resolution cryo-electron microscopy studies (cryo-EM) in order to study multi-component systems; some of their targets, like RNA polymerase or the exosome that degrades mRNA, have a well-established status in biology. Others have just been identified in systematic tandem-affinity purification/mass spectrometry studies. These complexes have no known function, but with yeast, a wealth of genetic and biochemical tools are available to characterise them while the structural analysis is ongoing. Atomic resolution may not be reachable for some of the targets, but useful models can be obtained by docking into the electron density of cryo-EM images, the high-resolution models obtained by X-ray crystallography on some of the components.

All these studies integrate the expertise acquired by labs that were part of the first round of Structural Genomics initiatives to which they owe many of their tools and first of all, efficient methods to produce and analyse recombinant proteins.[12] Figure 1.1 describes the standard procedure that was set up to express and prepare proteins of *Saccharomyces cerevisiae* during the four years of the Yeast Structural Genomics pilot-project.[13] It comprises three major steps:

1. *Cloning:* We use the PCR reaction to amplify the target sequence in genomic DNA (mostly intron-free in *S. cerevisiae*); the two primer oligonucleotides contain appropriate restriction sites and the 3'-primer codes for a six-histidine tag placed just after the last codon. The PCR products are purified, digested with restriction enzymes and inserted into an expression vector. Their DNA sequence is checked. In *E. coli*, we use vectors derived from the pET plasmid, which place the target gene under control of the highly efficient phage T7 promoter.

2. *Protein Production:* The level of gene expression and the solubility of the target protein are evaluated in small-scale cultures of several *E. coli* strains, each grown at four different temperatures. The conditions that yield the most soluble protein are retained for large-scale production in 1 litre flasks.

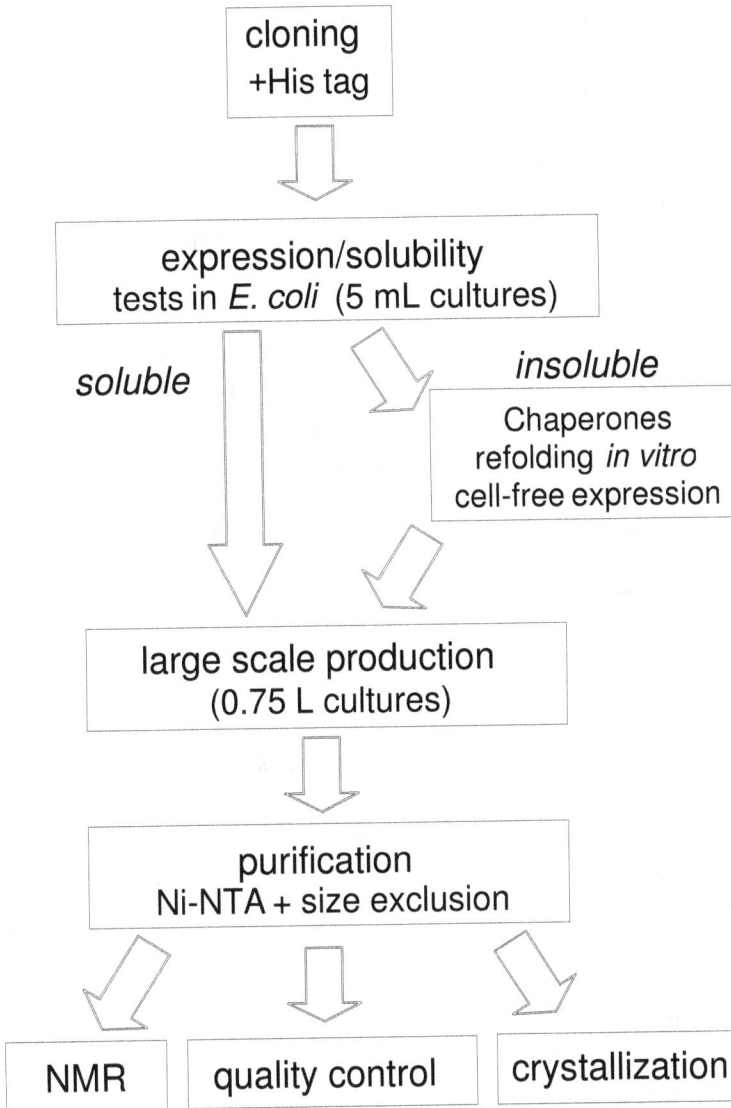

Fig. 1.1. Flowchart of the protein expression/purification procedure. During the Yeast Structural Genomics pilot-project, 250 *S. cerevisiae* genes were cloned and tagged in a standard protein preparation procedure. Expression in *E. coli* succeeded for 80% of the proteins with less than 350 residues. Soluble protein could be purified in two steps from the cell extract, and insoluble protein could be recovered in a number of cases from inclusion bodies (adapted from Ref. 13).

3. Protein Purification and Quality Controls: The His-tagged protein is purified on a Ni-NTA resin, concentrated and run on a size exclusion column. Its degree of purity (usually > 95%) is judged by electrophoresis on a SDS gel and its chemical integrity by mass spectrometry.

The cloning step was carried out on 250 *S. cerevisiae* target genes with a success rate above 90%. After optimization of the growth conditions, most of the cloned genes were highly expressed in *E. coli*; an overnight culture in a shaken flask yielded the target protein in milligram quantities. However, more than one-third of the constructions gave insoluble protein in inclusion bodies. About half of those could be recovered as soluble protein either by co-expressing bacterial chaperones, by solubilizing the inclusion bodies in 6 M guanidinium chloride and screening for refolding in a number of buffers,[14] or by using a cell-free expression system.[15]

Carrying out the whole procedure on all the targets was outside the scope of a pilot-project, and therefore, we focused our work on a subset of proteins of interest. Starting with 140 well-expressed yeast genes, we obtained 72 proteins purified to homogeneity in quantities of 0.5 to 10 mg that could be subjected to automated crystallization screens. A majority of the screens gave crystalline hits, not always of sufficient quality for structure determination, but some of these leads could be optimised as discussed below. Fourteen proteins had their X-ray structure determined to resolutions of 1.3 to 2.6 Å within the four-year course of the pilot-project[16] (*http://genomics.eu.org/spip/Overview*), and another ten during the two years after. Therefore, the goal of 20 new structures that we had initially fixed to the pilot-project had been reached by 2006, leaving the place for new projects mostly concerned with protein–protein complexes.

Other SG centres have had success rates similar to ours, often on a much larger scale.[17] The second generation programmes that opened in 2005 in the US and Japan, have built on that experience to set up high-throughput production chains for the structure determination of single gene products by both X-ray crystallography and NMR. Whereas most of the first-generation targets were from prokaryotes or yeast, more difficult targets from higher eukaryotes and including membrane proteins are now being addressed, albeit with a much lower throughput.[12,18]

1.3 Preparing Protein–Protein Complexes and Multi-component Assemblies

The preparation scheme of Fig. 1.1 has a success rate of 50% that may be considered as satisfactory on a target that is a single gene product. The same scheme can be used to produce multigenic protein assemblies by preparing each component separately. For a binary complex, the expected 25% yield makes it worth trying, but with more than two components the chance is poor that all the subunits can be prepared separately as soluble proteins that will self-assemble when mixed together. Nevertheless, the one-by-one approach has had some remarkable successes. For instance, the *Xenopus* genes coding for the four different histones that constitute the nucleosome core particle could be individually expressed in *E. coli*, and the core particle was reconstituted by mixing them together in appropriate proportions.[19] More frequently, some but not all of the components of a multi-component complex are obtained in soluble form. The complex itself cannot be reconstituted, but some of the soluble components form subcomplexes that can yield important information on the assembly, and they may be suitable for high-resolution structural studies complementing a cryo-EM analysis of the whole complex.

Figure 1.2 describes the strategy that we developed for preparing yeast protein–protein complexes. It offers several alternatives to the one-by-one gene expression approach (Pathway 3). One possible approach is to prepare the assembly directly from yeast extracts, either at its natural abundance (Pathway 1) or after over-expressing all its components (Pathway 2). Over-expression can also be attempted in *E. coli* (Pathway 4). Pathway 1 is the one that was used in the structural studies of bacterial ribosomes, and also of the yeast 20S proteasome.[20] The cells can be grown in large quantities, the ribosome and the proteasome are very abundant, and they can be purified by techniques that do not require affinity tags. In all other cases, the complexes must be over-expressed. A simple procedure would be to build an expression vector for each of the genes of interest, and introduce them into the same bacterial or yeast strain. However, it is difficult to maintain more than two plasmids in the same host, and even with a binary complex, the level of expression of two genes carried by different vectors is likely to be very unequal, compromising the formation of an assembly with a well-defined stochiometry. The approach that we and others favour is therefore to make operon-like genetic constructions, in which several genes of interest are placed next to each other.

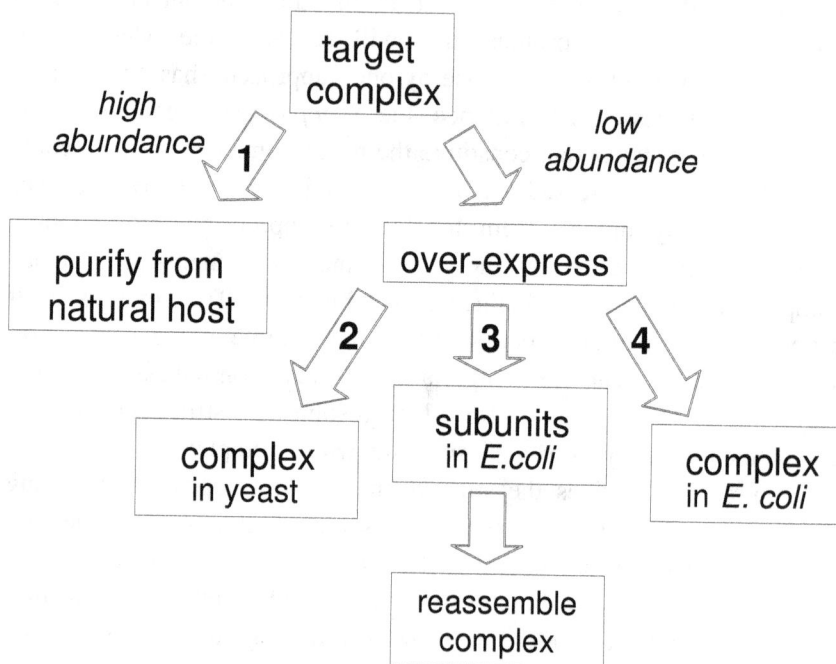

Fig. 1.2. Strategy for the purification of multi-subunit yeast complexes.

They form a single transcription unit under the control of the same promoter, and a ribosome binding site is placed between each stop and start codon.[21] In practice, five or more medium-size genes can be co-expressed in this way, one of them bearing an affinity purification tag. The construction can be facilitated by placing restriction sites at strategic locations, or dispensed of altogether by using synthetic DNA. The genes in the operon are transcribed into a single mRNA, they are translated at similar levels and their products are able to associate as they exit the ribosome. Thus, components that would be insoluble (or disordered and degraded) if expressed alone, can be rescued through their interaction with the partner chains. The procedure does not apply to systems such as the complexes of the respiratory chain, because their assembly requires specialized chaperones or cofactors. Still, the co-expression and self-assembly in *E. coli* of eukaryotic protein–protein complexes has had a remarkable success rate, and most of the methodological developments in progress follow Pathway 4.

1.4 Crystallization and X-ray Studies

Crystallization is a well-recognised bottleneck in structural studies. A number of new tools have been developed in recent years, mostly in SG labs. These techniques were designed primarily for single-gene products, but they work equally well for multi-component assemblies and play a key role in the present study. In spite of many attempts to make it rational, the crystallization of proteins, nucleic acids and their complexes still depends on testing hundreds of conditions that combine different precipitants, pHs and additives. One of the very first upshots of the SG initiatives, the one that spread the most quickly, was automatic crystallization. Unlike an attempt we had made[22] to use robotics in the early nineties, the devices and procedures that were developed ten years later in the framework of the SG centres immediately found industrial support and are now used routinely by the protein science community. Pipette robots and crystallization kits greatly facilitate the preparation of the precipitant solutions. Equally important, the amount of biological material required to do the tests has dropped by one or two orders of magnitude, thanks to liquid-dispensing robots that prepare arrays of nanodrops in 96-well plates.[23-24] A standard set of four plates can be prepared in a couple of hours with a minimum of human intervention, and it uses up only a milligram or two of pure protein material. Moreover, the success rate is remarkably high: in our hands, about half of proteins entering crystallization trials give crystals of some sort. As many are not suitable for diffraction experiments because of their size and shape, or of the low resolution of the X-ray diffraction pattern they give, the conditions under which they are obtained have to be optimised, and that again can be done by automatic procedures.[25]

The crystallization methods are essentially the same for single proteins and for complexes, although meeting the criterion of homogeneity may be more difficult in the latter case. The protein concentration in the crystallization mix – typically 5–10 mg/ml – allows the formation of low-affinity complexes with K_d values in the micromolar range by simple mixing of the components. But the stoichiometry of the mixture may not be exact and the components in excess can interfere with crystallization, or crystallize separately. It is possible to avoid this problem by purifying the complex on a size exclusion column, or by other chromatographic techniques before crystallization. However, such procedures are only applicable to stable assemblies.

Once diffracting crystals are obtained, the steps that follow usually require labeled material. Labelling is needed for structure solution by both crystallography and NMR. NMR makes extensive use of isotopic labels such as ^{15}N and ^{13}C (see also Chapter 7), and crystallography needs heavy atom labels for phase determination. In a recombinant protein, heavy atom labelling can be very efficiently achieved by incorporating selenomethionine in place of methionine.[26] The widespread utilization of the dispersive and anomalous signal of selenium at wavelengths near 0.98 Å has been the key to the development of high-throughput crystallographic methods in the last ten years. Synchrotron radiation centres make this wavelength easily available on experimental setups that allow a complete diffraction dataset to be recorded in a matter of minutes. This is achieved by making full use of the high beam intensity of synchrotron radiation, and by the efficiency of the X-ray detectors developed over the past ten years.

Biological crystallographers have at their disposal an extensive library of software that performs all the steps which follow the recording of X-ray patterns: data reduction, phase determination, model building and refinement.[27] Diffraction data taken at several wavelengths (in the Multiple Anomalous Diffraction or MAD method), or even at a single wavelength (in the Single Anomalous Diffraction or SAD method) on just one crystal, yield good quality phases and electron density maps in which a large part of the polypeptide chain can be traced automatically.[28] In the case of a complex, it may be sufficient to label one of the components if prepared separately. On the other hand, selenomethionine incorporation requires the protein to be expressed in bacteria or yeast grown on special media, or possibly *in vivo* in a cell-free system. While the method may be extended to other expression systems in the future, biological material extracted from a natural host cannot be labelled in this way. However, elements other than selenium give a dispersive and anomalous signal that can be used for phasing, for instance the metal ions present in metalloproteins, or just the sulfur atoms of cysteines in favourable cases.[29] If none of these methods are applicable, one must return to classical heavy atom labelling techniques; Hg reacting with cysteines for instance, has been used in the past for multiple isomorphous replacement, and is nowadays also suitable for MAD or SAD phasing.

1.5 The Geometric Analysis of Protein–Protein Interfaces

Structure determination by crystallography or NMR ends with the deposition of set atomic coordinates at the Protein Data Bank[30] (PDB) that makes it available to the community. The information present in a PDB entry is chemical: the nature of each atom; and geometric: the atomic positions. Its conversion into terms of physical and/or biological relevance is rarely straightforward, and specific tools have been developed for that purpose. Thus, at least three geometric tools are appropriate when defining the interface between two molecules or macromolecules A and B that form a complex AB:

1. *Distance:* atoms or chemical groups i of A and j of B are part of the A:B interface if they satisfy the condition $d_{ij} < d_0$, where d_0 depends on the atomic or group radii r_i and r_j, and on a cutoff value r_0 in the range 0.5–2 Å:

$$d_{ij} < d_0 = r_i + r_j + r_0 \qquad (1.1)$$

2. *Buried Surface:*[31] the A:B interface comprises all the points of the solvent accessible surface of A or B that do not belong to the accessible surface of AB.

3. *The Alpha-complex,*[32] a geometric construction related to the Voronoi diagram.

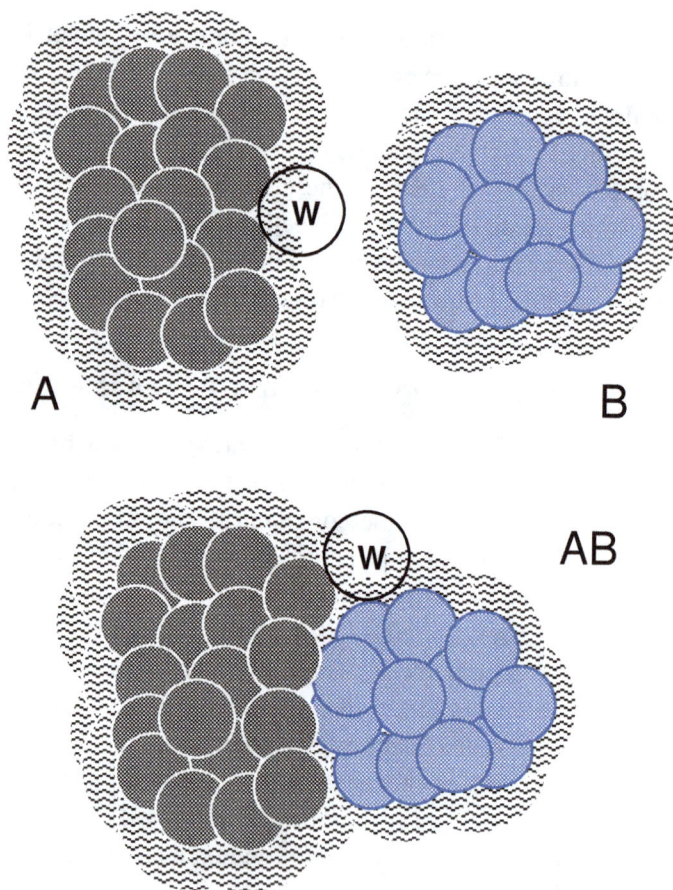

Fig. 1.3. Solvent accessible and buried protein surfaces. The centre of the water probe (W) defines the solvent accessible surface. The A:B interface comprises the points of the solvent accessible surface of A and B (top) that are no longer accessible in the AB complex (bottom).

The first definition is the simplest, but as it depends on an arbitrary cut-off, the second is used more commonly nowadays. In Fig. 1.3, the spheres represent atoms or chemical groups with radii that are augmented of the probe radius ($r_W = 1.4$–1.5 Å for a water probe) on the surface; the solvent accessible surface is the border of their union. Its construction may implement the rolling sphere algorithm[33] or an equivalent analytical algorithm. The buried surface area is a convenient measure of the interface size. It can be computed as:

$$BSA = ASA_A + ASA_B - ASA_{AB} \qquad (1.2)$$

where ASA_A, ASA_B and ASA_{AB} are the accessible surface areas of A, B and the AB pair. The atoms and residues that contribute to the BSA are part of the interface, and in practice the BSA is proportional to their number, and also to the number of atom pairs that satisfy Eq. 1 with $r_0 = 2r_W$. The alpha-complex on which the third definition is based is an extension of the Voronoi diagram, a geometric construction first applied to proteins by Richards.[34] That diagram associates to each atom its Voronoi cell, the convex polyhedron that contains all points of space closer to that atom than to any other atom. To account for the different sizes of the chemical groups, the Euclidean distance is commonly replaced by the power distance $p(\mathbf{x})$:

$$p(\mathbf{x}) = d^2 - r^2 \qquad (1.3)$$

Here, r is the radius of the sphere that represents the atom, and d the distance of point $\mathbf{x}$ to its centre. The Voronoi cell of an atom comprises all points $\mathbf{x}$ that have a power distance to that atom less than to any other atom.[35–36] Its facets belong to the radical plane, which contains the intersection of the spheres if there is one. The Voronoi or the power diagram offer a natural definition of contacts: two atoms are in contact if their Voronoi cells share a facet. The interface area may then be calculated as the sum of the areas of these 'bicolour' facets. In Fig. 1.4a, the blue and red circles that represent atoms of A and B (in two dimensions) have radii that are augmented of r_W as in Fig. 1.3 above. The blue lines are the facets shared by atoms of A, and the green lines are the bicolour facets shared between A and B. But atoms on the molecular surface always have unbounded facets, like the one between A_1 and A_3 in Fig. 1.4a. They raise a problem to which the alpha-complex gives an elegant solution.[32] It is built like a power diagram, except that one restricts the Voronoi cell of each atom to its associated ball and seeks intersections between these restricted regions. Thus, a facet between two atoms is not part of the alpha-complex if the associated spheres do not intersect, or if the facet lies outside the intersection. In Fig. 1.4a, the B_1 and A_3 balls intersect outside their Voronoi cells, due to the presence of A_2. The corresponding bicolour facet is dashed to indicate that it is not part of the alpha-complex, and the Voronoi interface comprises only the two facets of B_1 with A_1 and A_2.

An interface defined in this way may still contain a few unbounded facets that Ban *et al.*[37–38] remove through an iterative retraction procedure, and Cazals *et al.*[39] by testing appropriate geometric criteria. Figure 1.4c illustrates how the Voronoi interface defined by the Cazals procedure approximates the shape of the buried molecular surface in the protease–inhibitor complex of Fig. 1.4d. When the two procedures are applied to the set of protein–protein interfaces of Chakrabarti and Janin,[40] the Voronoi interface area calculated as the sum of the areas of the bicolour facets, correlates linearly with the BSA, but the correlation is much better with the Cazals than the Ban procedure ($R^2 = 0.98$ vs 0.85). A remarkable result of Cazals *et al.*[39] is that about 13% of the atoms that share bicolour facets do not contribute to the BSA, mostly because they are not solvent accessible to start with. An example is shown in Fig. 1.4b: on top, the blue and red atoms are shown to share a facet, but the bottom panel indicates that the red atom is not solvent accessible due to the presence of other atoms in molecule B. As a consequence, the Voronoi interface generally comprises significantly more atoms, and especially more main chain atoms, than the buried surface.

1.6 Types and Sizes of Protein–Protein Interfaces

The protein–protein interfaces in the PDB are of several types that represent different categories of interactions.[41,6–9] One may distinguish between the non-obligate interactions that occur when two preformed proteins form (non-covalent) complexes, and other interactions that are obligate and permanent.

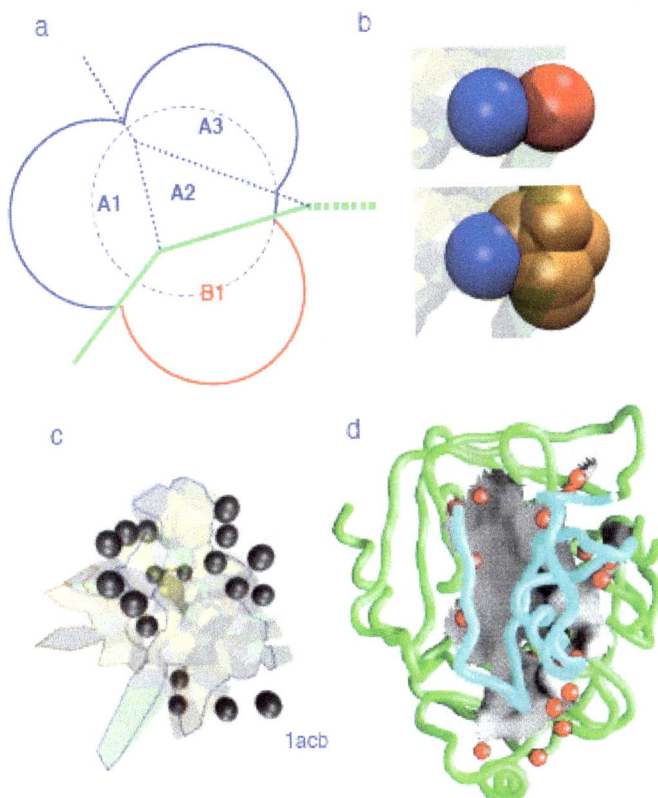

Fig. 1.4. The Voronoi model of protein–protein interfaces. (a) Voronoi interface in two dimensions: The blue and red circles represent atoms of A and B; their radii are augmented of the probe radius; the dashed blue lines are 'monocolour' Voronoi facets shared by atoms of A, the green lines are 'bicolour' facets shared by A and B. The A:B interface comprises the two facets drawn as full lines; they are part of the alpha-complex, but the B1:A3 facet (green dashes) is not. (b) A buried atom can be part of the Voronoi interface: the blue and red balls intersect, although other atoms of molecule B (in gold) make the red ball inaccessible to a solvent probe in the free molecule. (c) The Voronoi interface of a protease–inhibitor complex (1acb); the balls are interface water molecules. (d) The protease surface buried in contact with the inhibitor is viewed through the inhibitor backbone drawn as a blue tube; the green tube is the protease backbone. Panels a–c are adapted from Ref. 39.

The complexes of an antibody with the cognate antigen, of an enzyme with a protein inhibitor, or those that mediate signal transduction in cells, all illustrate non-obligate interactions. In contrast, the interactions between the subunits of an oligomeric protein usually form while their synthesis takes place on the ribosome, or soon after, and break only when the protein is denatured or degraded.

Obligate or not, they play major roles in the structure and function of the protein assemblies that they stabilise. The PDB also contains many examples of a third type of interactions: those that hold protein crystals together. Unlike those that stabilise functional assemblies, crystal packing contacts are unspecific and not subject to any biological selection. They are laboratory artefacts (with some interesting exceptions), yet they are of the same physico-chemical nature as the interactions that stabilise complexes of oligomeric proteins. Any geometric, chemical or physico-chemical feature is of interest if it is able to distinguish between the interfaces created by crystal packing contacts and those that reflect biological interactions, because such a feature may contribute to the specificity of recognition between the protein surfaces involved.

The size of the interface is the most obvious one. Figure 1.5 shows histograms of the BSA in sets of non-obligate protein–protein complexes and homodimeric proteins assembled by Chakrabarti and Janin[40] and Bahadur *et al.*,[42] and compares their interfaces with crystal packing interfaces. Mean values and standard deviations are cited in Table 1.1. Sets assembled by Jones *et al.*[5,43] yield similar values. With the complexes, the distribution peaks near 1,600 Å^2, and a majority of the interfaces buries less than 2,000 Å^2. With the homodimers, most of the interfaces are larger, and often much larger. On average, their BSA is twice that of the complexes: 3,900 Å^2 instead of 1,910 Å^2, with a large standard deviation which confirms that the sample is very heterogeneous in terms of interface size.

The crystal packing interfaces have a mean BSA of only 570 Å^2 and therefore, they should be easy to tell apart from the specific interfaces of the complexes and the homodimers. In most cases indeed, a visual inspection of the molecular contacts suffices to identify units of biologically relevance, which the PDB calls the 'biomolecules'.

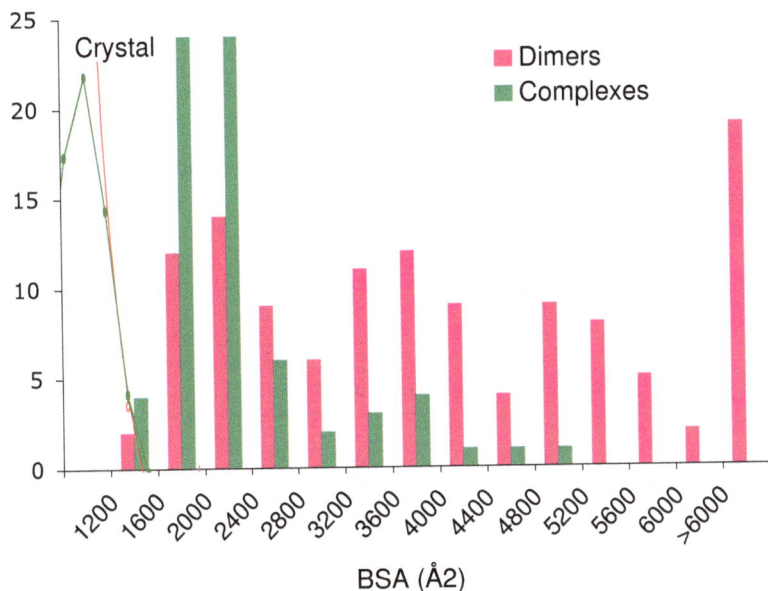

Fig. 1.5. Size distribution of the protein–protein interfaces. Histogram of the BSA in 122 homodimeric proteins, 70 protein–protein complexes, and 1320 crystal packing interfaces (the green line is drawn at a different vertical scale). The references are cited in Table 1.1.

However, what appears in a PDB entry is not the biomolecule, but the crystal asymmetric unit (ASU). The relation between the two is far from obvious: a monomeric protein can yield crystals with two or more chains in the ASU. An oligomeric protein may result in crystals with only one chain, in which case it must have crystal symmetries. A complex may even have subunits in several ASUs. The header of post-1999 PDB entries contains two records, REMARK 300 and REMARK 350, which relate the biomolecule to the content of the ASU. As it takes some effort to convert this information into a set of coordinates, the PDB created Biounit, a database accessible through its RCSB interface, in which the biomolecule is built on the basis of REMARK 300/350 or of supporting information from the authors. The reliability of this procedure can be assessed by comparing it with the composition reported in the biochemical literature. A wide, albeit incomplete, literature survey carried out by Lévy[44] indicates that the two disagree in about 15% of the PDB entries, and in up to 27% of the proteins with non-redundant sequences. The results of the search are accessible through the PiQSi (Protein Quaternary Structure Investigation) database, well worth checking each time there in case of doubt.

Table 1.1. Properties of protein–protein interfaces.

	Protein-protein complexes[a]	Homodimers[b]	Crystal packing[c]
Number in data set	70	122	188 (1,320)
BSA (Å^2)	1,910	3,900	1,510 (570)
s.d.	760	2,200	520
Composition (BSA %)			
non-polar	58	65	58
neutral polar	28	23	25
charged	14	12	17
Polar interactions[d]			
number of interface H-bonds	10	19	5
BSA per H-bond (Å^2)	190	210	280
Water molecules per 1,000 Å^2	10	11	15
Atomic packing[e]			
Fraction of buried atoms (f_{bu}, %)	34	36	21
Shape complementarity (S_c)	0.69	0.70	0.63
Gap volume index (I_{gap}, Å)	2.5	2.1	4.4
Packing index (L_D)	42	45	32

a. Data of Chakrabarti and Janin[40] on a subset of the complexes in Lo Conte *et al.*[49]
b. Data of Bahadur *et al.*[42]
c. Pairwise interfaces in crystals of monomeric proteins. The values in parentheses are for all the interfaces present in 152 crystal forms of monomeric proteins[48]. All other numbers are for the subset of interfaces with BSA > 800 Å^2 in Bahadur *et al.*[47]
d. The data on interface water are from Rodier *et al.*[61]
e. Data from Bahadur *et al.*[47]

1.7 Chemical and Physical Chemical Properties of the Interfaces

Twelve per cent of the crystal packing interfaces have a BSA above 800 Å^2, the size of the smallest interfaces seen in complexes and homodimers. Thus, they cannot be distinguished from biologically relevant interfaces on the basis of their size only.[45–47] Moreover, many of the larger crystal packing interfaces are associated with elements of 2-fold symmetry (crystallographic or local). They constitute 'crystal dimers' that may be mistaken for real homodimers when interpreting the X-ray structure. Other properties of the interfaces should be considered, their composition for instance. In a protein crystal, each molecule has many neighbours: eight on average in the set of Janin and Rodier;[48] even though each interface may be small, together they bury a large fraction of the protein surface. The chemical and amino acid composition of this surface must be like that of the solvent accessible surface, whereas the interfaces of complexes and oligomeric proteins are expected to be different. The chemical composition of the protein surface or an interface may be estimated as the fractional contribution of each atom type to the ASA or the BSA. On average, the non-polar (carbon containing) groups, which form 57% of the ASA, contribute marginally more to the BSA in complexes and crystal packing interfaces, but only the interfaces of homodimers are significantly more hydrophobic. The charged groups of Asp/Gly/Lys/Arg side chains are abundant (17–19%) on the protein surface and at crystal packing interfaces relative to the interfaces of complexes or homodimers (12–14%). On average, the crystal packing interfaces contain fewer H-bonds in proportion to their size than in complexes and homodimers: on average, one per 280 $Å^2$ of BSA instead of one per 190–210 $Å^2$. In contrast, they contain more residual hydration water (Table 1.1). However, the surface composition, the number of H-bonds and the hydration of the interfaces vary widely from a protein or a complex to another.[49] Thus, the differences between the mean values are always less than the standard deviations.

Still, the procedures that aim to distinguish between biological vs crystal packing interfaces often rely on criteria derived from the hydrophilic/hydrophobic character of the interfaces, in addition to their size. Examples are PQS (Probable Quaternary Structure)[50] and the related databases of the European Bioinformatics Institute (Hinxton, UK). PQS applies crystal symmetries to the molecules in the ASU, generates neighbours, scores each pairwise interface on the basis of the BSA and a solvation energy term in PQS, and builds molecular assemblies iteratively by retaining the interfaces that achieve high scores. PITA (Protein InTerfaces and Assemblies)[51] is the same, except that a statistical potential replaces the solvation energy. PISA (Protein Interfaces, Surfaces and Assemblies)[51] uses a graph exploration algorithm to survey all the assemblies that can be formed in the crystal, and an empirical energy is calculated for each. All these procedures ignore the REMARK 300/350 information, and PQS disagrees with it in 18% of the cases.[52] Moreover, PISA fails to recognise as stable assemblies some classical complexes, the D1.3 antibody–lysozyme complex (1vfb) for instance. Thus, the combined criteria of the interface size and its chemistry are not always sufficient to determine the nature of the biomolecule in a crystal.

Current approaches are based on statistical pairwise potentials and machine-learning procedures (reviewed in several chapters of this volume) that allow many other criteria to be taken into account. They should perform better than PQS or PISA, and in fact they can achieve a success rate close to 95% on test sets of limited size.[53] However, they have not yet been applied to the whole PDB, or if they have the results have yet to be made accessible like PQS's or PISA's. In the same way, methods based on phylogeny and sequence conservation (see Chapter 5) have proved their efficiency in a number of cases, but the field of application is still limited.

1.8 Atomic Packing and Interface Topology

Upon visual inspection, the crystal packing interfaces often seem poorly packed and split into small groups of atoms.[47] In contrast, protein–protein complexes have interfaces that are close-packed like the protein interior[49] and they form a single contiguous patch,[40] at least when their BSA is less than 2,000 Å^2. The quality of the atomic packing and the connectivity of an interface express the shape complementarity of the surfaces in contact. They are important characteristics that govern the energetics of the van der Waals and hydrophobic interactions, but they are not easy to quantify. The volumes of the Voronoi cells can be accurately measured to show that the packing density is the same within 1–2% inside globular proteins[54] and at the interfaces of protein–protein complexes[49] or oligomeric proteins.[55] But Voronoi volumes can only be estimated for buried atoms (atoms with zero ASA), and those are a minority at macromolecular interfaces: 34–36% on average in complexes and homodimers, 21% at crystal packing interfaces (Table 1.1).

Other descriptors of the geometric complementarity are the S_c index of Lawrence and Colman,[56] the gap volume index of Laskowski,[57] and the L_D index of Bahadur *et al.*[47] The mean values reported in Table 1.1 confirm that the complementarity is less good at crystal packing interfaces than in complexes or homodimers. In Figure 1.6, these values are normalised to 1 for the homodimer interfaces, and their standard deviations are marked. The three descriptors behave similarly, but the contrast is poor with S_c, and the gap volume index has a large standard deviation due in part to the strong edge effects that affect it when the interface is small or split.

A loosely packed interface with a large gap between the two surfaces in contact must bury few atoms in proportion to its size, and therefore the fraction of buried atoms (f_{bu}) is related to the packing. In practice, f_{bu} is at least a good criterion to distinguish specific from non-specific interfaces as S_c or the gap volume index.[8–9] Interfaces that are split into several regions also bury fewer atoms. Chakrabarti and Janin[40] applied a geometric clustering algorithm to the interface atoms in order to identify connected regions and define the interface topology. These regions, called recognition patches, generally occur in pairs, one on each protein. A majority of the protein–protein complexes has only one pair, the average number being 1.4. Homodimers, which have larger interfaces, contain more: 1.7 pairs on average, with still a majority of single-pair interfaces.[42] The algorithm gives unreliable results with crystal packing interfaces. On the other hand, the model based on the alpha-complex provides a straightforward definition of an interface topology: the set of Voronoi facets that constitute the interface can be split into connected components, subsets of facets that have an edge in common, and these subsets correlate well with the recognition patches defined by the clustering algorithm.[39]

Interfaces may be split in many other ways: along the amino acid sequence into interface chain segments[5,58] or secondary structure elements.[59] Chakrabarti and Janin[40] distinguish within each interface between a core made of the residues that contain atoms buried in the contact, and a rim in which all interface atoms are solvent accessible. In protein–protein complexes and homodimers, the interface rims have essentially the same amino acid composition as the solvent accessible protein surface, but the cores are significantly different; similarly, the interface core residues tend to be conserved in evolution, but not the rim residues.[60]

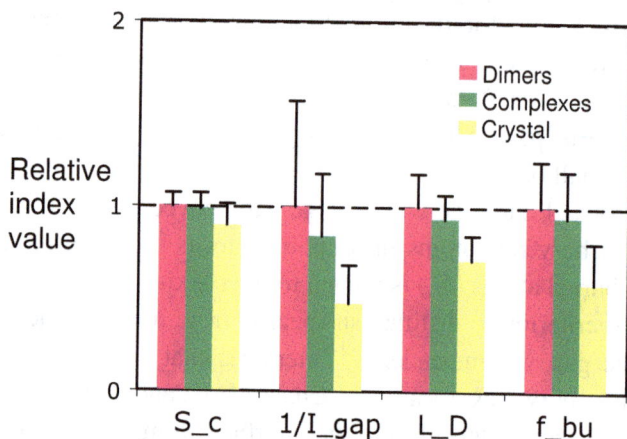

Fig. 1.6. Atomic packing and surface complementarity. Mean values and standard deviations (black bars) of the four parameters reported in Table 1.1 for 122 homodimeric proteins, 70 protein–protein complexes, and 188 crystal packing interfaces with a BSA > 800 Å^2. All have been scaled to 1 for the homodimer interfaces. Comparatively low values of the S_c shape complementarity index, of the reciprocal of the I_{gap} gap volume index, of the L_D packing index, and of the fbu fraction of buried atoms, all confirm that the atomic packing is less compact and the surface complementarity less good in crystal packing interfaces.

1.9 Conclusions and Outlook

The interfaces of binary assemblies such as the protein–protein complexes and the homodimeric proteins have properties that multi-component assemblies are likely to share, but only to some extent. In larger oligomeric proteins or in virus capsids,[62] some of the interfaces are comparable in size, chemical composition and atomic packing density to those of the homodimers. Others are much larger and may span more than two subunits, or they are much smaller and resemble crystal packing interfaces. Presumably, the larger interfaces play a greater role in the stability of the assembly, and are more subject to a selection pressure than small interfaces. The analysis of how different types of interfaces cooperate to stabilise large macromolecular system and contribute to their self-assembly will certainly be one of the most interesting aspects of the ambitious ongoing structural studies.

Acknowledgements

This work greatly benefited from the expertise of Dr A. Poupon, S. Quevillon-Chéruel and other members of the Yeast Structural Genomics team in Orsay, and of B. Séréphin (CNRS, Gif-sur-Yvette). Discussions with F. Cazals (INRIA, Sophia-Antipolis), S. Wodak (University of Toronto), P. Chakrabarti (Bose Institute, Calcutta) and R.P. Bahadur (Jacobs University Bremen), and support of the 3D-Repertoire and SPINE2-Complexes programmes of the European Union are gratefully acknowledged.

References

1. Sali A. (1998). *Nature Struct Biol* 5: 1,029–1,032.
2. Janin J., Séraphin B. (2003). *Curr Opin Struct Biol* 13: 383–388.
3. Devos D., Russell R.B. (2007). *Curr Opin Struct Biol* 17: 370–377.
4. Dutta S., Berman H.M. (2005). *Structure* 13: 381–388.
5. Jones S., Thornton J.M. (1996). *Proc Natl Acad Sci USA* 93: 13–20.
6. Nooren I.M. (2003). *EMBO J* 22: 3,486–3,492.
7. Wodak S.J., Janin J. (2002). *Adv Protein Chem* 61: 9–73.
8. Janin J., Rodier F., Chakrabarti P., Bahadur R.P. (2007). *Acta Crystallogr D Biol Crystallogr* 63: 1–8.
9. Janin J., Bahadur R.P., Chakrabarti P. (2008). *Quart Rev Biophysics* 41: 1–48.
10. Janin J. (2007). *Structure* 15: 1,347–1,349.
11. Vakser I.A. (2008). *Structure* 16: 1–3.
12. Structural Genomics Consortium *et al.* (2008). *Nature Methods* 5: 135–46.

13. Quevillon-Cheruel S., Collinet B., Trésaugues L., Minard P., Henckes G., Aufrère R., Blondeau K., Zhou C.Z., Liger D., Bettache N., Poupon A., Aboulfath I., Leulliot N., Janin J., Van Tilbeurgh H. (2007). Cloning, production, and purification of proteins for a medium-scale structural genomics project. *Macromolecular Crystallography Protocols*, Vol. 1, S. Doublié (ed.) *Methods Mol Biol* 363: 21–37.
14. Trésaugues L., Collinet B., Minard P., Henckes G., Aufrère R., Blondeau K., Liger D., Zhou C.Z., Janin J., Van Tilbeurgh H., Quevillon-Cheruel S. (2004). *J Struct Funct Genomics* 5: 195–204.
15. Kigawa T., Muto Y., Yokoyama S. (1995). *J Biomol NMR* 6: 129–34.
16. Quevillon-Cheruel S., Liger D., Leulliot N., Graille M., Poupon A., de La Sierra-Gallay I.L., Zhou C.Z., Collinet B., Janin J., Van Tilbeurgh H. (2004). *Biochimie* 86: 617–623.
17. Burley S. K., Joachimiak A., Montelione G.T., Wilson I.A. (2008). *Structure* 16: 5–11.
18. Sauder M.J., Rutter M.E., Bain K., Rooney I., Gheyi T., Atwell S., Thompson D.A., Emtage S., Burley S.K. (2008). *Methods Mol Biol* 426: 561–575.
19. Luger K., Rechsteiner T.J, Flaus A.J., Waye M.M., Richmond T.J. (1997). *J Mol Biol* 272: 301–311.
20. Groll M., Ditzel L., Löwe J., Stock D., Bochtler M., Bartunik H.D., Huber R. (1997). *Nature* 386: 463–471.
21. Tan S. (2001). *Protein Expression and Purification* 21: 224–234.
22. Sadaoui N., Janin J., Lewit-Bentley A. (1994). *J Appl Cryst* 27: 622–626.
23. Bodenstaff E.R., Hoedemaeker F.J., Kuil M.E., de Vrind H.P., Abrahams J.P. (2002). *Acta Crystallogr D Biol Crystallogr* 58: 1,901–1,906.
24. Sulzenbacher G., Gruez A., Roig-Zamboni V., Spinelli S., Valencia C., Pagot F., Vincentelli R., Bignon C., Salomoni A., Grisel S., Maurin D., Huyghe C., Johansson K., Grassick A., Roussel A., Bourne Y., Perrier S., Miallau L., Cantau P., Blanc E., Genevois M., Grossi A., Zenatti A., Campanacci V., Cambillau C. (2002). *Acta Crystallogr D Biol Crystallogr* 58: 2,109–2,115.
25. Leulliot N., Trésaugues L., Bremang M., Sorel I., Ulryck N., Graille M., Aboulfath I., Poupon A., Liger D., Quevillon-Cheruel S., Janin J., Van Tilbeurgh H. (2005). *Acta Crystallogr D Biol Crystallogr* 61: 664–670.
26. Doublié S. (1997). *Methods Enzymol* 276: 523–530.
27. Rossmann M.G., Arnold E. (eds) (2006). Crystallography of Biological Macromolecules. *International Tables for Crystallography* Vol. F.
28. Smith J.L., Hendrickson W.A., Terwilliger T.C., Berendzen J. (2006). Crystallography of Biological Macromolecules. Rossmann M.G. and Arnold E. (eds) *International Tables for Crystallography* Vol. F: 299–309.
29. Dauter Z., Dauter M., de La Fortelle E., Bricogne G., Sheldrick G.M. (1999). *J Mol Biol* 289: 83–92.
30. Berman H.M., Battistuz T., Bhat T.N., Bluhm W.F., Bourne P.E., Burkhardt K., Feng Z., Gilliland G.L., Iype L., Jain S., Fagan P., Marvin J., Padilla D.,

Ravichandran V., Schneider B., Thanki N., Weissig H., Westbrook J.D., Zardecki C. (2002). *Acta Crystallogr D Biol Crystallogr* 58: 899–907.

31. Chothia C., Janin J. (1975). *Nature* 256: 705–708.
32. Edelsbrunner H., Mucke E.P. (1994). *ACM Trans Graphics* 13: 43–72.
33. Lee B.K., Richards F.M., (1971). *J Mol Biol* 55: 379–400.
34. Richards F.M. (1974). *J Mol Biol* 82: 1–14.
35. Gellatly B.J., Finney J.L. (1982). *J Mol Biol* 161: 305–322.
36. Aurenhammer F. (1987). *SIAM J Computing* 16: 78–96.
37. Ban Y.E.A., Edelsbrunner H., Rudolph J. (2004). *RECOMB* 2: 205–212.
38. Headd J.J., Ban Y.E., Brown P., Edelsbrunner H., Vaidya M., Rudolph J. (2007). *J Proteome Res* 6: 2,576–2,586.
39. Cazals F., Proust F., Bahadur R.P., Janin J. (2006). *Protein Sci* 15: 2,082–2,092.
40. Chakrabarti P., Janin J. (2002). *Proteins* 47: 334–343.
41. Larsen T.A., Olson A.J., Goodsell D.S. (1998). *Structure* 6: 421–427.
42. Bahadur R.P., Chakrabarti P., Rodier F., Janin J. (2003). *Proteins* 53: 708–719.
43. Jones S., Thornton J.M. (1995). *Prog Biophys Mol Biol* 63: 31–65.
44. Lévy E.D. (2007). *Structure* 1: 1,364–1,375.
45. Ponstingl H., Henrick K., Thornton J.M. (2000). *Proteins* 41: 47–57.
46. Ponstingl H., Kabir T., Thornton J.M. (2003). *J Appl Cryst* 36: 1,116–1,122.
47. Bahadur R.P., Chakrabarti P., Rodier F., Janin J. (2004). *J Mol Biol* 336: 943–955.
48. Janin J., Rodier F. (1995). *Proteins* 23: 580–587.
49. Lo Conte L., Chothia C., Janin J. (1999). *J Mol Biol* 285: 2,177–2,198.
50. Henrick K., Thornton J.M. (1998). *Trends Biochem Sci* 23: 358–361.
51. Krissinel E., Henrick K. (2007). *J Mol Biol* 372: 774–797.
52. Xu Q., Canutescu A., Obradovic Z., Dunbrack R.L. Jr (2006). *Bioinformatics* 22: 2,876–2,882.
53. Bernauer J., Bahadur R.P., Rodier F., Janin J., Poupon A. (2008). *Bioinformatics* 24: 652–658.
54. Tsai J., Taylor R., Chothia C., Gerstein M. (1999). *J Mol Biol* 290: 253–266.
55. Ponstingl H., Kabir T., Gorse D., Thornton J.M. (2005). *Prog Biophys Mol Biol* 89: 9–35.
56. Lawrence M.C., Colman P.M. (1993). *J Mol Biol* 234: 946–950.
57. Laskowski R.A. (1995). *J Mol Graph* 13: 323–330.
58. Pal A., Chakrabarti P., Bahadur R., Rodier F., Janin J. (2007). *J Biosci* 32: 101–111.
59. Guharoy M., Chakrabarti P. (2007). *Bioinformatics* 15: 1,909–1,918.
60. Guharoy M., Chakrabarti P. (2005). *Proc Nat Acad Sci USA* 102: 15,447–15,452.
61. Rodier F., Bahadur R.P., Chakrabarti P., Janin J. (2005). *Proteins* 60: 36–45.
62. Bahadur R.P., Rodier F., Janin J. (2007). *J Mol Biol* 367: 574–590.

CHAPTER 2

A Structural Perspective on Protein–Protein Interactions in Macromolecular Assemblies

Ranjit P. Bahadur

School of Engineering and Science, Jacobs University Bremen,
Campus Ring 1, D-28759 Bremen, Germany
E-mail: ranjitp_bahadur@yahoo.com

Many cellular processes are carried out not only by a single biomolecule, but require an assembly consisting of several partners, which gives rise to macromolecular machines. Structural genomics initiatives aim at providing a repertoire of three-dimensional structures of such macromolecular assemblies. The analysis of the interactions between the biomolecules in an assembly can give valuable insight into the assembly process, and a better understanding of its function. Virus capsids are important examples of such assemblies consisting of many proteins and nuclei acid subunits. The structural information forms the basis for the analysis of the physical–chemical and structural properties of the protein–protein interfaces in icosahedral virus capsids. The results are compared to those found for binary protein–protein interfaces exemplified by homodimers, heterodimers and crystal-packing monomers. Finally, the findings are correlated with different experimental data available in the literature to give an overview on the capsid self-assembly procedure.

2.1 Introduction

Recent discoveries in molecular and structural biology lead us to believe that most, if not all, of the cellular functions and activities are organised and carried out not by single molecules, but by the orchestrated action of biomolecular complexes.[1] Proteins are often the major players in this orchestra and may assemble inside the cell giving rise to macromolecular machines. Indeed, the entire cell can be viewed as a factory that contains an elaborate network of such machines. These molecular machines can grossly be divided into two major categories. In the first class, they are involved largely in the overall architecture and mechanical function of the organisms, exemplified in virus capsids[2] or the components of the cellular cytoskeleton.[3] In the other class, they are essential for many biochemical reactions as found in the ribosome,[4] the protein synthesis unit of the cell, or in chaperones,[5] that guide the protein folding process. To understand the functional mechanism of these complex structures, it is important to better understand their folding and assembly pathways.

While binary protein complexes (see Chapter 1) consist of only two polypeptide chains, multi-subunit cellular machines involve several polypeptide and/or nucleotide chains, and often have a more complicated structure than binary complexes. So far the Protein Data Bank (PDB)[6] stores only a few of these complex structures at atomic detail, and among those assemblies it is the virus capsids and ribosomes that have been studied most extensively.[7] Structural details are also available for the 20S proteasome core particle which is involved in removing damaged and misfolded proteins from cells;[8] for the large chaperonin complexes GroEL and GroES;[9] and for several multi-enzyme complexes such as glutamine synthetase[10] illustrating their functional mechanism. While the virus capsids are symmetrical cellular machines, most of the others assemblies of known structure are asymmetric in nature. Here, we discuss the different physical–chemical and structural properties of the protein–protein interfaces in icosahedral virus capsids, which represent a well studied system for the analysis of multi-subunit protein assemblies. The conservation of the interfaces during evolution will also be analysed. Furthermore, the interfaces in virus capsids as examples of multi-protein assemblies will be compared with those found in binary protein–protein complexes in terms of several structural and physical–chemical properties. Finally, the implication of the observations for explaining the mechanism of the self-assembly process of viral capsids will be discussed.

2.2 The Icosahedral Viruses

Icosahedral virus capsids are the classical examples of multi-subunit assemblies with a large number of protein and nucleic acid chains assembled in a regular geodesic structure. The capsids are spherical in shape, which encapsulate and protect the viral genome. They are frequently implicated in the recognition and infection of target cells. Icosahedral viruses are capable to infect a broad spectrum of living cells, ranging from bacteria to humans, and this ability makes them an important target element for biochemical and structural studies. In 1956, Crick and Watson[11] first proposed that the capsids of many viruses have spherical symmetry. This property greatly helped in the structural analysis by X-ray crystallography so that the first atomic structure was solved almost three decades ago for the tomato bushy stunt virus[12] and the satellite tobacco necrosis virus.[13] At present, icosahedral virus capsids are the best representative large protein assemblies in the PDB, and the atomic detail structures help us to understand the mechanism of self-assembly that determines the functional role in the cell.

2.3 The Structure of the Icosahedral Virus Capsids

The structure of an icosahedron can be constructed with 20 equilateral triangles, which are composed of 12 vertices, 20 triangular facets and 30 edges (see Fig. 2.1). An icosahedron has six 5-fold axes of symmetry passing through its vertices, ten 3-fold axes extending through each faces and 15 2-fold axes passing through its edges (Fig. 2.1). In 1956, Crick and Watson[11] first proposed that the capsid of many viruses have cubic *I* (icosahedral) point group symmetry, which was confirmed by the electron microscope studies of Caspar and Klug (1962).[14] It is therefore necessary that each protein component should be present in 60 copies related by one of the 2-fold, 3-fold or 5-fold point group symmetry elements of the icosahedron to make a sphere like structure.

The icosahedrons made by 60 copies are the smallest in size, and to make a bigger capsid it has to accommodate more than 60 copies of the polypeptide chain. In 1962, Caspar and Klug[14] introduced the rule of quasi-equivalence and lattice 'Triangulation Number' (T-number), and demonstrated the increased volume of the capsids for packaging large viral genomes. This T-number represents the number of subunits present in an icosahedral asymmetric unit (IAU) and a whole capsid contains 60T subunits.

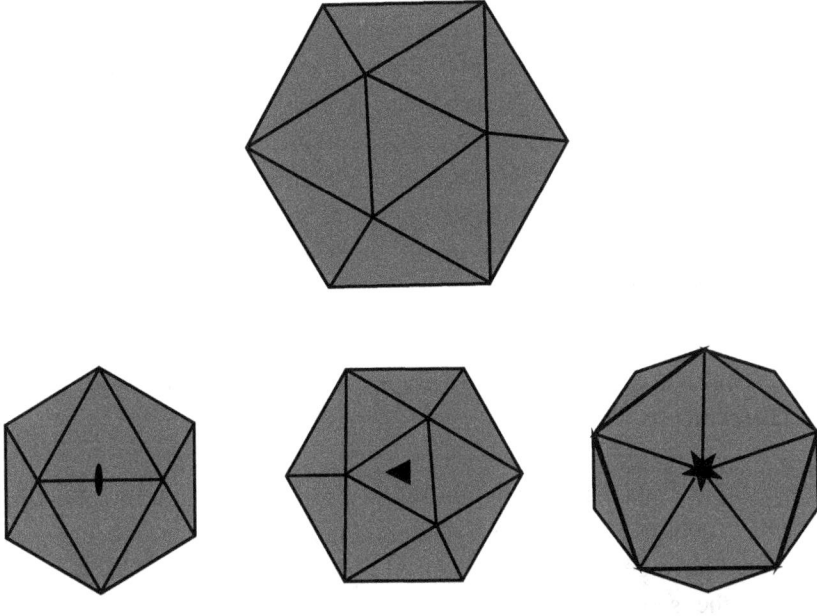

Fig. 2.1. Symmetries in an icosaherdron. An icosahedron consists of 20 equilateral triangles (upper panel). Different 2-fold, 3-fold and 5-fold symmetry axis (lower panel).

To retain the quasi-symmetry one has to restrict the values of T with the following equation,

$$T = (h^2 + hk + k^2) f^2$$

where h, k and f are integers. Caspar and Klug suggested quasi-equivalent environments for identical protein subunits.[14] However, after the structure determination of human rhinovirus 14 by Rossman et al.[15] in 1985, it became evident that these subunits are not necessarily always identical to form the predicted lattice, and it can be described as a pseudo-symmetry.[15] For T = 1, the corresponding capsid contains only one protein subunit in the IAU and a total of 60 in the whole capsid. This is for example the case for the satellite tobacco necrosis virus.[13] In capsids with T = 3, there are three identical subunits in the IAU and the whole capsid comprises of 180 subunits. They are larger in volume compared to the T = 1 and are more common in isometric plant viruses.[16] Most of the animal viruses have pseudo T3 (pT3) symmetry where the IAU contains three structurally identical subunits but with different amino acid sequences.[16] It has been assumed that a pT3 lattice has evolved from a T = 3 lattice by a triplication of the coat protein gene, followed by independent evolution of the three major subunits. All these structures with different lattice triangulation numbers survived due to the non-covalent interactions between all the protein and nucleotide subunits that build the capsid.

2.4 Structural and Chemical Features of the Protein–Protein Interfaces

The Protein Data Bank contains many examples of icosahedral virus capsid structures. The protein–protein interactions in 49 such capsid structures have been recently analysed by Bahadur *et al.*[17,18] Within this dataset, the sequence identity was less than 45% between any two polypeptide chains and the dataset contained 11 entries of T = 1 capsids, 17 with T = 3, and 10 with pseudo T = 3. All these structures obey the quasi-equivalence rule of Caspar and Klug. It also included a few entries that do not obey the quasi-equivalence rule; the capsid of the Blue Tongue virus[19] and bacteriophage phiX174[20] are among the most complex three-dimensional capsid structures known till today.

2.4.1 *Symmetry and Size of Interfaces*

For each capsid, Bahadur *et al.*[17] identified a unique set of pairwise interfaces from which all other interfaces between polypeptide chains in the whole capsid can be generated employing the icosahedral symmetry. These unique interfaces are repeated 60 times by the icosahedral symmetry except for those having 2-fold axes that are repeated only 30 times. The whole dataset contained 779 such interfaces, 16 per capsid, which corresponded to an average of 920 pairwise interfaces in each capsid. The interfaces were further divided into different symmetry classes and it was found that one-third occur within the IAU.

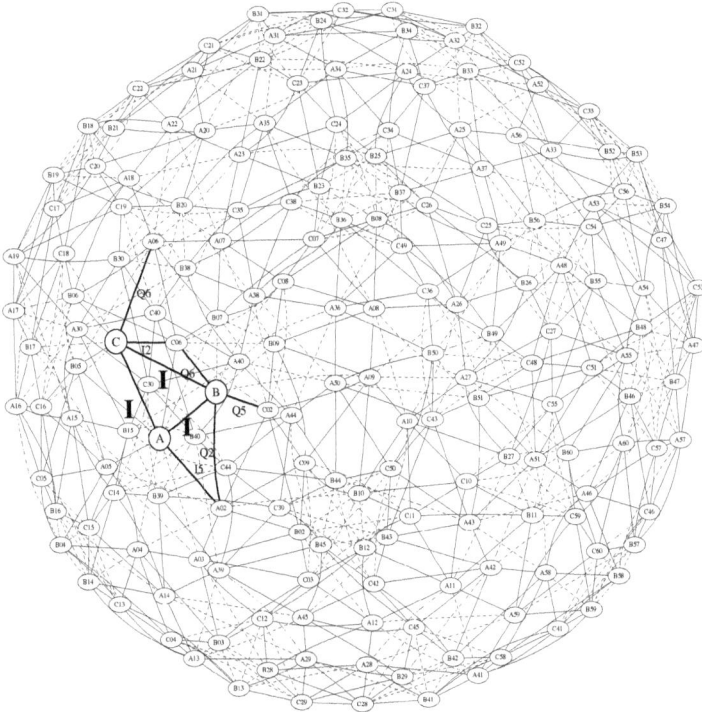

Fig. 2.2. A graphical representation of the protein–protein interaction network in the black beetle virus capsid (PDB code 2bbv). Nodes in the graph represent proteins, edges an interaction between them. Solid lines indicate interfaces with BSA > 800 Å^2, while smaller interfaces are represented by broken lines. The black beetle virus has a T3 capsid with three chains ABC in the icosahedral asymmetric unit (IAU). Chains ABC form nine unique interfaces with BSA > 800 Å^2 (thick lines) and five smaller interfaces. Symmetry-related subunits are labels Ni where 'N' stands for each polypeptide chain, and 'i' (2 to 60) for the relevant symmetry element. The unique interfaces are labelled according to symmetry type; IAU: icosahedral asymmetric unit; I2,I5: icosahedral 2- and 5-fold; Q2,Q5,Q6: quasi 2-, 5- and 6-fold.

The other two-thirds were almost equally distributed between icosahedral 2-, 3- or 5-fold symmetry, or quasi-symmetry. The range of the unique interfaces in T = 1 capsids was three to five – much less than the overall average of 16 – and contained at least one of each of the 2-, 3- and 5-fold icosahedral symmetry types. The number of unique interfaces in T = 3 capsids ranged from 8 to 17 in which 3 occur between the three polypeptide chains in IAU. Besides icosahedral symmetry, they also identified interfaces with non-icosahedral symmetry that derives from quasi-equivalence and is commonly called quasi-symmetry. A typical pT3 capsid contains 25 to 30 unique interfaces and these are related either by the symmetry in IAU, 2-fold, 5-fold or quasi 6-fold.

For example, in the black beetle virus (2bbv), the T = 3 capsid contains 14 unique interfaces and nine above 800 Å^2. Figure 2.2 represents the protein–protein interaction network in the black beetle virus capsid, where the nodes and the edges in the graph represent the protein subunits and interactions between them, respectively. Three of the unique interfaces occur within the IAU between three chains; one is the 2-fold marked I2 between the C chains, and one more is the 5-fold marked I5 between the A chains. The other four involve interfaces that are labelled Q2 for quasi 2-fold, Q5 for quasi 5-fold and Q6 for quasi 6-fold by analogy to the T = 3 capsids. In viral capsids, the average pairwise interface buries 1,750 Å^2 of BSA (Buried Surface Area, as defined in Chapter 1), and on each polypeptide chain it implicates 90 atoms and 25 residues (Table 2.1). Bahadur *et al.*[17] considered all the pairwise interfaces with BSA > 10 Å^2 and divided them into three size categories. The 'small' class, which buries less than 800 Å^2, makes up nearly 40% of the whole sample, but contributes only 7% of the overall BSA. On the other hand, the 'large' interfaces with BSA between 800 to 2,000 Å^2 make up only 28% of the sample, but contribute 68% of the BSA. They also identified very large interfaces with BSAs ranging from 5,000–10,000 Å^2. The majority of them occur within the IAU. The size distribution of these interfaces is shown in Fig. 2.3. Between different icosahedral symmetry related interfaces, on average 2-fold interfaces are the largest in size, and 5-fold interfaces which are related by 144° rotation are the smallest (Table 2.1).

R.P. Bahadur

Table 2.1. Structural and chemical properties of protein–protein interfaces in virus capsids

Parameters	All	Size category				Symmetry type					Capsid type			
		Small	Medium	Large	IAU	2-fold	3-fold	5-fold	5*-fold[a]	others	T=1	T=3	pT3	Misc
Unique interfaces	779	307	253	219	252	55	64	84	35	289	44	196	199	340
BSA (Å²)	1750	300	1360	4190	2450	2260	1480	1980	230	1180	2530	1490	1530	1890
Fraction of (%)														
interface no.	100	39	33	28	32	7	8	11	4	37	6	25	25	44
interface area	100	7	25	68	46	9	7	12	1	25	8	22	22	48
Fraction of (%)														
non-polar area (f_np)	63	61	61	64	63	65	63	63	63	62	61	64	64	62
buried atoms (f_bu)	29	11	23	33	30	30	32	28	10	25	34	28	29	27
core residues (f_core)	51	22	44	58	52	56	55	53	18	46				
H-bonds														
Number per interface	7.1	0.7	5.0	18.5	10.2	9.2	6.4	8.2	0.6	4.5	11.1	5.9	7.5	7.0
BSA per bond (Å²)	250	410	280	22.5	240	240	240	240	360	260	228	253	203	270
Packing index														
L_D	26	11	33	4.3	32	31	21	30	6	23	33	26	25	27
G_D	0.9	0.7	0.9	1.1	0.9	1.0	0.9	0.9	0.5	0.9	1.2	0.9	0.9	0.9
Patches														
Number per chain	2.3	1.3	2.2	4.2	3.1	2.9	1.9	2.4	1.3	1.9	2.8	2.1	2.2	2.6
BSA per pair (Å²)	720	220	620	1000	795	790	770	810	170	640	920	715	700	740
Segments														
Number per chain	3.9	1.8	3.7	7.1	5.0	4.4	3.2	4.4	1.5	3.1	4.7	3.6	3.1	4.5
Number per 1000Å²	4.5	1.2	5.4	3.4	4.2	3.9	4.3	4.4	13.0	5.3	3.7	4.9	4.1	4.7
Length	6.4	3.1	5.7	8.1	-	-	-	-	-	-	7.6	6.1	6.9	6.2
Mean normalized entropy <ε>														
Protein core	0.7										0.8	0.6	0.8	0.7
Protein surface	1.6										1.4	1.3	1.9	1.4
Interface	0.9	0.8	0.8	0.9	0.9	0.8	0.8	0.8		0.8	0.9	1.0	0.9	0.9
Interface core	0.8	0.9	0.7	0.8	0.9	0.7	0.8	0.7		0.7	0.8	0.9	0.8	0.8
Interface rim	1.2	0.8	0.9	1.0	1.0	0.8	0.9	1.0		0.9	1.1	1.2	1.2	1.1
Multi-interface	0.8										0.8	0.9	0.7	0.9
Mean entropy ratio														
Interface core/rim	0.7	1.1	0.8	0.8	0.9	0.9	0.9	0.7		0.8	0.7	0.8	0.7	0.8
Protein core/surface	0.5										0.6	0.5	0.4	0.6

[a]5-fold interfaces related to 144° rotation.

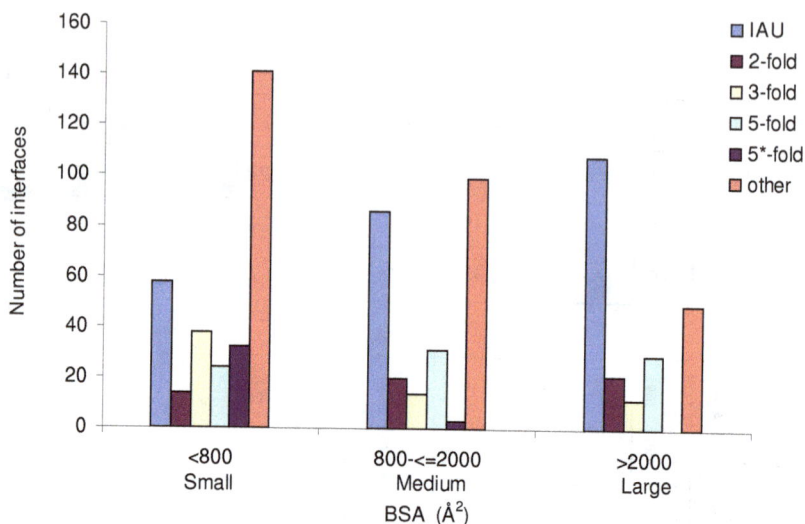

Fig. 2.3. Distribution of protein–protein interface sizes with different symmetries related to icosahedral virus capsids. The different size interface categories are: Small (BSA < 800 Å^2), Medium (800 < BSA < = 2,000 Å^2) and Large (BSA > 2,000 Å^2). Interfaces which are related by a 144° rotation are designated as 5*-fold interfaces.

In viral capsids, pairwise interfaces overlap and many residues are in contact with several adjacent polypeptide chains.[17] In the T = 1 and T = 3 capsids, each polypeptide chain is in contact on average with seven neighbouring chains that bury about 45% of its ASA. In pT3 capsids, this value increases to 12 and the total buried fraction to 60%. These multi-interfaces contain 15% atoms that are in contact with two neighbouring subunits and 2% with three. About 25% of the amino acid residues of these interfaces are in contact with two neighbours, 5% with three and only 2% with four neighbours.

2.4.2 *Chemical Composition and Hydrogen Bonds*

One can define the chemical composition of these interfaces at an atomic level of detail as well as on a residue level. At the atomic level, nitrogen, oxygen and sulfur can be considered as polar and all the carbon containing groups as non-polar. The fraction of area contributed by the non-polar group (f_np)[21] at the interface is about 63% of the total BSA, which is similar in different symmetry or quasi-symmetry related interfaces (Table 2.1). The large interfaces are slightly more non-polar than the average, whereas the f_np is highly variable in small interfaces with few atoms. At the residue level one can count the contribution of different interface amino acid residues to BSA and can also compare the contribution to the rest of the solvent accessible protein surface. In capsids, interface residue composition strongly differs from that of their surface. They are enriched with non-polar (Ile, Leu, Val) residues and depleted in charged residues (Asp, Glu, Lys) with only one exception in the case of Arg, which is a major contributor to the BSA of capsid interfaces.

Polar atoms (and the residues to which they belong) can make hydrogen bonds between the polypeptide chains at the interface. Bahadur et al.[17] have identified about seven hydrogen bonds per capsid interface, which corresponds to one H-bond per 250 Å^2 of BSA. It was found that the surface density of H-bonds is similar in different symmetry related interfaces, except those with the 5-fold related interfaces with 144° rotation, many of which are very small in size with no H-bonds at all.

2.4.3 *Atomic Packing of the Interfaces*

The fraction of atoms buried (f_bu defined in Ref. 21) at the protein–protein interfaces in viral capsids is 29% on average and depends on the size of the interfaces: large interfaces bury 33% compared to small which bury only 11% (Table 2.1). When one considers the whole assembly, f_bu increases to 36% (instead of 29% in pairwise contacts) in the presence of all neighbouring subunits. This is because atoms in contact with more than one neighbour can have zero ASA even though they are accessible in each of the pairwise interfaces. Buried atoms are part of the residues which form interface core[22,23] that covers half of the residues in pairwise interfaces, and which increases to two-thirds in the presence of all neighbouring subunits in the whole capsid.

The fraction of buried atoms is related to the packing density of the interface atoms and the shape complementarity of the protein surfaces in contact. A close-packed interface buries more atoms in proportion, and its presence implies that the two protein surfaces have complementary shapes. Two packing indices, L_D and G_D that measure the packing density of the interfaces,[21] and Table 2.1 show that large- and medium-sized capsid interfaces are well packed. In contrast, small interfaces are loosely packed and resemble crystal-packing interfaces (typical contacts of proteins in a protein crystal).

2.4.4 *Interface Patches and Segments*

According to Chakrabarti and Janin,[22] one can divide interface atoms into one contiguous region or several regions depending on the spatial distribution of the interface atoms. The authors used an average-linkage clustering algorithm with a cut-off value which depends on the type of the complexes. When applied to a virus capsid assembly with a threshold value of 15 Å, this algorithm yielded a single-patch in most of the small interfaces, 2–3 patches in medium size interfaces and multiple patches in large interfaces (see Table 2.1). Assignment of interface patches with a clustering method considers the three-dimensional distribution of the interface atoms. One can also distribute the interface residues into different segments depending upon the primary sequence of the polypeptide chain.[24,25] The average number of such segments is 3.9 per polypeptide chain, and each of these segments contains an average of 6.4 interface residues. The small interfaces tend to be fragmented into very short segments with only three interface residues per segment on average, while the large interfaces contain fewer segments and the segment length is quite large with an average of seven per polypeptide chain. Some of these long segments include up to 47 interface residues, which belong to the 'tail' region of the polypeptide chain as defined by Bahadur *et al.*[17] The large segments involving the tail region are frequently found in T = 3 and pT3 capsids. In T = 3 capsids, tails are preferentially involved in the 2-fold and quasi 6-fold interfaces, while in the pT3 capsids they contribute mainly to the interfaces in the IAU. Tails often play a dominant role during the process of a capsid assembly.

2.4.5 *Residue Conservation*

It is well established that residues which are involved in protein–protein interfaces are generally better conserved than other surface residues.[26–30] There are experimental studies such as alanine scanning mutagenesis,[31–33] as well as theoretical studies, that consider the sequence entropy[34,35] used to identify conserved residues at the binding site and also their contribution to the free energy of binding. In viral capsids, several polypeptide chains interacting with more than one partner are stabilised by non-covalent interactions. They should therefore impose stringent evolutionary constraints on the protein sequence.

Recently, Bahadur and Janin[18] analysed the residue conservation in viral capsids by using the 'Shannon Entropy',[36] which measures divergence at individual sequence positions in sequence alignments taken from the HSSP database.[37] As found in other binary complexes,[35] residues at the interface core in viral capsids are also more conserved than other interface residues. The novel finding of the work was that residues that are part of multiple interfaces are even more conserved than all other interfaces or surface residues; but the degree of conservation varies widely between different interfaces and also between different regions within each interface. This is evident in the human rhinovirus[38] (PDB code 1aym, a pT3 capsid) where the average normalised entropy $<s>$ is larger than one for non-interface residues, and close to one for residues in one interface (Figure 2.1b of Ref. 18). Residues that are present in two or more interfaces have $<s>$ in the range of 0.4–0.7. This value is significantly lower than 1.0, and cannot be explained solely by a preference of these residues for the interface core. There is no correlation found between the symmetry of the interfaces and their mean sequence entropy. Bahadur and Janin[18] also found that conservation of the capsid interface residues is generally not related to either the size of the interface or its structural and chemical properties, such as polarity or atomic packing. The sequence conservation can also differ significantly within a given interface. For example, in Fig. 2.4 the interface between A and C subunits in the human rhinovirus (1aym) is well conserved ($<s>$ = 0.9), but chain 'A' contains a highly divergent region at the C-terminus ('C-tail') with a very high entropy ($<s>$ = 2.0), and a highly conserved

N-terminus ('N-tail') with a very low entropy (<s> = 0.6), both of which contributes to the A:C interface (Figure 2.4).[18] In pT3 capsids, N-tails are more conserved than the C-tails in general, but in other capsids they are as conserved as other parts of the whole polypeptide chains that form the globular structure.

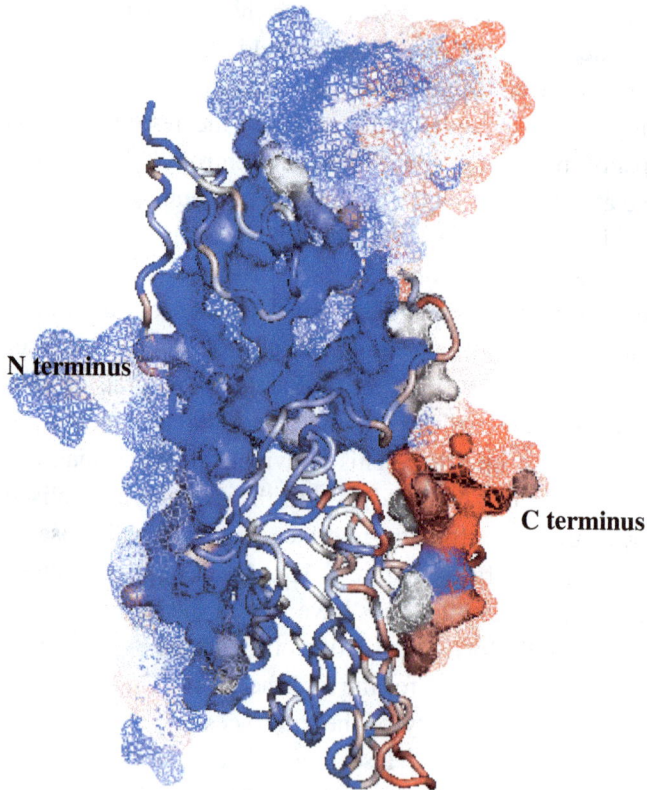

Fig. 2.4. Residue conservation at the interface between A and C subunits of the human rhinovirus (PDB code 1aym) capsid. Subunit residues of A are shown as molecular surface when they are in contact with chain C, otherwise they are drawn as mesh. C subunit is drawn in backbone tube. The gradient of colour is from blue (conserved) to red (divergent) according to the entropy of each residue. Chain A has a well conserved N-terminus and a poorly conserved C-terminus; both are in contact with the C subunit.

2.5 Comparison with Binary Interfaces

The pairwise interfaces in viral capsids can be directly compared with the interfaces in binary protein–protein complexes, homodimers, and crystal-packing interfaces of monomeric proteins analyzed in Chapter 1. The average size of the 'medium' capsid interfaces is comparable with the size of the interfaces in protein hetero complexes as analysed in,[22] which largely fall into the 'standard size' category as defined by Lo Conte *et al.*[39] The interfaces in the 'large' category in capsids are comparable to the size of the interfaces found in homodimers,[23] while the 'small' capsid interfaces are comparable with the crystal-packing interfaces in monomeric proteins as described in Ref. 21.

The area contributed by the non-polar groups at the large capsid interfaces is very similar to that of the interfaces found in homodimers or oligomeric proteins (see Table 2.1 in Chapter 1). On the other hand, small capsid interfaces are marginally more polar and very close to the interfaces found in protein–protein complexes and crystal-packing interfaces of monomeric proteins. Polar groups at the interface make hydrogen bonds. On average, the hydrogen bond density in capsid interfaces is lower than that of the other three types of interfaces. This parameter highly depends on the quality of the crystal structures. In high resolution capsid structures, the H-bond density is very close to other binary protein–protein interfaces.

Amino acid composition of the pairwise capsid interfaces is comparable to the other binary interfaces. Aliphatic residues like Ile, Leu, Val and Met contribute significantly to the interfaces in capsids and homodimers compared to the rest of the protein surface and crystal-packing interfaces (see Figure 2.5). Charged residues like Asp, Glu and Lys are abundant at the protein surfaces and crystal-packing interfaces but contribute less significantly to the interfaces either in capsids or homodimers. Arg behaves differently from other charged residues, and is uniformly present in all types of interfaces. Capsid interfaces however, differ in composition from homodimeric interfaces. They are depleted in aromatic residues (Phe, Tyr, Trp) and enriched in neutral polar residues (Ser, Thr, Asn, Gln) and Pro. These five residues contribute almost equally to the capsids and the crystal-packing interfaces, as well as surfaces of monomeric proteins. However, these residues contribute less significantly to the BSA of the homodimer interfaces.[17,21,23] This shows that the amino acid composition of the viral capsid interfaces is intermediate between homodimers and crystal-packing interfaces.[17] The pairwise capsid interfaces bury fewer atoms and have fewer core residues than in homodimers, but they become similar to those of homodimer interfaces when calculated in the presence of all neighbouring subunits. Small capsid interfaces are poorly packed with few buried atoms and core residues, and involve very short segments of the polypeptide chain. In contrast, the medium and large interfaces are well packed, and have comparatively larger chain segments at the interfaces. The average lengths of these segments are essentially the same in protein–protein complexes and homodimer interfaces.

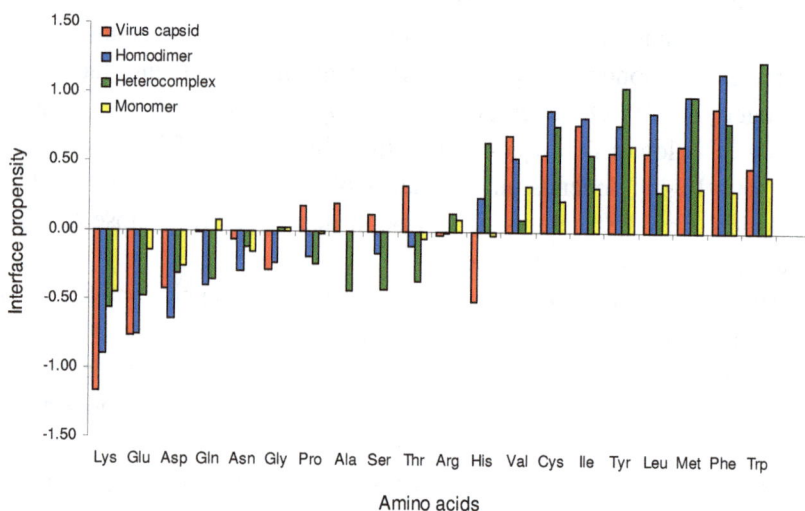

Fig. 2.5. The propensity of a residue to be part of the interface rather than the protein surface is given by ln(f/f') where f and f' are the percentage area contribution of each residue to the interface, or the rest of the solvent accessible protein surface respectively.[17] Values of protein–protein interfaces in viral capsids,[17] homodimers,[23] heterocomplexes,[39] and crystal-packing monomers[21] are shown in different colours. A positive value indicates the preference of a particular residue to be at the interface rather than the protein surface, while the negative value indicates just the opposite.

Homodimers or complexes differ significantly from viral capsids in terms of assemblies. The former group is a binary assembly with a single interface, while the latter is a multi-component assembly in which each polypeptide chain has several neighbours and forms several interfaces. The average number of neighbours in capsids and protein crystals is almost equal, and indicate that the former contain as many interfaces per polypeptide chain as the latter.[17,40] Capsid interfaces cover half or more of the protein surface also found in crystals of small proteins even though crystal-packing interfaces are comparatively smaller in size.[40] In binary assemblies, only one-fifth of the protein surface participates in subunit contacts; the remaining large part is accessible to solvent molecules,[41] whereas in viral capsids more than half of the protein surface is buried between the subunits. In relation to homodimers and complexes, viral capsids have fewer surface residues that are solvent accessible and many residues are part of several interfaces simultaneously.

2.6 Mechanism of the Capsid Assembly

The self-assembly of a virus capsid not only involves the folding of the subunits and their association, but also involves nucleic acids, accessory proteins, host chaperons, as well as processes like maturation through conformational changes and covalent modifications such as proteolysis.[42–44] Though it is a very complicated process, yet it is amazingly fast. For example, bacteriophage T4 goes through a complete cycle of infection, replication and lysis in 15 minutes; the capsid assembles in minutes despite the large number of polypeptide chains that constitute it.[45] Although the capsid assembly mechanism applies a high evolutionary pressure on the polypeptide chains, yet the interface amino acid composition is not markedly different from that of the protein–protein interfaces of simpler systems, suggesting that it goes through a series of low order – possibly binary – steps similar to the assembly of small oligomeric proteins.[17,18]

The intermediate oligomeric species in the assembly pathway are called capsomers. Bahadur et al.[17] postulated that 'large' and 'medium' size interfaces are mainly involved in building the capsomers, which leads ultimately to the formation of the capsid. In contrast, small interfaces are unlikely to take part in the assembly process and contribute to the stability of the assembly after its formation. Several experimental, as well as theoretical studies, have been carried out to understand the capsid self-assembly process.[42-44,46-53] Xie and Hendrix[48] demonstrated the capsid assembly of bacteriophage HK97 during *in vitro* studies. Reddy *et al.*[54] explained the assembly pathways of three icosahedral viruses with the help of van der Waals and electrostatic association energies calculated for each pairwise unique interface. Recently, Stockley *et al.*[55] detected assembly intermediates and investigated the formation of a whole T = 3 bacteriophage MS2 capsid by using the electrospray ionization–mass spectrometry.

In vitro studies of bacteriophage P22 and HK97 indicated that the capsid formation goes through several levels of oligomeric intermediates.[47,48] In solution, the HK97 subunits form pentamers and hexamers that interconvert slowly, and the procapsid assembly process is most efficient when the two species are in a ratio of 1:5, the same ratio as found in the capsid.[48] The pairwise interfaces identified by Bahadur *et al.*[17] explain the assembly process of HK97. They found 'large' 5-fold and 6-fold interfaces that resemble the properties of a stable homodimer interface.[23] They also identified 'medium' size interfaces formed between pentamers and hexamers, and also between pairs of hexamers which are related either by icosahedral or quasi 2- or 3-fold symmetry. These pairwise interfaces can stabilise the intermediate oligomeric species and generate the capsid by the stepwise addition to hexamers, one-by-one, in pairs or in triplets, to the pentamer. In another example, Stockley *et al.*[55] have recently studied the assembly of bacteriophage MS2 by electrospray ionization–mass spectrometry. The T = 3 capsid of MS2 dissociates into symmetric dimers, some of which become asymmetric upon addition of a RNA stem-loop fragment. Symmetric and asymmetric dimers then associate into dimers of hexamers, but no pentamer is formed.

These observations are compatible with the model of Bahadur *et al.*:[17] MS2 has large 2-fold and quasi-2-fold interfaces each with BSA ~ 6,400 $\mathring{A}^2$ that build, respectively, a symmetric and an asymmetric dimer; the

R.P. Bahadur

next largest interfaces are the quasi-3-fold interfaces in the dimeric hexamers; they are of medium size with BSA ranging from 800–1,100 $Å^2$ and larger than the 5-fold interface needed to build pentamers (see Fig. 2.6). Properties of these capsid interfaces (size as well as in physical–chemical and structural composition) are comparable with those found in stable protein–protein complexes.[56]

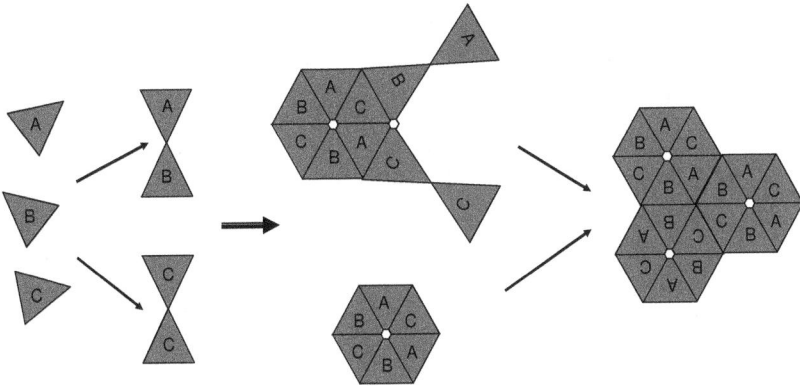

Fig. 2.6. Subunit interfaces and the mechanism of self-assembly of a T3 bacteriophage MS2 procapsid. The procapsid consists of 180 identical protein chains arranged in a a T = 3 lattice. The IAU includes three chains labelled ABC that form a cyclic hexamer. In solution, the subunits form symmetric (between C chains) and asymmetric (between chains A and B) dimers from which the assembly can proceed.[55] The two largest interfaces with a BSA of about 6,400 $Å^2$ occur between the symmetric and asymmetric dimers that stabilise the assembly. Medium-size (800–1,100 $Å^2$) interfaces may allow to first associate symmetric and asymmetric dimers into dimer hexamers, followed by higher-order assemblies and the stepwise addition of hexamers or pairs of hexamers.

The amino acid residues that constitute the interfaces are well conserved.[18] Thus, in viral capsids self-assembly must involve the stepwise formation of these different oligomeric interfaces, and the physical chemistry of proteins in solution may set a priority for the large interfaces to form first.

2.7 Conclusions and Outlook

Virus capsids are very large and symmetrical molecular assemblies that illustrate features which are likely to exist in other cellular machines. Hence, virus capsids are excellent model systems to study formation of multi-protein complexes. The subunits in the capsid are engaged in a large number of pairwise contacts simultaneously and form interfaces that have large differences in size. On the one hand, the large interfaces resemble those found in oligomeric proteins, while on the other hand the small interfaces display the properties of crystal-packing interfaces in monomeric proteins. Capsid assembly has often been compared to crystallization, but the comparison is somewhat misleading. In a protein crystal, the molecules are engaged in many interfaces but these are generally too small to stabilise a small intermediate assembly. Thus, the crystal grows by adding molecules, one-by-one, to a nucleus that is inherently unstable. In contrast, a viral capsid may assemble through a series of steps involving capsomers and oligomers of the subunits. At each step, one (or more) large interface is created yielding a species that, presumably, would be stable even if further growth was prevented. This mechanism is more efficient and more robust than nucleation–growth of protein crystals, and it has been retained by natural selection.

Acknowledgements

R. P. Bahadur is grateful to J. Janin, P. Chakrabarti, F. Rodier and M. Zacharias for valuable discussions.

References

1. Alberts B. (1998). *Cell* 92: 291–294.
2. Johnson J.E. (1996). *Proc Natl Acad Sci USA* 93: 27–33.
3. Welch M.D., Mullins R.D. (2002). *Ann Rev Cell Devl Biol* 18: 247–288.
4. Ramakrishnan V. (2002). *Cell* 108: 557–572.
5. Hartl F.U. (1996). *Nature* 381: 571–579.
6. Berman H.M., Battistuz T., Bhat T.N., Bluhm W.F., Bourne P.E., Burkhardt K., Feng Z., Gilliland G.L., Iype L., Jain S., Fagan P., Marvin J., Padilla D.,

Ravichandran V., Schneider B., Thanki N., Weissig H., Westbrook J.D., Zardecki C. (2002). *Acta Crystallogr D Biol Crystallogr* 58: 899–907.

7. Dutta S., Berman H.M. (2005). *Structure* 13: 381–388.
8. Whitby F.G., Masters E.I., Kramer L., Knowlton J.R., Yao Y., Wang C.C. (2000). *Nature* 408: 115–120.
9. Wang J., Boisvert D.C. (2003). *J Mol Biol* 327: 43–855.
10. Yamashita M.M., Almassy R.J., Janson C.A., Cascio D., Eisenberg D. (1989). *J Biol Chem* 264: 17,681–17,690.
11. Crick F.H., Watson J.D. (1956). *Nature* 177: 473–475.
12. Harrison S.C., Olson A.J., Schutt C.E., Winkler F.K., Bricogne G. (1978). *Nature* 276: 368–373.
13. Liljas L., Unge T., Jones T.A., Fridborg K., Lövgren S., Skoglund U., Strandberg B. (1982). *J Mol Biol* 159: 93–108.
14. Caspar D.L., Klug A. (1962). *Cold Spring Harbor Symp Quant Biol* 27: 1–24.
15. Rossmann M.G., Arnold E., Erickson J.W., Frankenberger E.A., Griffith J.P., Hecht H.J., Johnson J.E., Kamer G., Luo M., Mosser A.G., *et al.* (1985). *Nature* 317: 145–153.
16. Rossmann M.G., Johnson J.E. (1989). *Annu Rev Biochem* 58: 533–573.
17. Bahadur R.P., Rodier F., Janin J. (2007). *J Mol Biol* 367: 574–590.
18. Bahadur R.P., Janin J. (2008). *Proteins* 71: 407–414.
19. Grimes J.M., Burroughs J.N., Gouet P., Diprose J.M., Malby R., Ziéntara S., Mertens P.P., Stuart D.I. (1998). *Nature* 395: 470–478.
20. Dokland T., Bernal R.A., Burch A., Pletnev S., Fane B.A., Rossmann M.G. (1999). *J Mol Biol* 288: 595–608.
21. Bahadur R.P., Chakrabarti P., Rodier F., Janin J. (2004). *J Mol Biol* 336: 943–955.
22. Chakrabarti P., Janin J. (2002). *Proteins* 47: 334–343.
23. Bahadur R.P., Chakrabarti P., Rodier F., Janin J. (2003). *Proteins* 53: 708–719.
24. Jones S., Thornton J.M. (1997). *J Mol Biol* 272: 121–132.
25. Pal A., Chakrabarti P., Bahadur R., Rodier F., Janin J. (2007). *J Biosci* 32: 101–111.
26. Lichtarge O., Bourne H., Cohen F. (1996). *J Mol Biol* 257: 342–358.
27. Valdar W.S., Thornton J.M. (2001). *Proteins* 42: 108–124.
28. Kortemme T., Baker D. (2002). *Proc Natl Acad Sci USA* 99: 14,116–14,121.
29. Ma B., Elkayam T., Wolfson H., Nussinov R. (2003). *Proc Natl Acad Sci USA* 100: 5,772–5,777.
30. Mintseris J., Weng Z. (2005). *Proc Natl Acad Sci USA* 102: 10,930–10,935.
31. Wells J.A. (1991). *Methods Enzymol* 202: 390–411.
32. Clackson T., Wells J.A. (1995). *Science* 267: 383–386.
33. Bogan A.A., Thorn K.S. (1998). *J Mol Biol* 280: 1–9.
34. Elcock A., McCammon J. (2001). *Proc Natl Acad Sci USA* 98: 2,990–2,994.

35. Guharoy M., Chakrabarti P. (2005). *Proc Natl Acad Sci USA* 102: 15,447–15,452.
36. Shannon C.E. (July and October 1948). A Mathematical Theory of Communication. *Bell System Technical Journal* 27: 379–423 and 623–656.
37. Sander C., Schneider R. (1993). *Nucleic Acids Res* 21: 3,105–3109.
38. Hadfield A.T., Lee W., Zhao R., Oliveira M.A., Minor I., Rueckert R.R., Rossmann M.G. (1997). *Structure* 5: 427–441.
39. Lo Conte L., Chothia C., Janin J. (1999). *J Mol Biol* 285: 2,177–2,198.
40. Janin J., Rodier F. (1995). Protein–Protein Interaction at Crystal Contacts. *Proteins* 23: 580–587.
41. Bahadur R.P., Zacharias M. (2008). *Cell Mol Life Sci* 65: 1,059–1,072.
42. Liljas L. (1999). Virus assembly. *Curr Op Struct Biol* 9: 129–134.
43. Dokland T. (2000). *Structure* 8: R157–R162.
44. Steven A.C., Heymann J.B., Cheng N., Trus B.L., Conway J.F. (2005). *Curr Opin Struct Biol* 15: 227–236.
45. Leiman P.G., Kanamaru S., Mesyanzhinov V.V., Arisaka F., Rossmann M.G. (2003). *Cell Mol Life Sci* 60: 2,356–2,370.
46. Caspar D.L. (1980). *Biophys J* 10: 103–135.
47. Prevelige P.E. Jr, Thomas D., King J. (1993). *Biophys J* 64: 824–835.
48. Xie Z., Hendrix R.W. (1995). *J Mol Biol* 253: 74–85.
49. Johnson J.E., Speir J.A. (1997). *J Mol Biol* 269: 665–675.
50. Steven A.C., Trus B.L., Booy F.P., Cheng N., Zlotnick A., Caston J.R., Conway J.F. (1997). *FASEB J* 11: 733–742.
51. Zlotnick A. (2005). *J Mol Recognit* 18: 479–490.
52. Twarock R. (2006). *Phil Trans R Soc* A364: 3,357–3,373.
53. Nguyen H.D., Reddy V.S., Brooks C.L. 3rd. (2007). *Nano Lett* 7: 338–344.
54. Reddy V.S., Giesing H.A., Morton R.T., Kumar A., Post C.B., Brooks C.L. 3rd, Johnson J.E. (1998). *Biophys J* 74: 546–558.
55. Stockley P.G., Rolfsson O., Thompson G.S., Basnak G., Francese S., Stonehouse N.J., Homans S.W., Ashcroft A.E. (2007). *J Mol Biol* 369: 541–552.
56. Janin J., Bahadur R.P., Chakrabarti P. (2008). *Quart Rev Biophysics* 41: 1–48.

CHAPTER 3

Energetics of Protein–Protein Interactions

Ilian Jelesarov

*Department of Biochemistry, University of Zurich,
Winterthurerstr. 190, CH-8057 Zurich, Switzerland
E-mail: iljel@bioc.uzh.ch*

Proteins have evolved as versatile components of the molecular organisation of life. Formation of protein–protein complexes is indispensable in maintaining the organisation and integration of practically all biological processes. Not surprisingly, therefore, the identification and structural characterization of protein–protein interactions is a major line of research nowadays. This chapter focuses on the energetic aspects of the recognition process. Firstly, the thermodynamic formalism describing formation of binary complexes is outlined. A brief discussion of the merits and limitations of experimental methods facilitating quantification of binding reactions follows. The emphasis is on isothermal titration calorimetry. This technique provides direct and precise estimates of the magnitude of all relevant thermodynamic functions in an extended temperature range. The third part of the chapter is devoted to the obvious and widely anticipated, yet notoriously challenging problem: is it possible to predict binding affinity and to understand the magnitude of thermodynamic parameters from a structural perspective?

3.1 Introduction

Maintaining the molecular organisation of living matter is impossible without formation of highly-specific protein–protein complexes and protein networks. Protein–protein interactions are involved in the catalysis, coordination and structuring of all vital processes ranging from upkeep and realization of the genetic information to transformation and utilization of chemical and thermal energy, to signal transduction to integration and regulation of the cell as a whole. Not surprisingly, the identification and characterization of protein–protein interactions is an important line of research in modern biochemistry.

Present-day large-scale efforts aiming at establishing system-oriented approaches will possibly provide a global understanding of biological processes. However, many aspects of the intimate molecular mechanisms involved in biological function remain obscure. Protein–protein recognition is a typical example. Notwithstanding the serious progress that has been achieved over the past three decades, details about the mechanistic, energetic and kinetic principles of the binding process remain vaguely understood. The reasons are manifold. From a structural point of view, only recently the fast progress of molecular biology and structural methods facilitated the accumulation of high-resolution structures of protein complexes exhibiting architecture different from the architecture of the so far paradigmatic antibody-antigen and protease-inhibitor complexes. Similarly, the wider use of high-sensitivity biophysical instrumentation, especially titration calorimetry, has started to provide insights into the energetic principles of protein–protein recognition. However, the unification of the structural and energetic perspectives on the binding process is still a serious challenge in protein biophysics. Three main conceptual difficulties can be outlined:

1. 'Classical' physical–chemical concepts cannot be easily adapted to proteins given their size, structural inhomogeneity and ensemble nature.

2. Binding involves structural re-arrangements of the interacting partners, which cannot be (or only occasionally are) completely captured by structural methods.

3. Measured thermodynamic parameters mirror the energetic behaviour of the entire system under study, including, next to the energy of bonds accumulated at the binding interface, also re-arrangements of hydration shells and ions, and energetic expenditures caused by structural changes taking place upon binding. Big steps have been made towards energetic description of protein–protein complexes. Yet, admittedly, there is a long way ahead until the principles governing the strength and specificity of protein–protein recognition are rigorously understood, and

it is possible to rationally design or modulate protein–protein interactions for the purposes of biomedical and biotechnological applications.

It is impossible to cover all aspects of the energetics of protein–protein recognition in a single chapter. I will start with describing the thermodynamic formalism required for quantitative treatment of binding data. Next, a brief overview of the present-day methods measuring energetic parameters will be given. Finally, I will discuss some ideas about possible ways to predict binding affinity and to search for the yet elusive links between the energetics of protein–protein complexes and their structure.

3.2 Thermodynamic Formalism Describing the Energetics of Binding Reactions

There are many types of protein–protein binding reactions: association of two different proteins, self-association of a protein to form a dimer or higher-order oligomer, formation of oligomers from different types of subunits, etc. In fact, the interactions between independently folding domains of a single polypeptide chain can also be regarded as protein–protein binding. If one protein binds two or more partners, the individual binding events can be the same (*a priori* identical binding sites) or can differ (*a priori* non-identical binding sites). Furthermore, multiple binding sites can be independent from each other (no cooperativity), or occupation of one site can influence binding to another site (cooperativity). It is far beyond the scope of this section to treat all possible cases. In the following I delineate the relevant formalism for the most commonly occurring cases: formation of a 1:1 heterologous complex and formation of a self-associating dimer (1:1 homologous association). The interested reader can find extensive description of more complicated binding models and general treatment of binding phenomena in Refs 1 and 2.

A further level of discrimination is necessary. A survey of the literature identifies two general types of studies on the energetics of protein–protein interactions. Firstly, the prime interest is determination of the binding affinity. Usually, such studies are devoted to the mapping of binding sites and clarification of binding modes by measuring the affinity (dissociation constants) of wild type complexes and designed variants thereof. Secondly, the focus is on in-depth thermodynamic description of a particular complex, or of closely related complexes, with the idea to understand the enthalpic and entropic contributions to binding from a structural point of view. For the reader's easier navigation, I therefore split the following discussion into two sub-sections.

3.2.1 Determination of the Binding Affinity

3.2.1.1 Heterologous Binding

Experimental quantification of the affinity between two proteins, i.e. the strength with which they interact with each other, requires the design of what is commonly named a titration binding experiment. Typically, the two interacting molecules A and B are mixed at different molar ratios in a series of experiments, in which all other conditions are kept identical. Let us assume that increasing amounts of B are added to a fixed amount of A^a. The system equilibrates between the associated (bound) state AB and the dissociated (unbound) state $A + B$ according to:

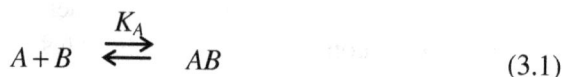

$$A + B \overset{K_A}{\rightleftharpoons} AB \tag{3.1}$$

The fractional population of the two states at equilibrium depends on the total concentrations of A and B and the magnitude of the equilibrium binding constant K_A, which is defined as (units of M^{-1}):

$$K_A = \frac{[AB]}{[A][B]} \tag{3.2a}$$

---

[a] Note that in the 1:1 binding mode the choice of the component that is held at fixed concentration is arbitrary. In most cases, the decision is guided by the solubility and the tendency for aggregation at high concentrations. Often, the production costs are also an important factor.

The square brackets indicate the *free* concentration of species at equilibrium. Biochemists often use the reciprocal value of K_A, which is called *dissociation constant* (units of M) and is defined as:

$$K_D = K_A^{-1} = \frac{[A][B]}{[AB]} \qquad (3.2b)$$

The normalised extent of reaction, typically called degree of saturation, Y ($0 \leq Y \leq 1$), represents the fractional occupancy of binding sites on A, and is related to the concentration of species and K_A by:

$$Y = \frac{[AB]}{[A]_{tot}} = \frac{[AB]}{[A]+[AB]} = \frac{K_A[B]}{1+K_A[B]} = \frac{[B]}{[B]+K_D} \qquad (3.3)$$

The function represented by Eq. 3.3 is called *binding isotherm* since the titration experiment is performed at constant temperature.[b] There are alternative ways to calculate K_A according to Eq. 3.3. Historically, the most popular way to plot binding data has been the Scatchard-Rosenthal transformation (see Fig. 3.1):[3,4]

$$\frac{[AB]}{[B]} = K_A[A]_{tot} - K_A[AB] \qquad (3.4)$$

The slope of this plot numerically equals $- K_A$. According to Klotz[5] the plot of Y versus log[B] is symmetrically sigmoidal, the inflection point (Y = 0.5) corresponding to $\log(1/K_A)$ (i.e. $\log K_D$; Fig. 3.1). Both methods suffer problems. In the Scatchard equation (Eq. 3.4) the independent variable ([AB]) and the dependent variable ([AB]/[B]) are statistically coupled, preventing calculation of error margins.

---

[b] Biochemical experiments are usually performed in open systems at constant atmospheric pressure. Otherwise, the pressure must also be controlled. K_A depends also on the ionic strength and can be altered if solvent components interfere with the binding process. Strictly speaking, calculation of K_A requires knowledge of the activity coefficients. However, in dilute (typically nanomolar to micromolar) solutions the activity coefficients are assumed to be 1 and the solution is treated as an ideal solution. Deviation from ideality is expected with highly concentrated protein solutions.

The Klotz plot requires the concentration of B to be varied in a very broad range in order to obtain data in the plateau regions. This is rarely possible with proteinous ligands, since in the low concentration regime the sensitivity of the assay is challenged, while in the high concentration regime problems with aggregation can appear. Eq. 3.3 describes a hyperbolic dependence of Y on $[B]$ (Fig. 3.1). $K_D = 1/K_A$ corresponds numerically to $[B]$ at half-saturation (Y = 0.5), and can be easily calculated by non-linear optimization.[c] All these methods, in fact, are variations of a common theme: the equilibrium concentrations of one of the participating species A, B and AB must be known. For example, if $[AB]$ is known, from the law of mass conservation $[B] = [B]_{tot} - [AB]$. If $[A]$ is known, $[AB] = [A]_{tot} - [A]$, etc. Unfortunately, physical separation of the free proteins and their complex without disturbing the equilibrium is typically impossible.

Often it is possible to indirectly determine the degree of saturation. (The methods are discussed in Section 3.3.) Plots of Y vs $[B]_{tot}$ are hyperbolic. As illustrated in Fig. 3.1, at half-saturation $[B]$ no longer corresponds to K_D in this case. Since $[A] = [A]_{tot} - [AB]$ and $[B] = [B]_{tot} - [AB]$ substitution in Eqs 3.2a or 3.2b and re-arrangement results in the following quadratic equation:

$$[AB]^2 - \left([A]_{tot} - [B]_{tot} - \frac{1}{K_A}\right)[AB] + [A]_{tot}[B]_{tot} = 0 \qquad (3.5)$$

Denoting the only physically meaningful root of Eq. 3.5 as R, for each combination of $[A]_{tot}$ and $[B]_{tot}$ the degree of saturation (Y = R/$[A]_{tot}$) is a function of $[B]_{tot}$.[d] Non-linear regression analysis of Y = f($[B]_{tot}$) yields the value of K_A. The binding isotherms depicted in Fig. 3.1 illustrate that the choice of concentrations is a very important aspect of the binding experiment.

---

[c] This is probably why the use of K_D instead of K_A is more intuitive. K_D is measured in units of concentration, and its numerical value helps to roughly estimate the position of the equilibrium when the total concentrations of the interacting proteins are known.

[d] Remember that arbitrarily we have defined A as the component whose concentration is kept constant throughout the titration experiment, while the concentration of B varies.

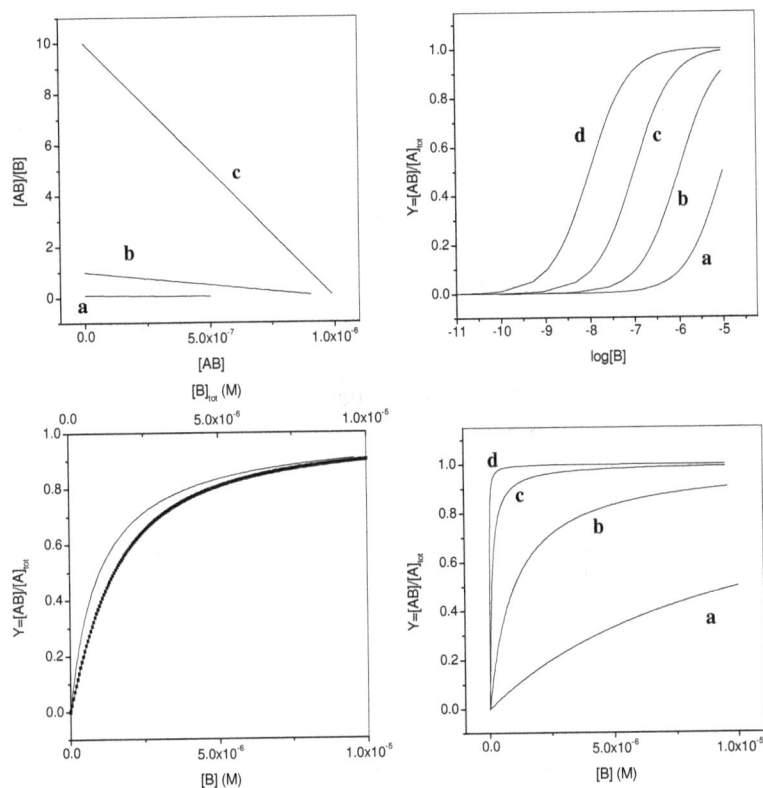

Fig. 3.1. Simulated binding plots. Molecule A ($[A]_{tot}$ = 1 µM) is titrated with molecule B ($[B]_{tot}$ = 0.1 nM to 10 µM). The binding constant, K_A, is 1×10^5 M^{-1} (a), 1×10^6 M^{-1} (b), 1×10^7 M^{-1} (c), and 1×10^8 M^{-1} (d). Panel A, Scatchard plots (Eq. 3.4). The y-axis intercept, the x-axis intercept and the slope correspond to $K_A[A]_{tot}$, $[A]_{tot}$ and $-K_A$, respectively. Panel B, Klotz plot. The logarithm of $[B]$ (*free B*) where Y = 0.5 equals to $1/K_A$. Panel C, degree of saturation as function of the *free* concentration of B (thin line; bottom x-axis) or the *total* concentration of B (thick line; top x-axis; Eq. 3.5). $K_D = 1/K_A$ equals $[B]$ at half saturation (Y = 0.5; Eq. 3.3). If $[B]_{tot}$ is the independent variable, K_A is calculated by non-linear optimisation according to Eq. 3.5. Panel D, shape of the binding isotherm as function of K_A at fixed $[A]_{tot}$. If $[A]_{tot} >> 1/K_A$, there is virtually no free B before saturation and no free A after saturation. In such case, the apparent K_A is only a lower estimate. If $[A]_{tot} << 1/K_A$ the degree of saturation increases little upon increasing the concentration of B. Panels A, B and D illustrate that high concentrations of B are required to obtain a complete binding isotherm if the affinity is low.

A smooth binding isotherm results only if sizeable amounts of free interacting species A and B, and their complex AB are populated at molar ratios between 0.5 and 1.5. If the total concentration of A is much higher than K_D almost all ligand binds to the receptor below the equivalence point (molar ratio 1:1). Above the equivalence point, there is almost no binding, because there are virtually no unoccupied binding sites. In other words, there is little free ligand below the equivalence point and little free receptor above. In the opposite case, $[A]_{tot} \ll K_D$, the concentration of the complex increases very slowly with the progress of titration (near to linear change of Y; Fig. 3.1). Full saturation is reached only at very high molar excess of ligand. These properties of the binding isotherm set practical limitations. If binding is very strong, very low concentrations of A are required, and the success of the experiment will depend on the physical limits of detection in the particular assay. If binding is weak, high concentrations of A and B are necessary to reach sufficient saturation and problems with low solubility and protein aggregation can occur.

3.2.1.2 Homologous Binding (Self-association)

Proteins are often biologically active as dimers or higher-order oligomer. Commonly, homodimers are formed according to the following reaction scheme:

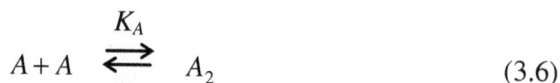

$$A + A \; \underset{}{\overset{K_A}{\rightleftarrows}} \; A_2 \qquad\qquad (3.6)$$

The equilibrium constant (units of M^{-1}) and the dissociation constant (units of M) are defined as, respectively:

$$K_A = \frac{[A_2]}{[A]^2} \qquad\qquad (3.7a)$$

$$K_D = \frac{[A]^2}{[A_2]} \qquad\qquad (3.7b)$$

In this case, the experiment is performed by variation of the total concentration of the protein, $[A]_{tot}$, as the independent variable. Again, determination of $[A]$ and $[A_2]$ is very difficult. Indirect measurements allow to determine the *fraction dimer, f_D* :

$$f_D = 1 - \frac{\sqrt{1 + 8K_A[A]_{tot}} - 1}{4K_A[A]_{tot}} \tag{3.8}$$

K_A is calculated by non-linear regression analysis of the data describing the changes of f_D as function of $[A]_{tot}$ (Fig. 3.2). As shown in the Figure, depending on the strength of association (K_A), $[A]_{tot}$ must be varied in a broad range in order to obtain data describing a portion of the isotherm sufficiently large to allow precise calculation of K_A.

3.2.2 *Free Energy, Enthalpy and Entropy of Association*

Once K_A is known, the free energy of association is calculated from:

$$\Delta G = -RT \ln K_A \tag{3.9}$$

R is the gas constant (8.314 J K^{-1} mol^{-1}) and T is the absolute temperature. Although ΔG in this equation is not superscripted by the usual superscript $°$ it is important to keep in mind that this is the *standard state free energy*.[e] $\Delta G = G_{AB} - G_{A+B}$ is the quantitative measure for the affinity between A and B. The larger the ΔG is, the larger the equilibrium population of the state that has the lowest energy. ΔG has an enthalpic and an entropic component according to:

$$\Delta G = \Delta H - T\Delta S \tag{3.10}$$

ΔH and $T\Delta S$, thus also ΔG, are sensitive to many factors: temperature, pressure, chemical composition of the system (concentration of protons, salts and other low-molecular substances).

---

[e] Equations 3.2 and 3.7 implicitly assume that the standard state is defined here as 1 mol L^{-1}. The choice of reference state will influence the numerical values of ΔG and the reaction entropy (ΔS), but not that of the reaction enthalpy (ΔH).

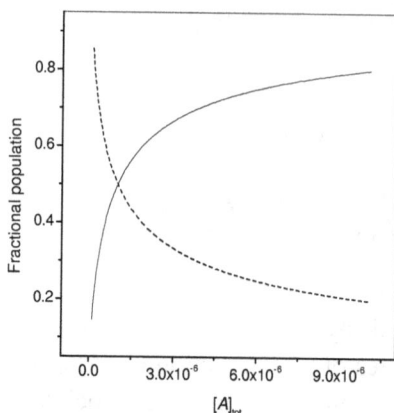

Fig. 3.2. Concentration dependence of the fractional populations of the dimeric state (continuous line) and the monomeric state (dashed line) in a self-associating dimeric system. The curves were simulated according to Eq. 3.8 with $K_A = 1 \times 10^6 \ M^{-1}$ and $[A]_{tot}$ varying between 0.1 and 10 μM.

Here, only the effect of the temperature is considered. The temperature dependence of ΔG is described by the Gibbs-Helmholtz equation:

$$\Delta G(T) = -RT \ln K_A = \Delta H(T_R) + \int_{T_R}^{T} \Delta C_p dT - T\left[\Delta S(T_R) + \int_{T_R}^{T} \Delta C_p d \ln T \right]$$

(3.11a)

T_R is an appropriate reference temperature used for integration. $\Delta H(T_R)$ and $\Delta S(T_R)$ are the enthalpy change and the entropy change at T_R, respectively. In the integrated form:

$$\Delta G(T) = \Delta H(T_R) + \Delta C_p (T - T_R) - T\left[\Delta S(T_R) + \Delta C_p \ln\frac{T}{T_R} \right]$$ (3.11b)

Since Eq. 3.10 is valid at any temperature, another popular notation of the Gibbs-Helmholtz equation is:

$$\Delta G(T) = \Delta G(T_R) + \Delta C_p \left(T - T_R - T \ln \frac{T}{T_R} \right) \qquad (3.11c)$$

The term ΔC_p in the above Eqs is the isobaric heat capacity change, which formally describes the temperature dependence of ΔH and ΔS according to:

$$\Delta C_p = \frac{d\Delta H}{dT} = T \frac{d\Delta S}{dT} \qquad (3.12)$$

For a full thermodynamic description of an association reaction the temperature dependence of ΔG, ΔH, and ΔS must be known.[f]

3.2.3 *Determination of Energetic Changes*

There are two practical approaches to access the energetic profile of the association reaction. First, let us consider the so-called van't Hoff approach. Combination of Eqs 3.9–3.11 yields the integrated form of the van't Hoff equation, which describes the temperature dependence of the equilibrium constant, and hence the temperature dependence of ΔG:

$$\frac{K_A(T)}{K_A(T_R)} = \exp\left[-\frac{\Delta H(T_R)}{R}\left(\frac{1}{T} - \frac{1}{T_R} \right) + \frac{\Delta C_p}{R}\left(\ln\frac{T}{T_R} + \frac{T_R}{T} - 1 \right) \right] \qquad (3.13)$$

T_R is arbitrarily chosen, typically within the temperature range of the measurements, so that $K_A(T_R)$ is known from a direct experiment. The binding constant, K_A is measured in a series of binding experiments performed at different temperatures and the data are plotted according to Eq. 3.13 in the form $\dfrac{K_A(T)}{K_A(T_R)} = f(1/T)$, or in a more popular way as

---

[f] ΔC_p itself is also a temperature-dependent parameter. However, its temperature variation between 0 and, say, 80 °C is small and can be neglected in most cases.

$ln\dfrac{K_A(T)}{K_A(T_R)}$ = f(1/T). Both plots are curvilinear with maximum at the temperature (T_H) where $\Delta H = 0$, meaning that binding is weaker at lower and higher temperatures. The data can be subjected to non-linear regression analysis to find the best fit in terms of ΔH and ΔC_p as fitting parameters in Eq. 3.13 (or any equivalent algebraic transformation). Theoretically, the slope of the tangent to both plots at any 1/T equals $-\Delta H(T)/R$. If it were possible to construct a secondary plot of $\Delta H(T)-$ vs $-T$, then ΔC_p could be calculated from Eq. 3.12. Having on hand high-precision K_A data as function of the temperature, $\Delta H(T)$ could be calculated from (see Eq. 3.13):

$$\Delta H\left(T\right) = RT^2 \frac{d\ln K_A}{dT} \qquad (3.14)$$

Unfortunately, binding experiments are burdened with many problems and typically yield estimates of K_A being (in favourable cases) within 10–15% of the statistical mean at a single temperature, thus seriously increasing the errors of ΔH determination (Fig. 3.3). Furthermore, collection of data in a broad temperature range (as to precisely describe the curvature of plots according to Eq. 3.13) is not possible either for technical reasons, or because the proteins undergo temperature-induced conformational changes, or simply because T_H is far away from the experimentally accessible region. The problem gets even more severe if one attempts to calculate ΔC_p and ΔS by taking temperature derivatives of ΔG:

$$\Delta C_p = -T\frac{\partial^2 \Delta G}{\partial T^2} \qquad (3.15)$$

$$\Delta S = -\frac{\partial \Delta G}{\partial T} \qquad (3.16)$$

Therefore, one should treat ΔH, ΔS and ΔC_p values obtained by van't Hoff analysis with caution, especially in long extrapolations outside the temperature interval of the actual measurements. Fortunately, the rapid advance in calorimetric instrumentation has provided a more direct and more precise determination of binding parameters. The principle and advantages of mixing calorimetry will be discussed in Section 3.5.

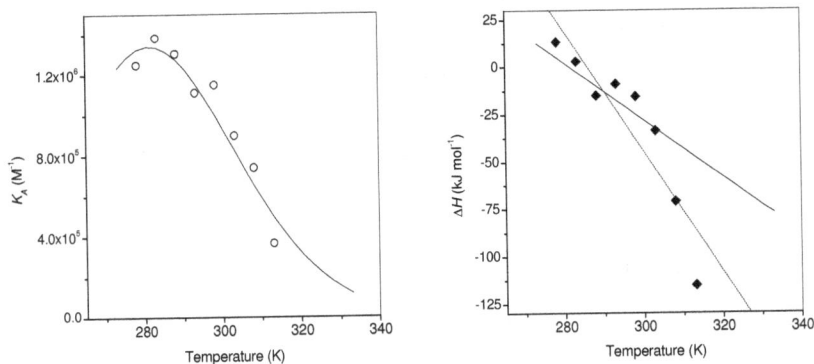

Fig. 3.3. Limitations of the van't Hoff analysis of binding data. The continuous line in Panel A represents the temperature dependence of K_A of a hypothetical binding reaction. The simulation was done according to Eq. 3.13 with the following parameters: $T_R = 25$ °C, $K_A(T_R) = 1 \times 10^6$ M^{-1}, $\Delta H(T_R) = -25$ kJ mol^{-1}, $\Delta C_p = -1.5$ kJ K^{-1} mol^{-1}. The entropy change is $\Delta S = 31$ J K^{-1} mol^{-1}. The symbols are simulated experimental data assuming that the random experimental error in K_A determination is modest: 10–20%. Eq. 3.14 was used to calculate the temperature dependence of ΔH from the temperature derivative of K_A (Panel B). The continuous line is the 'ideal' function $\Delta H(T) = \Delta H(T_R) + \Delta C_p(T-T_R)$. At 25 °C the van't Hoff analysis of the simulated 'experimental' data predicts $\Delta H = -16$ kJ mol^{-1} and $\Delta S = 62$ J K^{-1} mol^{-1} (the latter number was derived by Eq. 3.16). The slope of the dashed line equals the van't Hoff heat capacity change: $\Delta C_p = -3.2$ kJ K^{-1} mol^{-1}. All parameters are very different from the 'true' ones. Note that often K_A is determined with error as high as 50%. Moreover, temperature-dependent systematic errors can appear. In such cases the imprecision of the van't Hoff analysis will be substantially higher.

3.3 Experimental Methods to Measure the Energetics of Protein–Protein Association

A large variety of methods to quantitatively characterise the thermodynamics of protein–protein binding reactions has been developed over the years.[6] Because different assays are based on different physical principles, differ in sensitivity and cover different ranges of binding affinities amenable to determination, there is no universal method available. The selection of a particular experimental setup is dictated by the properties of the interacting proteins (like for example solubility, stability, presence of physical 'probes' to be followed, and in many cases by the availability of sufficient amounts of material). It is out of the scope of this section to provide a comprehensive compendium of available techniques and discuss at length their advantages and disadvantages. Merely, I would like to point to the diversity of the modern methodology in studying protein–protein interactions in quantitative terms.

3.3.1 *Methods utilising Physical Separation of Species*

It is clear from Eqs 3.2–3.4 that K_A (or K_D) can be readily calculated if the equilibrium concentrations [A], [B] and [AB] are known. This means that at equilibrium, A, B and AB have to be separated and their concentrations determined. Classical methods to characterise protein-small ligand binding such as equilibrium dialysis and Hummel-Dreyer size-exclusion chromatography are quite difficult to adapt to studies of protein–protein interactions, since the two proteins must significantly differ in size and suitable dialysis membranes, and chromatographic media are required. In principle it is possible to attempt separation of the reaction mixture either by precipitation of the complex at conditions where the components are still soluble, or else to recruit one species by sorption on a material, and wash out the non-sorbed molecules. For example, immunoprecipitation based on epitope–paratope interactions in a microtiter plate format can sometimes provide the solution. However, such 'direct separation' methods suffer problems with protein denaturation, non-specific binding/sorption and involve mandatory washing steps. Consequently, the equilibrium is disturbed and, therefore, the obtained binding constants should be treated only as crude estimates.

Electrophoretic techniques have potential in exploring the strength of protein–protein complexes. The requirement is that the free components and the complex possess different electrophoretic mobilities. The electrophoretic mobility shift assay (EMSA; sometimes called gel retardation assay) has found a broad application in studies of protein–DNA binding and can, in principle, be easily adapted to protein–protein interactions. Proteins A and B are pre-incubated in different molar ratios to allow equilibration. The samples are then transferred to a native gel and subjected to separation. In the electric field, A, B and AB will migrate to different positions. Visualization can be done by measuring radioactivity (although the production of proteins with high specific radioactivity may not be trivial) or fluorescence (native or caused by artificially introduced fluorescent markers; fusion to GFP or YFP is also an alternative). The intensity of the bands observed in separation of mixtures with differing A:B ratios can be related to the concentrations at equilibrium (after proper calibration). The caveats are that (i) this procedure will work quantitatively for slowly dissociating complexes, and (ii) interactions with the gel matrix and 'caging' can alter the apparent binding affinity. The latter problem is alleviated in capillary electrophoresis experiments, where the concentration of species can be determined by integration of the peaks moving along the capillary.

With the appearance of high-sensitivity instrumentation, analytical ultracentrifugation (AUC) has become a powerful tool to measure protein–protein binding constants for self-associating or heterologously-associating systems.[7] Let us consider for illustration the simple case of A:B (1:1) binding. In the sedimentation velocity regime, the weight-average sedimentation coefficient, s_w, is obtained by integration of the apparent sedimentation coefficient distribution. Since the individual sedimentation coefficients, s_A and s_B, can be measured separately, and considering the low of mass action K_A is calculated by regression analysis of the following equations:

$$s_w(A_{tot}, B_{tot}) = \frac{\varepsilon_A[A]s_A + \varepsilon_B[B]s_B + (\varepsilon_A + \varepsilon_B)K_A[A][B]s_{AB}}{\varepsilon_A A_{tot} + \varepsilon_B B_{tot}} \quad (3.17a)$$

$$A_{tot} = [A] + K_A[A][B] \quad and \quad B_{tot} = [B] + K_A[A][B] \quad (3.17b,c)$$

Where ε_i are the extinction coefficients and the square brackets indicate the equilibrium concentrations of A and B. The total concentration distribution observed for the A:B complex is:

$$a(r) = \varepsilon_A d [A]_{r0} e^{\Phi_A} + \varepsilon_B d [B]_{r0} e^{\Phi_B} + K_A (\varepsilon_A + \varepsilon_B) d [A]_{r0} [B]_{r0} e^{\Phi_{AB}}$$

(3.18)

The indices $r0$ denote the concentrations at the reference radius; d is the optical path length; the exponential factors Φ_i are functions of the buoyant masses, temperature and rotor speed.

In principle, electrospray ionization mass spectrometry (ESI–MS) is a fast and reliable method to study protein–protein interactions, without the need for derivatization and immobilization.[8] The relative ion abundances of the complexes and those of the unbound monomers have been shown to reflect the relative solution-phase concentrations of the respective species, thus allowing accurate determination of binding affinities. Clearly, the bound and unbound species must possess different mass-to-charge ratios, the resolution must be high enough to avoid overlaps and all charged species must be identified in the spectrum. Relatively few studies reporting binding constants that are in agreement with other methods have been published so far. It appears that ESI–MS is restricted to studies of high-affinity complexes, which remain sufficiently stable in the gas-phase. Alternatively, methods should be developed to introduce corrections for gas-phase dissociation. However, in cases of strong binding, reliable data can be obtained only at concentrations being close to the limit of detection.

All methods discussed so far (with the exception of AUC) cannot be easily performed in a broad temperature range, as to provide data to calculate not only K_A and ΔG but also other thermodynamic parameters according to Eqs 3.11–3.16.

3.3.2 *Indirect Spectroscopic Methods*

An indirect but versatile general approach to experimental determination of K_A is to monitor the value of a suitable spectroscopic signal σ that changes upon binding. Typically, the spectroscopic probe is located on

species *A*, whose concentration is kept constant in a titration experiment.[g]
Assume that the signal intensity is σ_0 in the unliganded *A*. If *B* is added
to *A*, the signal intensity changes to σ_i, depending on the molar ratio of *A*
and *B*. At full saturation ($[B]_{tot} \gg [A]_{tot}$, so that $[AB] \approx [A]_{tot}$), the
magnitude of the signal is $\sigma_i \approx \sigma_{sat}$. The normalised extent of reaction, Y
($0 \leq Y \leq 1$) is related to the concentration of species and K_A by (see also
Eq. 3.3):

$$Y = \frac{\Delta\sigma_i}{\Delta\sigma_{sat}} = \frac{\sigma_i - \sigma_0}{\Delta\sigma_{sat} - \sigma_0} = \frac{[AB]}{[A]+[AB]} = \frac{K_A[B]}{1+K_A[B]} \qquad (3.19a)$$

or

$$K_A = \frac{Y}{(1-Y)[B]} \qquad (3.19b)$$

The combined Eqs 3.5 and 3.19 can be fitted to the data ($\Delta\sigma_i =$
$f[B]_{tot}$; hyperbolic dependence; see Figs 3.1 and 3.4) to optimise the
values of $\Delta\sigma_{max}$ and K_A. In principle, if full saturation is reached, $\Delta\sigma_{sar}$
can be obtained directly from the data. However, if binding is weak, this
might be difficult (see Section 3.2.1.1).

Considering the physical nature of the signal σ, there are virtually no
limitations other than the sensitivity in the concentration range proper for
experimentation. Typically, UV absorbance between 200 and 300 nm
(caused by the peptide group and aromatic amino acid side chains)
changes little upon binding. Sometimes larger, conjugated double-bond
moieties being co-factors or prosthetic groups (heme, FAD, FMN, NAD,
etc.) are present near the binding site and their absorption 'feels' the
incoming ligand. CD spectroscopy is also somewhat limited in studies of
protein–protein binding reactions. It is uncommon for interacting
proteins to undergo large-scale secondary-structure conformational
transitions upon binding, so as to be monitored in the far-UV region.
Neither are significant changes in the environment of other optically
active groups ubiquitous, as to precisely monitor the binding process in

---

[g] The spectroscopic probe can be located also on *B*. However, in this case, and if the
experimental setup cannot be reversed (titration of *A* to fixed amounts of *B*), proper
controls are mandatory.

the near-UV and visible regions. Generally, ellipticity changes of proteins are of low intensity and CD measurements will be restricted to relatively weak complexes.

In many cases, changes of fluorescence as function of the partial saturation provides a sensitive probe to calculate the binding affinity. Binding may cause perturbation of the intrinsic fluorescence of chromophores located at or near the binding site. Typically, changes in tryptophan (or tyrosine) fluorescence are exploited. In a straightforward fashion, the experiment can be done by monitoring the changes of the fluorescence intensity (quantum yield) or the wavelength shift of the emission maximum, the reason being either the influence of a 'quencher' on the incoming binding partner, or changes in the hydration of the chromophore, or both. An example of a binding experiment followed by fluorescence is shown in Fig. 3.4.

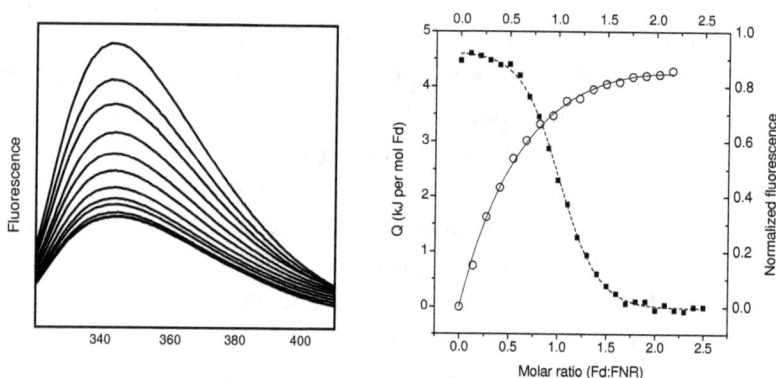

Fig. 3.4. An example of K_A determination by spectroscopy. Binding of spinach ferredoxin (Fd) to ferredoxin:NADP$^+$ oxidoreductase (FNR) was measured at 27 °C. Panel A, addition of Fd to FNR cause decrease of the Trp fluorescence of FNR in a concentration dependent manner. Fluorescent spectra recorded at Fd:FNR ratios ranging from 0 to ~ 2.5 are shown (top to bottom). Panel B, the normalised changes in fluorescence (see Eq. 3.19a) are plotted against the Fd:FNR molar ratio (open symbols; right y-axis). The associated continuous line visualises the results of non-linear regression analysis according to Eq. 3.3. The best fit was obtained with $K_A = 4.2\times10^6$ M^{-1}. For comparison, the binding isotherm measured by ITC (see Section 3) is shown (closed symbols and dashed line; left y-axis). From ITC data, $K_A = 6.5\times10^6$ M^{-1}. Data from Jelesarov and Bosshard (1994), *Biochemistry* 33: 13,321–13,328.

64   *I. Jelesarov*

In recent years, fluorescence anisotropy (or fluorescence polarization) has found a wide application in quantitative studies of protein–protein binding.[9] The underlying physical principle is the following. A linearly polarised laser excitation source preferentially excites fluorescent target molecules with transition moments aligned parallel to the incident polarization vector. The resultant fluorescence is directed into two channels that measure the intensity of the fluorescence polarised both parallel and perpendicular to that of the excitation beam. The observed changes in anisotropy/polarization at different *A:B* ratios reflects the changes in the degree of saturation.[h] The principal advantage of fluorescence-based techniques over other spectroscopic methods is their sensitivity, allowing characterization of strong binding at low concentrations.

NMR monitors the absorption of electromagnetic energy by magnetically active nuclei that are oriented in a strong magnetic field. The characteristic absorption frequencies of nuclei of the same type (same element) are strongly influenced by the chemical environment. Because of this property, the presence of a ligand will change the relative positions of the spectral lines of nuclei located at the binding site of a receptor. This so-called chemical shift sensitively reflects the relative population of bound and free receptor, i.e. reports on the degree of saturation in a titration experiment. NMR binding experiments can be performed in a variety of setups. Since individual nuclei are monitored, NMR is a powerful tool to delineate the boundaries of binding sites.[i] Furthermore, NMR binding experiments provide information about the overall dynamics of protein–protein complexes.[j] At present, the

---

[h] Polarization is defined as $P = \dfrac{I_{\updownarrow} - I_{\leftrightarrow}}{I_{\updownarrow} + I_{\leftrightarrow}}$ where the indices $\leftrightarrow$ and $\updownarrow$ denote the emission intensity parallel and perpendicular to the excitation light, respectively. The fluorescence anisotropy (A) is related to P by $A = 2P/(3\text{-}P)$. Both P and A provide the same information on the binding process.

[i] In practice, ^{15}N or ^{13}C labelling is required.

[j] In general, strong complexes (high K_A, low K_D) exhibit lifetimes longer than $1/(\nu_{bound}\text{-}\nu_{free})$, ν_{bound} and ν_{free} being the absorption frequencies of a nucleus in the complex and in the free protein, respectively. In this situation, two characteristic signals will appear in the NMR spectrum, corresponding to ν_{bound} and ν_{free}. Low-affinity binding is usually manifested by lifetimes of the complex being shorter than $1/(\nu_{bound}\text{-}\nu_{free})$. The signals are

application of NMR to extract K_A (and ΔG) is somewhat limited to studies of relatively weak complexes, since the experiments require high concentrations of reactants.[10]

3.3.3 *Methods Based on Refractive Phenomena*

I discuss two methods, which have a common physical foundation, separately, and illustrate how new technological developments widen the palette of experimental methods aimed at quantification of protein–protein interactions. The surface plasmon resonance (SPR) is a complicated phenomenon arising at the boundary between transparent media of different refractive index (for example glass and water), separated by a thin metal film.[11] Above a critical angle the light coming from the side of higher refractive index is totally reflected. At this condition the electromagnetic field component penetrates a short into the medium of a lower refractive index creating an evanescent wave. The intensity of the reflected light is reduced at a specific incident angle because of resonance energy coupling between the evanescent wave and the metal surface plasmons. It has been shown that the resonance condition (resonance angle) depends on the refractive index of the material adsorbed at the metal film, linearly changes with changes of the refractive index, and can be used to quantify changes in the mass concentration at the sensor chip surface. One binding partner is immobilised on the surface of a sensor chip in a flow cell. The other binding partner flows over the surface of the sensor chip and interacts with the immobilised molecules. The binding event on the surface of the sensor leads to a change in refractive index close to the surface of the sensor chip and is detected by the changes of the resonance angle – measured in arbitrary resonance units (RU). Starting from some value at zero saturation (no ligand present in the buffer) the RU signal increases with time when saturation increases (ligand present in the buffer). Since the process is dynamic (association and dissociation take place all the

---

broad, and the chemical shift represents a population-weighted average of ν_{bound} and ν_{free}. Complexes with intermediate behaviour are also known. Modern deconvolution techniques allow extraction of quantitative information in various situations.

time) some time elapses until the equilibrium is reached (constant RU signal). Therefore, the initial phase of the sensogram has the shape of a kinetic trace. Indeed, the association rate constant (k_{ass}) can be calculated from this part of the sensogram. In the region of the plateau the system is in the steady-state regime. At some point, the cell is flushed with buffer containing no ligand. The RU signal decreases exponentially since the complex dissociates. From this part of the sensogram the dissociation rate constant (k_{diss}) can be calculated. Binding affinities can be obtained either from rate constant measurements according to $K_A = k_{diss}/k_{ass}$ (see Section 3.3.4) or by measuring the steady state level of binding as a function of the sample concentration according to Eq. 3.19. The SPR approach in different implementations (the most popular and user-friendly being the BIAcore instrument) has been used in hundreds of studies of protein–protein interactions. Although problems might arise due to covalent immobilization of one of the binding partners (conformational changes, inaccessibility of a proportion of binding sites), the SPR technology yields K_A estimates in good or excellent agreement with truly in-solution methods.

Very recently, a truly in-solution assay has been proposed, which exploits a similar physical principle – back-scattering interferometry (BSI).[12] In a special optical arrangement, the interaction of laser light with the medium in a small reaction chamber produces scattered light with sharp interference fringes. Any change in the refractive index of the medium causes displacement of the interference pattern. The idea is that mixing the reactants in a stopped-flow-like setup will allow following the time course of BSI pattern shift. The binding affinity is calculated from a series of experiments at varying concentration of one of the reactants by either by analysis of the kinetic traces or by plotting the end-points of the BSI signal as function of the varying total concentration to produce the familiar hyperbolic saturation curve. The correspondence between results from BSI experiments and other type of binding experiments is impressive. The main advantage is that a wide range of K_A can be measured with very little material. However, the BSI method is still in the infant stage and extensive experimentation should delineate the limits of applicability.

3.3.4 *Kinetic Approaches*

The equilibrium between species A, B and AB is dynamic – binding and unbinding (association and dissociation) occur steadily:

$$A+B \underset{k_{diss}}{\overset{k_{ass}}{\rightleftharpoons}} AB \tag{3.20}$$

The microscopic rate constants for association and dissociation, k_{ass} and k_{diss}, respectively, are linked to the equilibrium constant by:

$$K_A = \frac{k_{diss}}{k_{ass}} \tag{3.21}$$

It follows, that another way to calculate K_A is to measure k_{ass} and k_{diss} in some type of experiment. As mentioned in Section 3.3.3 kinetic information can be obtained by SPR and BSI experiments. Also, time-resolved measurements of any spectroscopic probe can be exploited. If A and B are mixed in 1:1 molar and volume ratio, the disappearance of unbound A (or unbound B) can be described by:

$$-\frac{d[A]}{dt} = -\frac{d[B]}{dt} = k_{ass}[A][B] - k_{diss}[AB] = k_{ass}[A]^2 - k_{diss}[AB] \tag{3.22}$$

At time zero $[A] = [B] = [A_0]$. After substitution of $[AB]$ by $[A_0]$ $[A]$ and rearranging, the decrease of $[A]$ is given by:

$$-\frac{d[A]}{dt} = k_{ass}[A_0]\left(\frac{[A]^2}{[A_0]} + \frac{k_{diss}[A]}{k_{ass}[A_0]} - \frac{k_{diss}}{k_{ass}}\right) \tag{3.23}$$

and after integration,

$$[A] = \left(\frac{zb+zs-b+s}{1-z}\right)\frac{[A_0]}{2} \tag{3.24}$$

I. Jelesarov

where:

$$b = \frac{k_{diss}}{k_{ass}[A_0]} \qquad (3.25a)$$

$$s = \sqrt{4b + b^2} \qquad (3.25b)$$

$$z = \left(\frac{2 + b - s}{2 + b + s}\right) \exp\left(-k_{on}[A_0]\right) \qquad (3.25c)$$

The time course of the signal change is described by:

$$\sigma(t) = \sigma_0 + \Delta\sigma\left(1 - \frac{[A]}{[A_0]}\right) \qquad (3.26)$$

where σ_0 and $\Delta\sigma = \sigma_{eq} - \sigma_0$ are the signal at time zero and the maximum change of the signal when equilibrium is reached, respectively. Figure 3.5 presents an example.

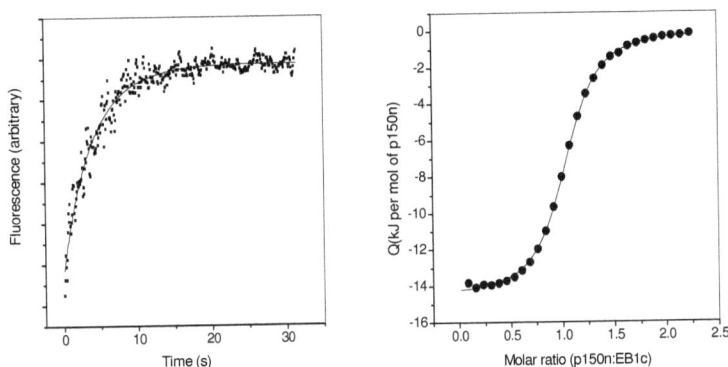

Fig. 3.5. An example of K_A determination by kinetics. Binding of the CAP-Gly domain of dynactin (p150n) to the C-terminal domain of End-binding protein 1 (EB1c) was measured at 25 °C. Panel A, time course of the fluorescence changes detected upon mixing of equimolar concentrations of the two proteins (20 μM). The continuous line is the best fit according to Eqs 3.20–3.26 to the data (symbols). The calculated rate constants for association and dissociation are $k_{ass} = 6.9\times10^4$ M^{-1} s^{-1} and $k_{diss} = 7\times10^{-2}$ s^{-1}, respectively. Thus, K_A determined by kinetics is 9.8×10^5 M^{-1}. In Panel B the equilibrium binding isotherm obtained by ITC is shown (see Section 3.3). From ITC data, $K_A = 3.5\times10^5$ M^{-1}. Data from Honnappa *et al.* (2006), *Mol. Cell* 23: 663–671, and unpublished results.

3.3.5 *Isothermal Titration Calorimetry (ITC)*

Finally, the principle and merits of ITC are described in some detail. The unique feature of the method is that the experimental signal is the heat released or absorbed upon interaction between molecules *A* and *B*. Since the experiment is performed at constant temperature and pressure in a quasi-adiabatic chamber, the measured heat values can be related to the molar enthalpy of the binding reaction. Therefore, even few ITC experiments provide immediate access to and precise estimates of all relevant thermodynamic functions describing the stability of a receptor–ligand complex.

Molecule *A* dissolved in the appropriate buffer is placed in the sample cell.[k] Molecule *B* dissolved in the same buffer is added stepwise in aliquots from a computer-controlled injection syringe, which rotates to effect rapid mixing of the reactants. If binding takes place, heat is released or absorbed upon each addition of *B* to *A*. In power compensation instruments the heat change is expressed as the electrical power applied by the feedback network to maintain a small temperature difference between the sample and the reference cell.

Each injection of *B* into *A* produces a deflection of the differential power signal (μcal s^{-1}) from the thermal baseline connecting the signal immediately before the injection and after re-equilibration (Fig. 3.6). Integration of the differential power with respect to time yields the heat change Δq (μcal) between injections i and i-1:[l]

$$\Delta q = q_i - q_{i-1} \tag{3.27}$$

---

[k] The reference cell is filled with high-purity, degassed water unless the heat capacity of the buffer is very different from the heat capacity of water.

[l] Δq is caused not only by the shift of the chemical equilibrium in the cell but contains contributions from the heat values of dilution and unspecific effects not pertaining to the binding event *per se*. Therefore, corrections have to be introduced by performing control experiments.

Δq is proportional to the volume of the calorimetric cell, which is known, to the molar enthalpy of binding, which is a constant at constant temperature and pressure, and to the change in the concentration of the complex, $\Delta[AB]$ (or bound B, $\Delta[B]_{bound}$), between injections i and $i-1$:

$$\Delta q = V_{cell}\Delta H \left([AB]_i - [AB]_{i-1}\right) \tag{3.28}$$

Using the known total concentration of A in the cell, $[A]_t$, and the degree of saturation Y, defined by Eqs 3.3a and 3.12 can be re-written as:

$$\Delta q = V_{cell}\Delta H [A]_t \left(Y_i - Y_{i-1}\right) \tag{3.29}$$

The experimental data can be plotted in two ways. The familiar sigmoidal titration curve, which is most frequently seen in the biochemical literature and shown in Fig. 3.6, represents the derivative of Eq. 3.29 with respect to the total concentration of B added, $[B]_t$:

$$\frac{d\Delta q}{d[B]_t} = V_{cell}\Delta H \left(\frac{1}{2} + \frac{1 - \dfrac{[B]_t}{n[A]_t} - \dfrac{1}{n[A]_t K_A}}{2\sqrt{\left(\dfrac{[B]_t}{n[A]_t} + \dfrac{1}{n[A]_t K_A} + 1\right)^2 - 4\dfrac{[B]_t}{n[A]_t}}} \right) \tag{3.30}$$

According to Eq. 3.29, the heat Q accumulated up to injection i is given by:

$$\sum_1^i \Delta q = Q = V_{cell}\Delta H [A]_t Y_i \tag{3.31}$$

Q can be expressed in terms of $[A]_t$ and $[B]_t$:

$$Q = V_{cell}\Delta H \left(\frac{1 + n[A]_t K_A + [B]_t K_A - \sqrt{\left(1 + n[A]_t K_A + [B]_t K_A\right)^2 - 4n[A]_t[B]_t K_A^2}}{2K_A} \right) \tag{3.32}$$

The data transformation according to Eq. 3.32 is illustrated in Fig. 3.6. Note that Eqs 3.30 and 3.32 include the parameter n, which represents the number of binding sites on molecule A. Obviously, in the simplest case of 1:1 interaction $n = 1$. Regression analysis of the data allows simultaneous determination of K_A, ΔH and n. [m]

According to Eqs 3.9 and 3.10, a single ITC experiment provides ΔG, ΔH and $\Delta S = (\Delta H - \Delta G)/T$ at the experimental temperature. From a couple of experiments performed at different temperatures ΔC_p can also be calculated according to Eq. 3.12. Hence, a series of ITC experiments, which can be performed within two–three days, yields high-precision estimates of all relevant energetic terms, alleviating the inherent problems of van't Hoff analysis of binding data (see Eqs 3.14–3.16). Figure 3.6 shows the complete thermodynamic profile of a protein–protein complex. For a more detailed discussion of the principle, experimental design, data handling and data interpretation Refs 13–19 are recommended.

3.4 Energetics of Protein–Protein Interactions

Calculation/prediction of thermodynamic quantities that characterise binding reactions is a major challenge in modern computational chemistry. Apart from the fundamental importance of understanding the energetic principles of molecular recognition, the interest to the problem in the field of biomedical research is fueled by the ever growing needs to design compounds that either mimic or interfere with biologically-relevant interactions between proteins and their ligands, as a way to prevent or cure diseases. Initial screening for and rational optimization of

---

[m] Note that the total volume increases in the course of the titration. Therefore, the actual concentrations of A and B after injection i are $[A]_{t,i} = [A]_{t,0}\left(1 - \dfrac{dV}{V_{cell}}\right)^i$ and $[B]_{t,i} = [B]_{t,0}\left(1 - \left(1 - \dfrac{dV}{V_{cell}}\right)^i\right)$, where dV is the injection volume. These corrections have to be introduced in Eqs 3.30–3.32.

lead compounds towards higher/desired affinity by purely experimental strategies has proved to be a tedious and time-consuming task.

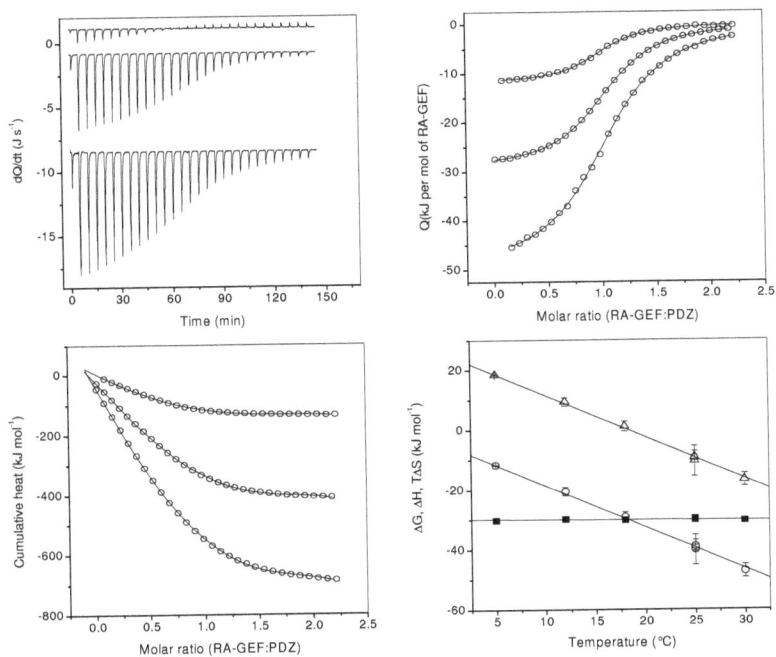

Fig. 3.6. ITC binding experiments. As an illustration of the informational content of ITC, shown are binding experiments with the PDZ2 domain of human protein tyrosine phosphatase 1E and the C-terminal pentadecapeptide of the RA-GEF protein. This reaction represents the biologically relevant binding mode. The peptide was titrated to PDZ at temperatures of 5 to 30 °C. Panel A, raw calorimetric output from experiments at 5, 18 and 30 °C (top to bottom). The traces are shifted on the y-axis for clarity. Panel B, the heat values released at each titration step were calculated by integration of the corresponding differential power peak, and plotted vs the molar ratio RA-GEF:PDZ2 (symbols). The lines visualise the best non-linear regression fits of Eq. 3.30 to the data. Panel C, the same data plotted in the cumulative mode. Best fits according to Eqs 3.31 and 3.32 are shown with lines. From both analyses, K_A, ΔH, and the stoichiometry of interaction (n) can be calculated. Panels A–C illustrate that the reaction becomes increasingly more exothermic at higher temperatures. Panel D, thermodynamic profile of the RA-GEF:PDZ2 complex. The temperature variation of ΔG (squares), ΔH (circles) and $T\Delta S$ (triangles) is shown. The slope of the line associated with ΔH data is the heat capacity change, ΔC_p. Note that the reaction is driven by both ΔH and $T\Delta S$ below 18 °C. Above that temperature, the entropic term becomes unfavourable. Since the temperature dependencies of ΔH and $T\Delta S$ are very similar, the binding affinity (ΔG) is nearly constant in the studied temperature interval. Data from Milev *et al.* (2007), *Biochemistry* 46: 1,064–1,078.

Therefore, computer-based and/or empirical predictions of binding affinities are highly desirable. The problem has two interlinked sides. On the one hand, the number of high-resolution structures of bio-macromolecular complexes appears to catch up with and even outrun the number of thoroughly characterised binary complexes. On the other hand, even the most precise measurement of any binding parameter yields just a number, which itself contains no descriptive power if there is no (at least vague) understanding of the underlying molecular process. Here, I first will very briefly sketch some approaches aimed at predicting binding affinities. The reader is advised to consult the original publications, as well as chapters on protein modelling from this volume for a thorough description. I will provide more details about the state of the currently popular field, sometimes also known as 'structural thermodynamics', where the efforts are to find correlations between experimentally measured thermodynamic parameters and the physical–chemical properties of the interacting molecular surfaces.

3.4.1 *Calculation of Binding Affinities*

Arguably, the statistical thermodynamics treatment represents the most rigorous way to calculate the standard free energy change of a binding reaction. Since at equilibrium the chemical potentials of the associated (AB) and dissociated ($A+B$) states of the system are equal, i.e. $\mu_{AB} = \mu_A + \mu_B$, the free energy is given by $\Delta G = \mu_{AB}^0 - \mu_A^0 - \mu_B^0$ where the superscript 0 indicates the standard chemical potentials. For each solute i the standard chemical potential μ_i^0 is proportional to the ratio of the canonical partition functions of the solute placed in the solvent and of the pure solvent, the volume of the system(s) at the standard pressure, and the partial specific volume of the solute. In principle, these quantities can be evaluated by MD simulations using multiconfiguration integrals over suitably defined vectors of momenta and atomic coordinates, so that the free energy of binding is given by:

$$\Delta G_{bind} = -RT \ln\left(\frac{C^\circ}{8\pi^2}\frac{\sigma_A\sigma_B}{\sigma_{AB}}\frac{Z_{AB}}{Z_A Z_B}\right) + P^\circ \Delta \bar{V}_{AB} \qquad (3.33)$$

$C°$ is the standard concentration; σ_i are the symmetry numbers (the number of identical orientations of the molecules); Z_i are the integrals; $P°$ is the standard pressure; $\overline{\Delta V}_{AB}$ is the change of the partial specific volume upon association. Any change of the system is divided in consecutive small perturbations and the calculation yields the work involved at each step. The sum of these stepwise work terms yields the free energy change. Probably the most popular implementation is the 'double-annihilation' method, where the ligand is first 'annihilated' from the solvated complex and is then 'annihilated' from the solvent in a second step. Although this methodology is certainly the most robust approach, it is clear that it is based on the assumption of ergodicity of the system(s) which cannot be proved. For an excellent discussion of this type of calculation see Ref. 20. The method has been devised and developed in studies of binding small ligands to proteins. However, the formulations are general and can be applied in principle in studies of protein–protein association, if accurate sampling of the relevant configurations of the solutes and solvent are available. Nonetheless, this methodology is at its best in studies of changes in ΔG_{bind} in response to mutations (*relative* binding affinities).

Calculation of *absolute* binding affinities by equilibrium sampling along a quasi-continuous transformation path connecting the initial and the final states is a computationally demanding task. Therefore, end-point protocols have been devised, where only the initial and final states of the system are evaluated. One such method combining the advantages of conformational sampling with the computational efficacy of continuum calculations is the Molecular Mechanics Poisson-Boltzmann (Generalized Born) Surface Area approach (MM-PB(GB)SA).[21] The binding free energy is expressed as the sum of energetic contributions:

$$\Delta G_{bind} = \langle \Delta E_{int} \rangle + \Delta G_{pol} + \Delta G_{np} - T\Delta S_{other} \qquad (3.34)$$

All these terms represent differences between state AB and state $A+B$.

$<E_{int}>$ is the internal (bond, angle, torsion) energy. ΔG_{pol} represents the sum of $<E_{el}>$ (the gas-phase electrostatic energy) and ΔG_{pol}^{solv} (the energy associated with solvation of polar groups). Likewise, ΔG_{np} can be partitioned into $<E_{vdW}>$ (the gas-phase van der Waals energy) and ΔG_{np}^{solv} (the energy associated with solvation of non-polar groups). All bracketed terms are evaluated as ensemble averages by molecular mechanics forcefield(s). ΔG_{pol}^{solv} is extracted by applying either the Poisson-Boltzmann finite-difference method or the generalised Born approach. The energetic effect of association of non-polar surfaces is assumed to be proportional to the solvent accessible surface area (ASA) according to $\Delta G_{np}^{solv} = \gamma ASA + b$, the values of the coefficients γ and b being different in different implementations. Finally, $T\Delta S_{other}$ in Eq. 3.34 is the entropic contribution from the changes in rotational, translational and vibrational degrees of freedom (see below). Solvent-related entropic effects are implicitly included in ΔG_{pol}^{solv} and ΔG_{np}^{solv}. Although the MM-PB(GB)SA method includes continuum approximations, it has been argued that it has a clear connection to statistical thermodynamics.[22]

The linear interaction energy method (LIE) also takes advantage of using sampling of ensembles generated by MD or MC.[23] Simulations are performed for the ligand in the free state and in the bound state to obtain the average energy of intermolecular van der Waals (vdW) and electrostatic (el) interactions (denoted by $<U>$). The binding free energy is then derived using the following formula:

$$\Delta G_{bind} = \alpha \langle \Delta U_{vdW} \rangle + \beta \langle \Delta U_{el} \rangle \qquad (3.35)$$

The delta sign indicates the change in energy in going from the free to the bound states, that is $\Delta U = U_{bound} - U_{free}$. The coefficient α is derived by linear regression fit to a set of binding energies and $\beta = 0.5$ as predicted by the linear response approximation for ionic interactions, but can be scaled accordingly for neutral polar ligands.

Another class of methods consists of empirical or semi-empirical approaches. They implement master equations derived by regression within a training set of known interactions (and affinities). ΔG_{bind} is expressed as the sum of different energetic terms representing either energetic benefits in formation of intermolecular contacts or energetic penalties (loss of degrees of freedom). Just as an example, I show here (in slightly different notation) the equation discussed in detail in Ref. 24:

$$\Delta G_{bind} = \sum \Delta G_{nHB} + \sum \Delta G_{iHB} + HSA\Delta G_{HE} + n\Delta G_{rot} + \Delta G_{t+r} \quad (3.36)$$

Binding is expected to be favoured by polar interactions, that is, formation of neutral hydrogen bonds (*nHB*) and ionic hydrogen bonds (*iHB*). The contribution of the hydrophobic effect scales with the amount of hydrocarbon surface (*HSA*) buried at the interface, ΔG_{HE} being the contribution per Å^2. The entropic losses due to freezing of *n* internal rotations (*rot*) and restrictions of translational and rotational degrees of freedom (*t+r*) are considered by the last two terms of the above equation, respectively. The various terms do not have a sound statistical thermodynamics or molecular mechanics foundation but describe the available structural information and binding data in a statistical manner. Clearly, the essential advantage of empirical methods is their efficacy in terms of speed. The main limitation, however, is that the precision critically depend on the size and quality of the training database.

There is a large variety of 'hybrid' methodologies available, combining physically rigorous approaches with empirical treatments.[25-28]

3.4.2 *Structure-based Prediction of Binding Parameters*

Real progress in understanding the mechanisms of macromolecular association can be achieved only if an intimate link is established between energy and structure. Since non-polar and polar contacts exhibit different thermodynamic signatures, the intuitive and relatively simple approach which has been pursued, is to parameterize the observed quantities in terms of the amount and type of surface buried in a complex.[29,30] The idea has been adopted from studies of protein energetics; namely that ΔH, ΔS, and ΔC_p of protein unfolding could be predicted with relatively good precision from the increase in surface exposure upon disruption of the compact native state (Ref. 31 and references therein). It is debatable whether the surface-energy correlations found for proteins really take into account the physics of the underlying process, or whether they simply represent a useful empirical observation. Nonetheless, if such correlations could be established for binding reactions, a deeper understanding of the forces holding the binding partners together would result and *predictions* of the binding profile may become possible.

The method follows the typical end-point strategy. The amount of surface buried in the interface is calculated by taking the difference between the solvent accessible surfaces (ΔASA) of the bimolecular complex and the isolated binding partners.[n] Using proper definitions, the total ΔASA can be partitioned into polar (ΔASA_{pol}) and non-polar (ΔASA_{np}) components. The central issue in the prediction of binding parameters from structural information only is the calculation of ΔC_p. The hydration heat capacity of polar and non-polar substances is positive and negative, respectively, and the elementary contributions ($c_{pol} < 0$ and $c_{np} > 0$; per unit of surface area) are known from pure-phase-to-water transfer studies with model compounds. The binding heat capacity change is calculated by the following empirical equation:[o]

$$\Delta C_p = \Delta C_{p,np} + \Delta C_{p,pol} = c_{np}\Delta ASA_{np} + c_{pol}\Delta ASA_{pol} \qquad (3.37)$$

Analytical expressions describing the intrinsic change of c_{pol} and c_{np} with temperature can be plugged into Eq. 3.37, yet the temperature dependence is weak and can be neglected below, say 80 °C.

Similarly, the binding enthalpy can be calculated by summing up contributions from polar and non-polar interactions, whose overall strength scales with the amount of buried surface of the corresponding type.

---

[n] Albeit the basic algorithm for calculation of ASA is universal – rolling a probe (a sphere representing the water molecule) on the van der Waals envelope of the protein – different radii can be assigned to both the probe and the van der Waals radii of protein atoms. Therefore, care should be taken to perform the calculations of thermodynamic parameters using ASA and elementary contributions per unit surface that are consistent with each other.

[o] All equations in this section consider the dissociated state ($A+B$) as the reference state. Therefore, each parameter ΔX represents $X_{AB} - X_{A+B}$.

$$\Delta H (60°C) = h_{pol} \Delta ASA_{pol} + h_{np} \Delta ASA_{np} \qquad (3.38)$$

The coefficients $h_{np} < 0$ and $h_{pol} > 0$ have been derived by regression analysis of the protein thermodynamic database. The assumption is that the average packing density at protein–protein interfaces approximates the average packing density of proteins. Therefore, the enthalpic term combining the enthalpy of formation of polar interactions and the concomitant changes in hydration enthalpy, is similar for protein unfolding and dissociation of protein–protein complexes. A more precise analysis requires calculation of the *actual* packing density and considering the presence of mixed polar/non-polar contacts.[29,32] At any other temperature T:

$$\Delta H (T) = \Delta H (60°C) + \Delta C_p (T - 60) \qquad (3.39)$$

In the absence of proton transfer (see below), the entropy of dissociation can be partitioned into contributions arising from solvent re-organisation, ΔS_{solv}, conformational effects, ΔS_{conf} and changes in the vibrational, translational and rotational spectra, ΔS_{other}:

$$\Delta S (T) = \Delta S_{solv} + \Delta S_{conf} + \Delta S_{other} \qquad (3.40)$$

Using the temperature(s) where the entropy of hydration of non-polar groups $(T_{S,np})$ and polar groups $(T_{S,pol})$ is zero, the solvent-related part is calculated according to:

$$\Delta S_{solv} (T) = \Delta C_{p,np} \ln \left(\frac{T}{T_{S,np}} \right) + \Delta C_{p,pol} \ln \left(\frac{T}{T_{S,pol}} \right) \qquad (3.41a)$$

According to Eq. 3.37,

$$\Delta S_{solv} (T) = c_{np} \Delta ASA_{np} \ln \left(\frac{T}{T_{S,np}} \right) - c_{pol} \Delta ASA_{pol} \ln \left(\frac{T}{T_{S,pol}} \right) \qquad (3.41b)$$

$T_{S,np}$ is known to be close to 385 K.[33,34] There are two conflicting estimates of $T_{S,pol}$: 385 K[34] or 333 K.[35]

In many cases, the conformational term (ΔS_{conf}) is dominated by the immobilization of side chains due to formation of intermolecular contacts (but see Section 3.4.3). The loss of entropy of side chains in going from the completely solvent-exposed state to the completely buried, immobilised state ($\Delta S_{SC}*$) has been estimated by statistical thermodynamic methods.[36,37] Assuming that partial burial causes partial loss of entropy, the total, side chain-related entropy effect can be calculated according to:

$$\Delta S_{SC} = \sum_i \frac{\Delta ASA_i}{ASA_i} \Delta S_{SC} * \tag{3.42}$$

Where ΔASA_i is the change in ASA of the side chain of residue i and ASA_i is the ASA of the corresponding side chain in the fully exposed state.

Perhaps the most problematic issue in binding entropy calculations is the magnitude of ΔS_{other} (Eqs 3.34, 3.36 and 3.40). Typically, vibrational contributions are neglected. When hydration entropy is calculated by semiempirical parameterization schemes based on organic compound transfer thermodynamic data, it necessarily includes – at least to some extent – vibrational contributions as well, which are thought to be relatively small in general. The estimates of the magnitude of ΔS_{other} (rotation/translation) are largely discrepant. It seems now that values stemming from gas-phase statistical mechanics derivations[38] largely overestimate this contribution.[39–41] Experimental studies suggest a contribution in the order of −40 J mol^{-1} K^{-1} for a bimolecular protein–protein association ($T\Delta S_{other} \sim$ −10 to −12 kJ mol^{-1} at 25 °C).[42–43] This is numerically close to the 'cratic entropy', $S_{cratic} = R\ln(1/55)$ ($T\Delta S_{cratic} \sim$ −10 kJ mol^{-1}).[44] A recent computational study yields $\Delta S_{other} \sim$ −100 J mol^{-1} K^{-1} ($T\Delta S_{other} \sim$ −30 kJ mol^{-1} at 25 °C), which perhaps can be regarded as an upper estimate.[45] It has been argued that ΔS_{other} will depend on the nature and strength of association.[45]

The binding free energy is $\Delta G(T) = \Delta H(T) - T\Delta S(T)$ with $\Delta H(T)$ and $\Delta S(T)$ calculated according to Eqs 3.39 and 3.40, respectively.

3.4.3 Understanding Binding: Are there Structure-Energy Relationships?

Protein–protein and protein–ligand binding involves formation of large intermolecular interfaces, which can be regarded as complementary arrays of highly cooperative, non-covalent bonding networks. In such systems, analysis of the changes in thermodynamic functions in terms of additive contributions from distinct physical forces and effects is indeed problematic on theoretical grounds.[20,46] Notwithstanding the coarse oversimplification of the physical reality, calculations using the formalism presented in the foregoing sections (especially Section 3.4.2) and comparisons with experimental data, provide a unique opportunity to take a glimpse into the energetic principles that govern association. In this respect, ITC data are regarded as the 'golden standard' because of the unprecedented accuracy of ΔH, ΔS, and ΔC_p determination.

The straightforward applications of Eqs 3.37–3.42 assume that binding is treated as 'lock-and-key' interaction. In this approximation, the shape (van der Waals) complementarity of the binding surfaces is close to optimal. The binding interface is completely dehydrated. The interaction partners are held together by non-covalent bonds (polar and non-polar). The enthalpic gain of intermolecular hydrogen, dipole and dispersion bonds realised in the low dielectric permittivity of the interface is partially offset by the loss of enthalpic interactions between the participating solute groups and the solvent in the unbound state. The principal gain in entropy comes from the increase in entropy upon release of water molecules bond to the dissociated, 'empty' binding sites into the bulk (ΔS_{solv}; hydrophobic force). The principle loss of entropy is caused by immobilization of rotors when bonds are formed (ΔS_{SC}). Entropic penalty is paid for the formation of one kinetic unit out of two (ΔS_{other}). This model can be accepted if there is numerical agreement between predicted and experimentally measured ΔH and ΔS (consequently also ΔG), within of course the uncertainties of both experimental data and parameterization coefficients (including the calculation of buried molecular surface). The success of the parameterization scheme(s) is sometimes impressive.[47,48]

Every so often, however, reports about significant discrepancies between calculation and experiment appear. In many cases the predicted numbers are not only far off the measured values but are even of opposite sign. This is an indication that additional processes and effects accompany binding. Here, only the most commonly encountered 'complications' are discussed.

Calculation of ΔH according to Eqs 3.38 and 3.39 yields the '*genuine*' enthalpy change, i.e. the change in ΔH caused by formation of contacts and dehydration of the interface. If association causes pK_a shifts of ionizable groups, protons will be released from or will bind to the complex. Necessarily, the corresponding number of protons are taken up into or released from the buffer compound, respectively. Since the ionization enthalpies of many commonly used buffers are large, the apparent binding heat values in the ITC experiment will contain a contribution from the buffer ionization heat. Due to thermodynamic linkage relationships and the intrinsic temperature dependence of pK_a shifts, the observed magnitude of all thermodynamic parameters, and their apparent temperature dependencies, will be influenced by proton transfer. The traditional way to detect proton exchange is to perform a series of experiments at the same pH in solutions buffered with compounds having different ionization enthalpies. Dozens of studies have demonstrated that plots of the observed enthalpy, ΔH_{obs}, as function of the buffer ionization heat, ΔH_b, are linear. Formally, the data can be described by the equation $\Delta H_{obs} = \Delta H_{b,0} + n_{H+}\Delta H_b$, where the slope, n_{H+}, and the y-axis intercept, $\Delta H_{b,0}$, quantify the number of transferred protons and the enthalpy of association in a (hypothetical) buffer with zero ionization enthalpy, respectively. Usually, $\Delta H_{b,0}$ is interpreted as representing the intrinsic (genuine) binding enthalpy in the absence of proton transfer effects.[p] Because the magnitude of pK_a shifts and ΔH_b both depend on the temperature, proton transfer will influence the apparent ΔC_p and ΔS values. Hence, *ASA*-based predictions are valid only if proton transfer effects are taken into consideration.

---

[p] This is strictly correct only at a given temperature and a given pH. See Ref. 49 for discussion.

Typically, the structures of the unbound components A and B are not known and calculations of ΔASA are performed with the structure of the complex AB. This is faulty in all cases where the binding partners experience structural re-arrangements in going from the free to the bound state. Conformational adaptation on a global or local scale is a ubiquitous phenomenon, yet its thermodynamic consequences are hard to grasp. In principle, it is expected that transition from a more unfolded and solvent-exposed free state to a more compact, more solvent-inaccessible bound state will contribute negatively to ΔH (because packing interactions are formed) and ΔC_p (because predominantly hydrophobic surface becomes newly buried). In analogy to protein folding, the entropic effect of refolding is expected to be unfavourable, since conformational rigidification generally over-compensates the hydrophobic effect in terms of entropy. Next to ΔS_{SC} (Eq. 3.42), additional entropic expenses have to be considered in calculation of ΔS_{conf}. They are linked to immobilization of the backbone when secondary structure elements are formed, and fixing side chains residing in such elements into the binding pocket. These contributions have been estimated by computational methods per each amino acid residue.[35,37] In favourable cases, the structures of the free components are available and they can be used in surface-accessibility calculations. Alternatively, one can attempt to simulate the free-to-bound structural transition by MD simulation starting with the structure(s) of the bound component(s). It should be noted, that ASA-based predictions made with the structure of a complex will be problematic, if binding proceeds *via* a conformational selection mechanism.[50] According to this model, one of the components exists in two (or more) alternative conformations, only one of them being capable to bind the partner molecule. Binding induces shift in the equilibrium distribution of binding incompetent and binding competent states. Therefore, the experimental data will contain ΔH, ΔC_p and ΔS contributions from this linked equilibrium.[51, 52]

The shape complementarity of binding surfaces is rarely perfect. Water molecules are often trapped in cavities and clefts at the binding interface. They contribute to the energetics of binding and must be taken into account when interpreting ASA-based calculations. Significantly, buried waters ($ASA < 10$ Å^2;) are considered as belonging to one of the binding partners in *ASA* calculation (see Ref. 53 for detailed discussion). On the practical side, the problem is that the resolution of X-ray structures may not be high enough to detect interpretable electron density in cavities and clefts, and NMR is unable to 'see' waters. Again, MD simulations may come to help in identification of long-lived trapped water molecules. It is expected that 'structural' waters will increase the favourable enthalpy of binding by improving the packing density, forming hydrogen bonds and alleviating the enthalpic penalty linked to charge burial. Also the unfavourable enthalpic effect of complete dehydration of non-polar groups is diminished. According to some estimates, the transfer of a water molecule from the bulk solvent and including it into an interface can be an exothermic process adding as much as -16 kJ mol^{-1} to ΔH.[54,55] At the same time, immobilised waters destabilise the complex entropically. The exact magnitude of the effect is not precisely known, but the upper limit has been estimated as ~ -30 J K^{-1} (mol of water)$^{-1}$ ($T\Delta S \sim 9$ kJ K^{-1} (mol of water)$^{-1}$ at 25 °C).[56] Water burial will decrease ΔC_p by 25–40 J K^{-1} (mol of water)$^{-1}$.[56]

In the framework of the discussed methodology, the sign and rough magnitude of calculated parameters is indicative of the balance of forces stabilising a given complex, even if the derived numbers are far off the experimental data. Large negative ΔH with small or medium temperature dependence (ΔC_p) is typical for formation of strong hydrogen bonds. Large positive ΔS and large negative ΔC_p with small negative or positive ΔH mark the dominant role of the hydrophobic effect. Large negative ΔS and ΔC_p can be attributed to either significant restriction of degrees of freedom in rigid-body type binding, or pronounced association-induced refolding reaction, or entrapment of many water molecules at the binding interface.

Words of caution are mandatory. First, *ASA*-based calculations assume that the structures of the free components and their complex are invariant in an extended temperature interval. This is not granted. Even if there are no large-scale conformational changes, protein molecules are 'soft' and flexible, and experience thermal fluctuations. Credible structure-oriented comparisons require correction of ΔH, ΔC_p and ΔS determined by ITC for any temperature-induced contributions from minor partial refolding and conformational flexibility. The correction of

ΔH and ΔC_p can be done by integrating the heat capacity differences between the associated and dissociated state of the system as function of the temperature.[57] It should be noted that the corrections are not a matter of 'cosmetics'. The temperature dependence of the corrected ΔH can be dramatically different from the experimentally observed one; sometimes even the sign in a given temperature interval can change. Secondly, the mere idea that ΔC_p reflects changes in surface hydration has been challenged. It has been argued that large ΔC_p is the consequence of cooperative formation of many weak non-covalent bonds and from very tight packing interactions.[58,59] Thirdly, the correspondence between predictions and experimental results is sometimes poor, even with carefully selected, high-resolution structures and high-precision data. The likely reason is energetic propagation. Strong *inter*molecular bonds can modulate the strength and vibrational content of *intra*molecular bonds in the vicinity of the binding sites.[24, 60–62] The effect can lead to 'extra' contributions to all thermodynamic parameters derived by experiment, yet cannot be captured by structural methods. Notwithstanding these problems, structure-based predictions of thermodynamic quantities have fruited a deeper understanding of the energetic principles of association reactions and remain an active field of research.

3.5 Conclusions and Outlook

Protein–protein interactions coordinate and structure virtually all processes in the living cell. New protein–protein interactions are being identified on a daily basis. Protein binding sites can have arbitrary geometry and there are only few unifying principles from a structural point of view. Binding interfaces span hundreds and thousands of square angstroms, yet point mutations can severely impair binding affinity. Structurally, unrelated proteins can sometimes effectively compete for the same binding site. It is still very difficult to achieve by rational design and optimization, high affinity and specificity of a molecule designed to compete with protein–protein binding. This is why methodologically-rigorous biophysical studies of diverse protein–protein complexes are central to the efforts towards better understanding of biological functions. The past decades have brought about an increasing interest in deciphering the thermodynamic principles of protein–protein recognition.

The thermodynamic formalism for extracting quantitative information on complex formation is long known. The free energy change (ΔG) which quantifies the relative population of the bound state (complex), can be calculated if the equilibrium binding constant (K_A) is known. There are a large variety of experimental methods available to determine binding constants. Some methods are 'direct' in the sense that the free and bound species can be separated and their equilibrium concentration measured. Other methods are based on following the apparent changes of spectroscopic, hydrodynamic, etc. properties reporting the degree of partial saturation at given total concentration of species. There is no universally applicable method to study protein–protein binding equilibria. Once K_A and ΔG are known, the enthalpy change (ΔH), entropy change (ΔS) and heat capacity change (ΔC_p) of the binding reaction can be derived by taking temperature derivatives of K_A and ΔG. Unfortunately, the precision of such estimates is low. The roll-out of the new generation titration calorimeters has ignited a real explosion in the field of macromolecular recognition. The main advantage of ITC is the possibility of a direct and very precise measurement of K_A and ΔH in a single experiment. In principle, a highly reliable thermodynamic profile of a protein–protein complex can be constructed in a few days.

The ultimate goal is to find the intimate relationships between molecular structure, energetics and dynamics, and to discover 'rules' guiding predictions of the energetic response of a particular complex to structural changes in the binding partners. In such research programmes theoretical approaches are indispensable. Predictions of the binding affinity and other thermodynamic quantities can be achieved by different means: statistical thermodynamics treatment, combination of molecular mechanics with continuum calculations, use of master equations derived by 'training', semi-empirical calculations, etc. One popular way to deal with the problem is to estimate ΔH and the hydration-related part of ΔS from the amount and type of surface buried in a complex. Since estimates for the magnitude of the conformation-related part of ΔS have been derived, the total predicted ΔH and ΔS terms can be tested against data obtained by calorimetry. In spite of the underlying coarse simplifications, the approach provides useful insights into the balance of forces stabilising protein–protein complexes, and sometimes achieves remarkable correspondence with experiments. Although not covered in this chapter, there are numerous new theoretical developments providing ways to analyse complicated binding equilibria and heterotropic effects. In the future, the task will be more and more to look *critically* at the accumulated results from theoretical perspective. Furthermore, it will be important to collect in a systematic way high-precision data on carefully selected, high-resolution systems. Arguably, there is a long way ahead which will lead to the elucidation of rigorous links between experimental thermodynamics and molecular structure.

Acknowledgements

Experimental work in my laboratory, results of which are cited in this chapter, was supported in part by the Swiss National Science Foundation and the Canton of Zurich.

References

1. Wyman J., Gill S.J. (1990). *Binding and Linkage. Functional Chemistry of Biological Macromolecules*, University Science Books, Mill Valley, California.
2. Klotz I.M. (1997) *Ligand-receptor Energetics: A Guide for the Perplexed*, Wiley, New York.
3. Scatchard G. (1949). *Ann N Y Acad Sci* 51: 660–672.
4. Rosenthal H.E. (1967). *Anal Biochem* 20: 525–532.
5. Klotz I.M. (1946). *Arch Biochem* 9: 109–116.
6. Fu H. (ed.) (1994). *Protein–Protein Interactions: Methods and Applications*, Humana Press, Totowa, New Jersey.
7. Scott D.J., Schuck P. (2005). In: Scott D.J., Harding S.E. & Rowe A.J. (eds). *Modern Analytical Ultracentrifugation: Techniques and Methods*, pp. 1–25, The Royal Society of Chemistry, Cambridge.
8. Krishnaswamy S.R., Williams E.R., Kirsch J.F. (2006). *Protein Sci* 15: 1,465–1,475.
9. Park S.H., Raines R.T. (1994). In: Fu H. (ed.) *Protein–Protein Interactions: Methods and Applications*, pp. 161–167, Humana press, Totowa, New Jersey.

10. Fielding L. (2003). *Curr Top Med Chem* 3: 39–53.
11. Schuck P. (1997). *Ann Rev Biophys Miomol Struct* 26: 541–566.
12. Bornhop D.J., Latham J.C., Kussrow A., Markov D.A., Jones R.D., Sørensen H.S. (2006). *Science* 317: 1,732–1,736.
13. Wiseman T., Williston S., Brandts J.F., Lin L.N. (1989). *Anal Biochem* 179: 131–137.
14. Freire E., Mayorga O.L., Starume M. (1990). *Anal Chem* 62: A950–A959.
15. Cooper A. (1998). In: Ladbury J.E. & Chowdry B.Z. (eds) *Biocalorimetry: Applications of Calorimetry in the Biological Sciences*, pp. 103–111, John Wiley & Sons, Chichester, U K.
16. Indyk L., Fisher H.F. (1998). *Meth Enzymol* 295: 350–364.
17. Jelesarov I., Bosshard H.R. (1999). *J Mol Recognit* 12: 3–18.
18. O'Brien R., Haq I. (2004). In: Ladbury J.E. & Doyle M.L. (eds) *BiocalorimetryII: Applications of Calorimetry in the Biological Sciences*, pp. 3–34, John Wiley & Sons, Chichester, UK.
19. Thomson J.A., Ladbury J.E. (2004). In: Ladbury J.E. & Doyle M.L. (eds) *BiocalorimetryII: Applications of Calorimetry in the Biological Sciences*, pp. 37–58, John Wiley & Sons, Chichester, UK.
20. Gilson M.K., Given J.A., Bush B.L., McCammon J.A. (1997). *Biophys J* 72: 1,047–1,069.
21. Massova I., Kollman P.A. (2000). *Perspectives in Drug Discovery and Design* 18: 113–135.
22. Swanson J.M.J., Henchman R.H. and McCammon J.A. (2004). *Biophys J* 86: 67–74.
23. Åqvist J., Luzhkov V.B., Brandsdal B.O. (2002). *Acc Chem Res* 35: 358–365.
24. Williams D.H., Stephens E., O'Brian D.P., Zhou M. (2004). *Angew Chem Int Ed* 43: 6,596–6,616.
25. Novotny J., Bruccoleri R.E., Saul F.A. (1989). *Biochemistry* 28: 4,735–4,749.
26. Horton N., Lewis M. (1992). *Protein Sci* 1: 169–181.
27. Novotny J., Bruccoleri R.E., Davis M., Sharp K.A. (1997). *J Mol Biol* 268: 401–411.
28. Olson M.A. (1998). *Biophys Chem* 75: 115–128.
29. Luque I., Freire E. (1995). *Meth Enzymol* 295: 100–127.
30. Baker B.M., Murphy K.P. (1995). *Meth Enzymol* 295: 294–315.
31. Robertson A.D., Murphy K.P. (1997). *Chem Rev* 97: 1251–1267.
32. Hilser V.J., Gomez J., Freire E. (1996). *Proteins* 26: 123–133.
33. Baldwin R.L. (1986). *Proc Natl Acad Sci USA* 83: 8,069–8,072.
34. Murphy K.P., Peivalov P.L., Gill S.J. (1990). *Science* 247: 559–561.
35. D'Aquino J.A., Gomez J., Hisler V.J., Lee K.H., Amzel L.M, Freire, E. (1996). *Proteins* 25: 143–156.
36. Creamer T.P., Rose G.D. (1994). *Proteins* 19: 85–97.
37. Lee K.H., Xie D., Freire E., Amzel L.M. (1994). *Proteins* 20: 68–84.

38. Finkelstein A.V., Janin J. (1989). *Protein Eng* 1: 1–3.
39. Amzel L.M. (1997). *Proteins* 28: 144–149.
40. Yu Y.B., Privalov P.L., Hodges R.S. (2001). *Biophys J* 81: 1,632–1,642.
41. Vinals J., Kolinski A., Skolnik J. (2002). *Biophys J* 83: 2,801–2,811.
42. Tamura A., Privalov P.L. (1997). *J Mol Biol* 273: 1,948–1,060.
43. Yu Y., Lavigne P., Kay C.M., Hodges R.S., Privalov P.L. (1998). *J Phys Chem B* 103: 2,270–2,278.
44. Kauzmann W. (1959). *Adv Prot Chem* 14: 1–63.
45. Luo H., Sharp K.A. (2002). *Proc Natl Acad Sci USA* 99: 10,399–10,404.
46. Mark A.E., van Gunsteren W.F. (1994). *J Mol Biol* 240: 167–176.
47. Gomez J., Freire E. (1995). *J Mol Biol* 252: 337–350.
48. Baker B.M., Murphy K.P. (1997). *J Mol Biol* 268: 557–569.
49. Armstrong K.M., Baker B.M. (2007). *Biophys J* 93: 597–609.
50. Bosshard H.R. (2001). *News Physiol Sci* 16: 171–172.
51. Bruzzese F.J., Connelly P.R. (1997). *Biochemistry* 36: 10,428–10,438.
52. Ferrari M., Lohman T.M. (1994). *Biochemistry* 33: 12,896–12,910.
53. Luque I., Freire E. (2002). *Proteins* 49: 171–180.
54. Dunitz J.D. (1995). *Chemistry & Biology* 2: 709–712.
55. Ladbury J.E. (1996). *Chemistry & Biology* 3: 973–980.
56. Dunitz J.D. (1994). *Science* 264: 670.
57. Privalov G.P., Privalov P.L. (2000). *Meth Enzymol* 323: 31–62.
58. Cooper A. (2000). *Biophys Chem* 85: 25–39.
59. Cooper A., Johnson C.M., Lakey J.H., Nöllmann M. (2001). *Biophys Chem* 93: 215–230.
60. Cooper A., Dryden D.T.F. (1984). *Eur Buophys J.* 11: 103–109.
61. Williams D.H., Westwell M.S. (1998). *Chem Soc Rev* 27: 57–63.
62. Hunter C.A., Tomas S. (2003). *Chemistry & Biology* 10: 1023–1032.

CHAPTER 4

Kinetics of Biomacromolecular Complex Formation: Theory and Experiment

Georgi V. Pachov*, Razif R. Gabdoulline*[γ], Rebecca C. Wade*

*Molecular and Cellular Modeling Group, EML Research gGmbH,
Schloss-Wolfsbrunnenweg 33, 69118 Heidelberg, Germany
*[γ]BioQuant, University of Heidelberg, Im Neuenheimer Feld 269,
69120 Heidelberg, Germany
E-mail: Rebecca.Wade@eml-r.villa-bosch.de

The kinetics of association contributes to the biological function of macro-molecular complexes. Here, we first discuss key features and determinants of bimolecular association kinetics. Then we give an overview of contemporary experimental techniques for measuring kinetic properties and for gaining associated structural and dynamic data. We then describe theoretical and computational approaches to calculating kinetic properties. We end this chapter by discussing recent computational advances with selected examples of protein–nucleic acid and protein–protein complexation.

4.1 Introduction

The formation of biological complexes between proteins, proteins and small molecules, and proteins and nucleic acids plays a part in many biological processes, including gene transcription, cell signalling, enzyme catalysis and the immune response. Molecular association is governed by both the kinetic and the thermodynamic properties of the molecules and the medium involved. Inside cells, the medium is crowded with a variety of different macromolecules. Biomacromolecular complexes vary widely in their affinities and lifetimes, ranging from obligate and permanent to transient and short-lived complexes. Here, we will consider bimolecular association to form a transient complex. Complexation is usually characterised in terms of affinity, as weak or tight and strong. The affinity is often largely determined by the dissociation rate. However, association rates can vary over many orders of magnitude between complexes and be critical in the biological context. For example, the snake toxin fasciculin must not only strongly inhibit acetylcholinesterase (which is critical to neural transmission) but also reach its target quickly.[1] Similarly, the intracellular inhibitor barstar protects the bacterium *Bacillus amyloquefaciens* from the enzyme barnase, which it excretes to act extracellularly as a ribonuclease.[2] The protein interleukin 4 (IL-4) forms a complex with its cellular receptor and the time of this process is a measure for the regulation of the immune system.[3] The speed at which the *lac* repressor binds to its chromosomal *lac* operator regulates gene expression in the living cell.[4]

In this chapter, we will address the problem of understanding how a biomolecular complex forms and what the macromolecular interactions involved are. Some important parameters for describing the kinetics of molecular association will be introduced. In addition, we will give an overview of recent experimental techniques and methods for measuring these kinetic parameters and investigating macromolecular interactions. We will then focus on theoretical and computational approaches for calculating association rates and will discuss the current limitations of these approaches. Finally, the most recent computational advances in studying protein–protein and protein–nucleic acid association will be reviewed.

4.1.1 *Bimolecular Association*

In the cellular environment, two biomolecules can bind by diffusing towards each other. Active transport processes may also contribute to binding but these will not be discussed here. Association can be considered to consist of two steps: first an intermediate is formed by diffusion; this is called a 'diffusional encounter complex'. Then this intermediate evolves to form a tightly bound complex. Association is diffusion-controlled when the first step is rate-limiting, and it is reaction-controlled when the second step determines the rate of the association process.

4.1.1.1 Diffusional Encounter Complex

The characterization of the diffusional encounter complex is of high importance for protein and nucleic acid design studies aimed at altering the association kinetics. In the case of diffusion-controlled processes, formation of the encounter complex determines the bimolecular diffusional rate constant. The rate of diffusional association has an upper limit: a reaction between two molecules cannot be faster than their rate of collision. This limit is around 10^9 $M^{-1}s^{-1}$ for uniformly reactive spheres of typical macromolecular size[5] in aqueous solvent with no forces between them. As a rule, a random collision of two molecules does not result in binding – a freely diffusing molecule X must come close to its binding patch on a target molecule Y in order to form a diffusional encounter complex. Geometrically, this complex can be viewed as an ensemble of configurations able to evolve to the bound state. During a single encounter, the two molecules have time to undergo substantial rotational reorientation while remaining trapped in the vicinity of each other and undergoing multiple collisions. This effect is known as a 'diffusive entrapment'. A Brownian Dynamics (BD) study[6] of two non-interacting spheres the size of small proteins showed about 400 times larger association rate ($2 \cdot 10^6$ $M^{-1}s^{-1}$), attributed to the diffusive entrapment effect, than the rate calculated by a simple geometric correction of the Smoluchowski rate considering two contacts as the criterion for binding ($1 \cdot 10^4$ $M^{-1}s^{-1}$). An association rate constant of about 10^6 $M^{-1}s^{-1}$ is typical of protein–protein pairs that bind without strong electrostatic interactions. Attractive electrostatic forces can lead to enhancement of the rates to values very close to the Smoluchowski rate.

4.1.1.2 Bound Complex

Once the encounter complex has formed, the biomolecules must reorient with respect to each other to form a fully bound complex. They may also undergo changes in conformation and induced-fit in order to achieve a bound complex. Within the complex, the biomolecules are held together by close-range non-covalent interactions such as salt bridges, hydrogen bonds and van der Waals interactions. These interactions depend on the chemical nature of the interacting groups of both molecules as well as on their spatial arrangement and can be mediated by individual water molecules. A biomolecule can have one or several binding sites stabilising the complex.

A subtle change in the binding site can change the binding mode significantly. Therefore, biological associations are dependent on the structure of both molecules and can be highly specific.

4.1.2 *Molecular Transport*

In the living organism, the biomolecules move in a fluid environment. Molecular transport depends on both the macroscopic and the microscopic properties of the surroundings.

4.1.2.1 *Diffusion*

Diffusion is the spontaneous motion of solute particles through a solution caused by collisions between the solute particles and the solvent molecules. For a particle undergoing normal diffusion, the average value of the square displacement along one spatial dimension is proportional to the time t elapsed during the process,

$$<r^2> = 2nD,   \qquad (4.1)$$

where D is the diffusion coefficient and n is the number of dimensions. In some studies of molecular diffusion in cells and nuclei, anomalous diffusion has been observed with the displacement showing a smaller or larger dependence on time corresponding, respectively, to subdiffusion or superdiffusion.[7] The flux of particles J across a certain area can be expressed through the concentration gradient in one-dimension by Fick's first law:

$$J = -D\frac{\partial C}{\partial x}   \qquad (4.2)$$

Many transport phenomena are described by the well-known continuity equation,

$$divJ + \frac{\partial C}{\partial t} = 0, \tag{4.3}$$

which describes the conservation of matter. From Eq. 4.2 and Eq. 4.3, Fick's second law can be derived,

$$\frac{\partial C}{\partial t} = D \frac{\partial^2 C}{\partial x^2}, \tag{4.4}$$

which is also known as the diffusion equation.

4.1.2.2 *Viscosity*

When a particle moves in a fluid it experiences friction to an extent depending on the properties of the fluid. The macroscopic quantity describing the internal resistance to flow is called viscosity η. For a moving sphere with radius r, it is inversely related to the diffusion coefficient D through the Stokes-Einstein formula.

$$\eta = \frac{k_B T}{6\pi r D} \tag{4.5}$$

Here, k_B is the Boltzmann constant and T is the temperature. The crowded cytoplasmic and nuclear environments have been observed to result in diffusion of small proteins such as GFP (green fluorescent protein) that is slower by a factor of about 4 than observed in aqueous solution.[7] The cellular environment is heterogeneous and thus it is a simplification to describe it by a macroscopic viscosity. Indeed, the crowded intracellular environment can result, depending on the solute size, in subdiffusion.[7]

4.1.3 *Molecular Interactions*

The interactions between biomolecules vary in strength, type and source. Therefore, a wide spectrum of different forces contribute to complex formation.[8] Here, we will discuss only the electrostatic, hydrodynamic and hydrophobic interactions since their contribution to the kinetics of bimolecular association is shown to be considerable. We describe well-established methods for modelling these interactions.

4.1.3.1 *Electrostatics*

Electrostatic interactions are important for bimolecular association because they are relatively long-range interactions and may therefore guide the association process by means of attractive and repulsive interactions. Their importance is shown by the dependence of association rate on ionic strength and the generally much greater influence on the association rate of mutations of charged than of neutral residues. The biological entities in the cell are surrounded by ions, which screen the electrostatic interactions between the species. One way to account for the ions is to compute the molecular electrostatic potential $\varphi(r)$ using the nonlinear Poisson-Boltzmann equation,

$$-\nabla \varepsilon(r)\nabla \varphi(r) = \rho(r) + \sum_i q_i n_i e^{\frac{q_i \varphi}{k_B T}} \qquad (4.6)$$

where $\varepsilon(r)$ is the position dependent dielectric permittivity, $\rho(r)$ is the molecular charge density, q_i and n_i are the charge and the concentration of the ith ionic species in the bulk, respectively. The above equation can be approximated by the linear Poisson-Boltzmann equation if the exponential is expanded as a Taylor series,

$$-\nabla \varepsilon(r)\nabla \varphi(r) + \varepsilon \kappa^2 \varphi = \rho(r) \qquad (4.7)$$

where κ is the Debye-Hückel screening length. Equations 4.6 and 4.7 are used in studies of interactions between macromolecules in continuous media, i.e. media for which water molecules and ions are not modelled explicitly. When two molecules approach each other in an aqueous solvent, an electrostatic desolvation effect arises due to the lower dielectric constant of the solute compared to that of the solvent. Charges located on the bimolecular complex interface become desolvated upon complex formation and this results in unfavourable electrostatic energy changes. This desolvation effect becomes significant at short distances and is mainly dependent on the location and magnitude of the charged groups.

4.1.3.2 *Hydrodynamics*

Hydrodynamic interactions between molecules are caused by the flow of solvent due to their mutual motions. Depending on the structure and shape of the molecules, these interactions can be either attractive or repulsive. To model them, one usually represents the diffusion coefficient as a tensor describing the properties of the solute and the media.[9] Another representation uses a mean field for hydrodynamic interactions with the diffusion coefficient depending on a local volume fraction of a system mimicking macromolecular crowding.[10]

4.1.3.3 *Hydrophobicity*

The hydrophobic effect results in favourable interactions between two macromolecules in aqueous solvent. The reason for this is that the nonpolar groups on the surface of the molecules can avoid interaction with polar groups or molecules (water) by forming a complex. This is caused by the fact that the water molecules are orientationally restricted by the presence of the nonpolar species, and this leads to an entropy decrease. Furthermore, the presence of the nonpolar species affects the ability of the water molecules to make hydrogen bonds. The hydrophobic force is temperature dependent and it is entropy driven only at low to room temperatures. At higher temperatures, the enthalpic contributions become significant.[11,12] Although, a generalised theory describing the hydrophobic forces does not exist, hydrophobic interactions can be modelled by incorporating a solvent accessible surface area term in the Gibb's free energy[13] or using a hydrophobic potential of mean force.[14]

4.1.4 *Reaction Rates*

If a molecule of type X forms a complex Z with a molecule of type Y, then the rate of formation of species Z and the rate of depletion of X are given by,

$$V_Z = \frac{dC_Z}{dt}, \qquad V_Z = \frac{dC_X}{dt} \qquad (4.8)$$

where C_Z and C_X are the concentrations of the Z and X species, respectively. This reaction process depends on the association and dissociation rate constants k_{on} and k_{off}, respectively.

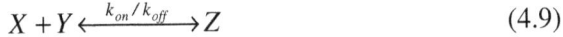

$$X + Y \xleftarrow{\quad k_{on}/k_{off} \quad} Z \qquad (4.9)$$

The reaction flux J can be expressed via the rate constant by the equation

$$J = kC^m \qquad (4.10)$$

where m denotes the order of a reaction. For our case we have,

$$J_Z = k_{off} C_Z \qquad (4.11)$$

$$J_X = J_Y = k_{on} C_X C_Y \qquad (4.12)$$

where Eq. 4.11 describes a first-order reaction whereas binding is a second-order process (see Eq. 4.12). The rate constants can be related to an equilibrium association constant:

$$K_a = \frac{k_{on}}{k_{off}} \qquad (4.13)$$

The reciprocal of K_a is the equilibrium dissociation constant K_d. An analytical solution of the diffusion-controlled association constant kon can be obtained for uniform spheres reacting at a distance r^5

$$k_{on} = 4\pi r (D_X + D_Y) \qquad (4.14)$$

where D_X and D_Y are the diffusion constants for species X and Y, respectively. Equation 4.14 is valid when there are no forces between the spheres.

In the case of interacting spheres, k_{on} is given by,[15]

$$k_{on} = \frac{4\pi\,(D_X + D_Y)}{\int\limits_r^\infty \dfrac{e^{U(r)/k_B T}}{r^2}\,dr} \tag{4.15}$$

where $U(r)$ is a centrosymmetric interaction potential between the spheres. For more complicated geometries and interaction forces, numerical approaches are necessary to compute association rates (see Section 4.3 below).

4.2 Experimental Techniques

There is a vast variety of experimental techniques and methods for measuring the properties affecting the kinetics of formation of biomolecular complexes. We will give an overview of some of them, focusing on the information they provide about binding kinetics.

4.2.1 Crystallography

X-ray and electron (EM)[16] crystallography can be used to determine the spatial atomic coordinates of biomolecules and provide some information on their dynamics. Time-resolved techniques can be used to induce a reaction process in the crystal and a time-resolved Laue diffraction pattern[17,18] can reveal structural re-arrangements in a time range from 100 ps to several seconds (see Fig. 4.1a). This technique was successfully applied to study the functional dynamics of hemoproteins, particularly the pathways of CO in myoglobin.[19,20] However, for a complete picture of the biomolecular dynamics the time-resolved experiments should be combined with complementary spectroscopic and computational techniques. In addition, the limitation of this technique is its dependence on the quality of the crystal,[18] which, for example, was shown to be affected by the association dynamics of the wild type and mutant forms of Ynd1p proteins.[21]

4.2.2 *Nuclear Magnetic Resonance (NMR)*

In contrast to X-ray crystallography, NMR analysis is applied to biomacro-molecules in solution. It can be used to investigate different structural and kinetic aspects of protein–protein and protein–nucleic acid complex formation via investigating the relaxation and chemical shift behaviour of the nuclear spin in the system.

A powerful method that has recently been introduced for detecting and visualising low population, weak and transient encounter complexes in biomolecular interactions is paramagnetic relaxation enhancement (PRE).[22] One of the molecules under investigation is labelled with a paramagnetic centre (e.g. Mn^{2+}), which has unpaired electrons with a large magnetic moment (see Fig. 4.1b). This results in magnetic dipolar interactions with the backbone amide protons of the other biomolecule. Since the transverse relaxation rate R2 depends on the distance r between the unpaired electron on the labelled centre and the proton as $< r^{-6} >$, the mutual position and orientation can be determined from the magnitude of the rate. In such an experiment, the transverse PRE rate Γ_2 can be measured from two time-points T_a and T_b and is given by:[22]

$$\Gamma_2 = R_2^{para} - R_2^{dia} = \frac{1}{T_b - T_a} ln \frac{I_{dia}(T_b) I_{para}(T_a)}{I_{dia}(T_a) I_{para}(T_b)} \qquad (4.16)$$

where R_2^{para}, I_{para} and R_2^{dia}, I_{dia} are the rates and the peak intensities in the paramagnetic and diamagnetic states, respectively. Apart from structural information, the PRE can provide information about the binding mechanism in an equilibrium system of two interacting biomolecules. However, to distinguish the PRE contribution of the less populated states from the resonance of the highly occupied ones, the rate of exchange between the states should be fast (> 50 s^{-1}, see Fig. 4.1b). In this way, low population transient intermediates can be studied in systems with relatively weak interactions. The PRE technique was applied successfully to protein–DNA, as well as protein–protein complex formation.[22] Nonspecific encounter complexes depending on the ionic strength were identified in both cases suggesting that the interactions facilitating the encounter formation are electrostatic in nature: hydrophobic interactions were also found to be important for the ultra-weak (Kd > 15 mM) self-association of the histidine-containing phosphocarrier protein (HPr) of *Escherichia coli*.[23]

4.2.3 *Stopped-flow Methods (SF)*

The stopped-flow (SF) technique is an experimental method for studying the kinetics of two or more solutions. Each of them is placed in a syringe and simultaneously the molecules are rapidly inserted into a mixing chamber (Fig. 4.1c). This chamber plays a role as an observation box, where the interactions between the biomolecules can be investigated via different experimental techniques. Most commonly, the SF experiments rely on fluorescence methods for reporting the progress of the binding reaction. Some of these methods will be discussed in the next sub-sections.

4.2.4 *Fluorescence Recovery After Photobleaching (FRAP)*

The fluorescence recovery after photobleaching (FRAP) technique is carried out in living cells in order to study biomolecular dynamics.[24,25] The molecules under interest are labelled with a fluorescent protein (e.g. green fluorescent protein GFP) and, after photobleaching with a high-intensity laser, the fluorescence of the GFP is destroyed within the area of the laser focus, whose radius is ω (see Fig. 4.1d).

The fluorescence recovers with time due to the movement of the surrounding molecules into the bleached region. If there is an immobile fraction of molecules in the photobleached region, then a partial recovery is observed. Hence, the mobile fraction M_f as well as the effective diffusion coefficient D_{eff} can be determined via

$$M_f = \frac{F_\infty - F_0}{F - F_0}, \qquad \tau_D \approx \frac{\omega^2}{4D_{eff}}, \tag{4.17}$$

where the Fs are the fluorescence intensities given in Fig. 4.1d and τ_D is the diffusion time. The FRAP curve reflects binding processes and binding information can be extracted from it.[26–28]

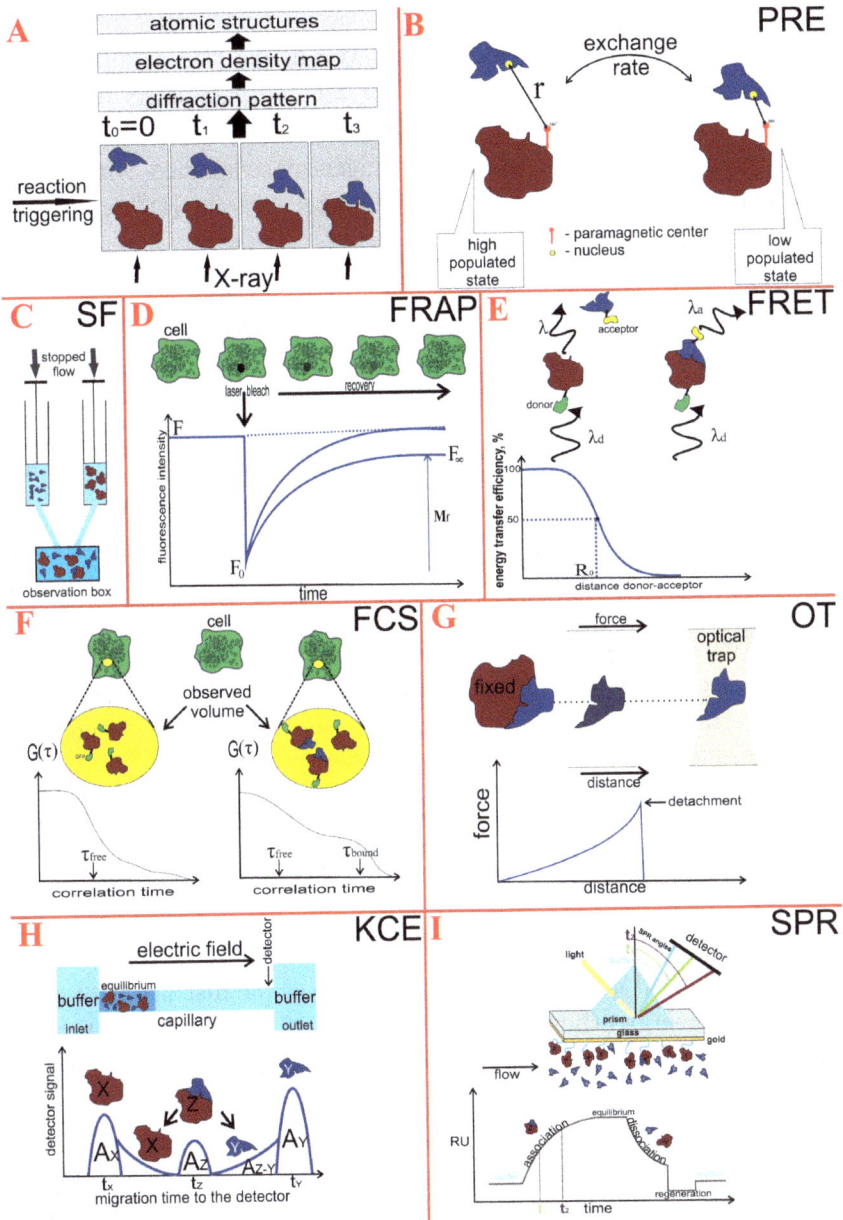

Fig. 4.1. Schematic figures of the experimental techniques (see the text).

Moreover, a 3D FRAP model taking into account the spatial localization of the binding sites within the cell, as well as the role of diffusion in binding, gives a better estimation of the binding parameters for DNA-transcription factor glucocorticoid receptor interactions.[29]

In another study,[30] systematic mutagenesis combined with FRAP revealed the binding geometry of the linker histone H1^0 within the nucleosome in chromatin fibre.

4.2.5 *Fluorescence Resonance Energy Transfer (FRET)*

The concept of fluorescence resonance energy transfer (FRET) is, as for FRAP, based on the use of fluorescent molecules for detecting the interactions between the biomolecules. However, in FRET the labelling of the species is done with different fluorophores exhibiting diverse excitation and emission spectra. In such a way, after illumination, the probe (donor-fluorophore) absorbs the light and either emits it to the surroundings or transfers it to another probe (acceptor-fluorophore) in close proximity (Fig. 4.1e). This energy transfer leads to an emission from the second probe with its characteristic wavelength λ. The energy transfer efficiency E depends on the distance between the donor and acceptor r via,

$$E = \frac{1}{1 + \left(\dfrac{r}{R_0}\right)^6},$$

(4.18)

where R_0 is called a Förster distance[31] and gives the distance between a pair at 50% energy efficiency (Fig. 4.1e). FRET has a wide application area in the study of protein–nucleic acid, protein–protein and protein ligand interactions.[32] For example, measurements of the end-to-end distance of the DNA on the nucleosome revealed its decrease in the presence of a linker histone H1, i.e. compact nucleosome complex forms.[33] In a related FRET study[34] on a trinucleosome system, it was shown that high salt concentration affects in a similar way the end-to-end distance while histone acetylation has a reverse effect. An investigation of DNA-integration host factor (IHF) interactions by time-resolved

FRET revealed a two-step process: firstly, IHF binds quickly to the DNA and secondly the DNA bends slowly, the latter step being the rate-limiting step at IHF concentrations over 150 nM.[35,36] In addition, it has been found that the binding constant of DNA wrapping around the IHF protein is strongly dependent on the salt and the anion identity.[37]

4.2.6 *Fluorescence Correlation Spectroscopy (FCS)*

Fluorescence correlation spectroscopy (FCS) examines biomolecular interactions *in vivo* and *in vitro* within a defined volume (Fig. 4.1f). The fluorescently labelled molecules under interest diffuse through the volume and the photon fluctuations arising from the fluorophores are measured.[38,39] This signal can be autocorrelated and, via the amplitude of the autocorrelation function $G(\tau)$, different parameters like concentrations, diffusion constant and correlation times can be determined.[40,41] When the tagged molecule binds a binding partner, the diffusion time is increased and it can be detected on the autocorrelation plot (see Fig. 4.1f). However, this process has a significant impact on the slope only when the size of the binding target is large. Using different measuring techniques, the FCS can reveal important information on the binding mechanism and the dynamic behaviour of biological systems.[40,41]

4.2.7 *Force Probe Methods*

There exist several well-established experimental methods for measuring forces between macromolecules like optical tweezers (OT), magnetic tweezers (MT) and atomic force microscopy (AFM). The technique is based on attaching one of the biomolecules under investigation to a surface and the other one to a force sensor, which can be a bead or a cantilever. The OT use light (a laser beam), which is a momentum and electric field carrier and can induce forces and torques on the biomolecule and, thus, the latter can be trapped in a potential well[42] (Fig. 4.1g). In contrast to the OT, the magnetic tweezers use a magnetic field gradient for trapping.[32] In this way, the molecule can be gradually detached from the binding species and the forces generated measured. The force acting on the object due to the laser beam increases with the distance between both macromolecules until the detachment takes place.[43] In applications of OT generally, this dependence is monitored and insights into the dynamical and mechanical properties of the biological macromolecules can be gained. Stretching nucleosomes in chromatin showed a relatively small dissociation constant implying a stable nucleosome complex.[44,45] The histone-like protein TmHU binding to the DNA displayed different reaction rates to the closely related HU protein which can be attributed to the different temperatures at which they function.[46]

The AFM uses a cantilever with a sharp tip that scans over the surface of the specimen.[47] The forces arising from the tip-surface interaction cause a detection of the cantilever. This detection leads to a change of the rejected laser beam from the cantilever, which is detected by photodiodes. The AFM can provide knowledge about the forces between the biomolecules and therefore permit determination of some kinetic parameters like the dissociation constant.[48]

Another recently developed force method is the nanopore force spectroscopy (NFS).[49] In contrast to OT, MT and AFM, the NFS can determine the association and dissociation constants by analysing many species in a short time. The technique is based on the insertion of a nanopore in a planar lipid bilayer and measuring the ionic current through it after applying a voltage.[49] When a molecule (ssDNA) goes through the nanopore, the current is partially blocked and the measured translocation time increased.

However, when a bound pair tries to pass through the nanopore, a very long blockage takes place because of the small pore diameter. To free the channel the voltage is reversed. For a weakly bound protein–DNA complex the measured translocation time is slightly longer than for the DNA itself. The association and dissociation rates for the ssDNA–Exonuclease I complex have been obtained in this way.[49]

4.2.8 *Electrophoresis*

Electrophoresis is a technique based on the movement of particles in solvent due to an external electric field. If a mixture of two species is placed in a capillary and an electric field is applied, the molecules will move depending on their charge, mass and the solvent viscosity (Fig. 4.1h). A method called kinetic capillary electrophoresis (KCE) can be used to investigate the interactions between biomolecules.[50] Using different variations of boundary and initial conditions, the technique can provide a quantitative description of the complex formation and dissociation processes. For example, in the non-equilibrium capillary electrophoresis of equilibrium mixtures (NECEEM) method, the biomolecules X, Y and their complex Z are equilibrated and inserted in the inlet of the capillary (Fig. 4.1h). Upon application of an electric field, the mixture separates and the components move towards the outlet with different velocities. At the end of the capillary, the concentrations are measured with a detector and an electropherogram is obtained (Fig. 4.1h). Knowing the migration times and the peak areas, the dissociation constant can be determined

$$K_d = \frac{X_0 \left(1 + \dfrac{A_Y}{A_Z + A_{Z \to Y}}\right) - Y_0}{1 + \dfrac{A_Z + A_{Z \to Y}}{A_Y}} \qquad k_{off} = ln\left(\frac{A_Z + A_{Z \to Y}}{A_Z}\right)\frac{1}{t_Z} \qquad (4.19)$$

where the areas A_Y, A_Z and $A_{Z \to Y}$ are depicted in Fig. 4.1h, X_0 and Y_0 are the total concentrations in the equilibrium mixture and t_Z is the migration time of the complex.[50] The ssDNA-single strand binding protein (SSB) interactions have been investigated using 6 KCE experimental methods and the results revealed two types of interaction between the biomolecules: specific and non-specific.[50] Gel electrophoresis experiments combined with fluorescence techniques have been used to determine the kinetic constants of formation of a chromatosome particle mediated by a histone chaperone NAP1[51] and of a DNA–Endonuclease IV complex.[52]

4.2.9 *Surface Plasmon Resonance (SPR) Biosensor*

A direct way to measure the association and dissociation rate constants of a biomolecular complex can be provided by a surface plasmon resonance (SPR) biosensor technology.[53] The detection is achieved via a biosensor chip consisting of a thin gold layer mounted on glass (Fig. 4.1i). The target molecules X are attached on the gold surface and a flow of binding partners Y is injected across the surface. A beam of light is passed through a prism linked to the glass surface and the refraction angles are measured with a detector. At a certain incident angle and wavelength, the electrons in the gold film are excited and a surface plasmon wave (resonance) is generated. This leads to a reduction of the light intensity. When the molecules Y bind to the target X, the concentration of the molecules, i.e. the mass, on the gold surface is increased. Hence, the SPR refractive angle changes with time and this is recorded as resonance units (RU), which depend on the mass of the complex (Fig. 4.1i).[54] Association will lead to an RU increase, dissociation to a decrease. Equilibrium can be reached when the reaction rates are fast.

In order to remove the bound molecules Y from X, a regeneration buffer can be inserted giving the initial RU state (Fig. 4.1i). With the SPR technology, one can measure the kinetics of protein–peptide,[55] enzyme–DNA[56] and protein–membrane[57] interactions as well as design drugs.[53,54]

4.3   Theoretical and Computational Approaches

There are two methodologically distinct approaches to computing bimolecular association rate constants for biomacromolecules, which we will describe in the next two sections. In the first approach, the absolute rate constants are computed from a model aimed at providing a complete description of the relevant properties of the interacting molecules and the forces between them. Diffusional motion is simulated by particle-based Brownian dynamics (BD) simulations,[6] or a density-distribution-based analytical or finite difference solution of the partial differential diffusion equation is made.[58]

In the second approach, the aim is the estimation of the relative rates of association assuming that the rates are scaled by a factor depending on interaction energy.

4.3.1 *Computation of Bimolecular Rate Constants*

The most commonly used formalism for calculating bimolecular association rate constants is known as the Northrup-Allison-McCammon (NAM) method.[59] It is assumed that the rate constant for the approach of two molecules to a separation b at which the forces between them are negligible or centrosymmetric is given by the analytical Smoluchowski expression, Eq. 4.14 or Eq. 4.15 with $r = b$. BD simulations are then used to compute the probability that, having reached separation b, the molecules go on to form a diffusional encounter complex. Thousands of trajectories are simulated and the fraction recorded in which criteria are satisfied for diffusional encounter complex formation.[5] The two binding partners can be modelled in atomic detail.

The first applications of this method were to studies of superoxide dismutase[60–62] and it has been applied to many diffusion-influenced enzymes.[63,64] With this method, it was possible to predict generic protein–protein association rates by monitoring the formation of native polar contacts observed in the bound complex.[65–67] The process of protein–protein complex formation was quantified for different systems ranging from glycolytic enzymes interacting with actin filaments to antigen–antibody interactions.[68–74]

Weighted Ensemble Brownian (WEB) dynamics was introduced in Ref. 75. Rather than simulate the motion of a single molecule as in the NAM method, one molecule is replaced by an ensemble of pseudoparticles or weighted probability packets. These occupy bins along the intermolecular reaction coordinate that are equally sampled through splitting and combining the weighted pseudo-particles.[75] It was shown that the WEB method can be efficient in calculating the rates in the presence of large free energy barriers to association.[76]

Lee and Karplus[77] proposed an alternative BD method based on calculating the time-dependent probability that a pair of reactant molecules, having started in the reaction zone, will be found in this zone again in the absence of reaction. Zhou developed an efficient algorithm to calculate the time-dependent rate coefficient via the survival probability of the pair of reactants started in the reaction zone.[78]

Continuum models solve the diffusion equation under appropriate boundary conditions. The flux of particle density through a reactive boundary gives the association rate.[79,80] Applied to enzyme–substrate association, this approach can be computationally more efficient than BD simulations; a disadvantage is that the atomic-detail properties of the molecular species that is treated as a particle density cannot be taken into account.

4.3.2 *Estimation of Rate Enhancements due to Electrostatic Interactions*

It was shown that the variation in bimolecular rate constant can be correlated with the variation of the electrostatic interaction energy of proteins in transient intermediate configurations.[81] In the PARE (Predicting Association Rate Enhancement) approach, these configurations are approximated by the bound state of two proteins.[82] In this approach, only the rate enhancement can be predicted and the rate in the absence of electrostatic interactions (the basal rate) should be estimated by a different method or derived from experiments. This approach can be used for rapid, structure-based calculation of the electrostatic attraction between two proteins in the complex.[83] The average Boltzmann factor near the active site is seen to be a good descriptor of rate enhancement[81] as well.

This approach can be applied to design proteins to bind faster and tighter to their protein–complex partner by optimization of electrostatic interactions between two proteins[83] and to the design of fast enzymes by optimising the interaction potential in the active site.[84]

4.4   Recent Advances in Computational Approaches

4.4.1 *Protein–Protein Interactions*

4.4.1.1 *Computation of Rates*

Rate enhancement due to electrostatic interactions was given an extensive analysis in the publications of the last years.[85] The Poisson-Boltzmann(PB) description was applied to modelling of the association of various proteins (E9:Im9, Bn:Bs, AChE:Fas and IL4:IL4BP),[86,87] and it was found that suitably parameterised PB calculations yield accurate predictions of association rates. The results were found to be particularly sensitive to the definition of the solute–solvent dielectric boundary and, in some cases, to the choice of linear or non-linear PB equation[86–88] Computational tools were developed to design proteins that bind faster and tighter to their protein partners by optimization of the electrostatic interactions between two proteins.[83] Calculation of electrostatic steering was applied to a large number of structurally characterised protein complexes,[89] and it was found that electrostatic steering may result in an increase of over 100-fold in k_{on} for about 25% out of 68 transient hetero–protein–protein complexes.

4.4.1.2 *Determinants of Binding*

Binding affinity is determined by both association and dissociation rates. It has been shown that it is possible to design mutants that change the binding affinity by changing only the association rates for a number of different protein complexes.[83,90] The association of Cdc25B phosphatase with its Cdk2-pTpY/CycA protein substrate was found to be governed to a significant extent by the interactions of the remote hotspot residues, whereas dissociation was governed by interactions at the active site of phosphatase.[91] Single residue mutations can result in large (>100 fold) changes in association rates, significantly modulating binding affinity.[91] Mutants of the Ras effector protein Ral, a guanine nucleotide dissociation stimulator (RalGDS), to optimise electrostatic steering to

Ras were investigated in Ref. 92. It was possible to correctly predict and design a triple mutant that associates 14 times faster with the Ras protein, with binding properties being close to Raf, another Ras effector.

The speed of interaction between CheA, an autophosphorylating protein histidine kinase (PHK), and CheY, a phospho-accepting response regulator protein (RR), appears to be to a large extent regulated by the rapid association of the P2 domain of CheA with CheY.[93] This indicates the importance of the association process (together with fast His-Asp phospho-transfer within the respective PHK–RR complexes) in fast response times in the chemotaxis system of *Escherichia coli*.[94]

4.4.1.3 *Encounter Complex Quantification*

When two proteins diffuse together to form a bound complex, an intermediate is formed at the end-point of diffusional association which is called the diffusional encounter complex. Its characteristics are important in determining association rates, yet its structure cannot be directly observed experimentally.[95]

The encounter complex is an ensemble of target positions for BD simulations, and achieving this ensemble ensures further binding of molecules when the association is diffusion-controlled. The nature of this intermediate state for the association of barnase and barstar was investigated by double-mutant cycle experiments.[96] Evidence for contacts between charged residues in this intermediate state was found. The activation entropy of the transition state was found to be small, indicating a small degree of desolvation. This is all consistent with the models of encounter complexes generated by BD.[65] The residue–residue contacts maintained in the transition state differ at low and high ionic strength, indicating that the structure of the intermediate state changes with changing solvent conditions.

The encounter complex was also quantified by introducing mutations that alter association rates and modelling the structures of bimolecular configurations that fit experimental data.[97,98] However, it was shown that mutations alter the encounter complex[99] and therefore quantifying the encounter complex using mutational data is not straightforward. Very recently, it has become possible to quantify such transient intermediate complexes using long-range distance restraints derived from paramagnetic NMR methods[100,101] and this is expected to shed more light on the nature of encounter complexes.

4.4.1.4 *Induced Fit Phenomena*

It was shown that there is more than one intermediate state in the association process[102] because protein–protein binding in general consists of multiple steps: diffusion, conformer selection, and refolding or induced fit. It is, however, not simple to quantify all intermediates experimentally, although it can be shown in some cases that a one-step model of association is not sufficient.[103]

An extreme case of induced fit occurs when the protein folds or refolds upon binding to its partner.[104] It was shown[105] that this may be followed by a fly-casting effect coupled to electrostatic steering for the Ets domain of SAP-1 protein binding to its specific DNA sequence. A significant induced fit was found in the case of fasciculin 2 (Fas2) binding to acetylcholinesterase (AChE) indicating that the conformation of Fas2 able to bind AChE is not stable in the unbound form of Fas2 and the association process should follow a conformational change of a stable form of Fas2 that is not complementary to AChE.[106,107]

4.4.1.5 *Crowding Phenomena*

The influence of crowding agents cannot be explained simply as exertion of obstacles, volume exclusion or the change in the solvent viscosity, because there is a complex dependence of the molecular dynamics and reactions in crowded solutions on the properties of the molecular interactions in the system. An inverse linear relation was found between translational diffusion of proteins and viscosity in almost all solutions tested, in accordance with the Stokes–Einstein relation. Conversely, no simple relation was found between either rotational diffusion rates or association rates (k_{on}) and viscosity.[108] In all crowded solutions, the measured absolute k_{on}, but not k_{off} values, were found to be slower as compared to buffer. In the presence of low mass crowding agents k_{on} depends inversely on the solution viscosity.

In high mass polymer solutions, k_{on} changes only slightly, even at viscosities 12-fold higher than in water.[109] See also a recent review on this topic.[110]

4.4.2 *Protein–nucleic Acid Interactions*

4.4.2.1 *Computation of Rates*

Nucleic acids like DNA and RNA are highly negatively charged biomolecules, preferentially attracting proteins with positive binding sites. The PARE approach for computing protein–protein association rates[87] has been successfully applied to an atomistic model of protein–RNA (U1A–U1SLII) interactions.[111] The effects of salt concentration and mutations on the association and dissociation rates agreed well with the experimental data.[111] A theoretical model verified by BD simulation showed that nonspecific binding to DNA leads to a rate enhancement.[112] Moreover, a slightly higher association constant was obtained for a linear DNA than for a circular DNA, which was attributed to the more open conformation of the former.[112] The influence of the electrostatic and hydrodynamic interactions on the association rate was analysed by BD simulations for the translation protein eIF4E binding to five analogous mRNA cap-molecules.[113] For all five complexes, a very good reproduction of experimentally obtained kinetic rates was obtained for a two-step mechanism accounting for the existence of a diffusional encounter complex.[113] Another study[114] proposed a quantitative model for a reversible two-step binding mechanism between the DNA and a bacterial RNA polymerase, where the formation of an open complex from the closed one was investigated. It was shown that the rate of open complex formation depends on the interaction energies of the closed and opened complexes as well as on the DNA duplex melting energy.[114]

4.4.2.2 *Specificity and Nonspecificity*

Nucleic acid binding proteins play a crucial role in gene regulation. Therefore, a detailed picture of the searching mechanism and location of the target molecule is necessary. Some of these proteins are known to bind DNA with association rates exceeding the Smoluchowski rate for three-dimensional diffusion, suggesting that one-dimensional diffusion of the protein along the DNA plays a role in the binding process. The first quantitative investigation of specific and nonspecific binding of proteins to DNA suggested that a protein should spend half of its time in a 3D diffusional search and the other half in 1D sliding on the DNA to find its correct binding site.[115]

Subsequent Monte Carlo lattice simulations revealed that transcription factors (TF) need only spend about 15% of their diffusional search time free in solution in order to locate the target DNA molecules; this result is in agreement with relevant experiments.[116] A simple model including the conformational fluctuations of the TFs during the diffusional search, as well as the sliding, aimed at finding out the shortest binding time to the DNA consistent with thermodynamics.[117] Many DNA binding proteins have multiple binding sites[30] and they can transiently bind to two or more DNA binding sites. This facilitates 'intersegment transfer' in which a protein can transfer from one DNA segment to another without having to fully dissociate.

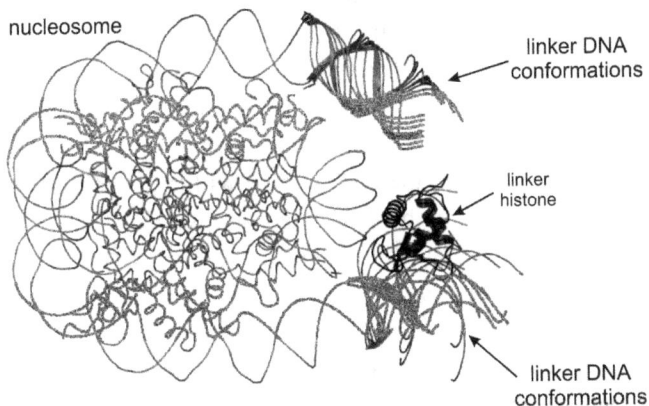

Fig. 4.2. Position of the linker histone GH5 (black) between different conformations of the linker DNAs (grey).

This process can contribute to enhancement of the rate of protein binding to specific sites on DNA binding.[118] In addition, lattice simulations[119] revealed that inter-segment transfer may increase protein diffusion while increasing the nucleic acid chain density. Considering only 1D sliding on the DNA chain, the diffusion coefficient was found to be reciprocal to the chain density.[119] A generalised theory has been developed for site-specific DNA–protein interactions arising from one-dimensional diffusion after the protein binds nonspecifically to DNA by three-dimensional diffusion.[120]

In summary, several factors contribute to the quick rates of association of nucleic acid binding proteins, such as transcription factors, and DNA; these include one-dimensional diffusion, intersegment transfer and conformational changes (upon binding).

4.4.2.3 *Chromatin Models*

In the cell nucleus, the DNA molecules compact to highly ordered chromatin structures assembling a biological network. Within this network, the DNA combines with other proteins and together they form complexes called nucleosomes, which in turn interact with each other. Several coarse-grained chromatin models[121–123] as well as atomistic ones[124] have been used to try to elucidate the important interactions within the chromatin fibre. The binding dynamics of the linker histone to the DNA was investigated both experimentally[30] and theoretically.[125] Recently, all-atom BD simulations revealed a unique binding mode of the linker histone GH5 to the nucleosome within a certain range of linker DNA conformations (unpublished data, see Fig. 4.2).

Moreover, a systematic mutagenesis identified two binding sites in agreement with experiments.[30] Since chromatin participates in the gene expression, the kinetics of its components must be clearly regulated – the DNA replication, transcription and repair are processes involving proteins, which usually have to rapidly identify the target DNA. The time of such initial target location can be directly affected by facilitated diffusion as well as, for example, by the level of DNA exposure and histone tail acetylation on the nucleosome.[126] Generally, all these studies contribute to the understanding of the chromatin fibre structure and function; some suggest the existence of different nucleosome repeat lengths in nature leading to different topological and mechanical properties of the fibre, while others primarily focus on the protein–DNA and DNA–DNA interactions resulting in diverse chromatin models.

4.5   Conclusions and Outlook

In this chapter, we have discussed characterization of the kinetics of biomacromolecular complex formation from theoretical and experimental points of view. New experimental techniques and methods are being developed to study the interactions between biomolecules over different time and length scales. However, these techniques are still insufficient to precisely describe and quantify the detailed dynamics of associating molecules. The computational approaches can provide a detailed description of the association process, but they are not able to simulate this process without approximations such as neglecting molecular flexibility. In addition, establishing the effects on macromolecular association of the crowded and heterogeneous cellular environment is a challenge for both computational and experimental approaches. To overcome this problem it is necessary to develop multi-scale and coarse-grained models with more accurate molecular interaction force fields and new supercomputers and highly parallelized software, allowing detailed simulations over several orders of time and length scales.

Acknowledgements

We thank the Klaus Tschira Foundation (KTS), the German Research Foundation (DFG) and the Center for Modelling and Simulation in the Biosciences (BIOMS). The authors are thankful to Prof. Dr Jörg Langowski for helpful discussion, and Anna Feldman-Salit and Paolo Mereghetti for reading the manuscript.

References

1. Quinn D.M. (1987). *Chem Rev* 87: 955–979.

2. Jucovic M., Hartley R.W. (1996). *Proc Natl Acad Sci USA* 93: 2,343–2,347.
3. Wang Y., Shen B.J., Sebald W. (1997). *Proc Natl Acad Sci USA* 94: 1,657–1,662.
4. Elf J., Li G.W., Xie X.S. (2007). *Science* 316: 1,191–1,194.
5. Smoluchowski M.V. (1917). *Z Phys Chem* 92: 129–168.
6. Northrup S.H., Erickson H.P. (1992). *Proc Natl Acad Sci USA* 89: 3,338–3,342.
7. Dix J.A., Verkman A.S. (2008). *Annu Rev Biophys* 37: 247–263.
8. Motiejunas D., Wade R.C. (2007). Structural, Energetic, and Dynamic Aspects of Ligand Receptor Interactions. *Comprehensive Medicinal Chemistry II* Vol. 4, Elsevier, Oxford.
9. Ermak D., McCammon J.A. (1978). *J Chem Phys* 69: 1,352–1,360.
10. Sun J., Weinstein H. (2007). *J Chem Phys* 127: 155,105.
11. Dill K.A., Truskett T.M., Vlachy V., Hribar-Lee B. (2005). *Annu Rev Biophys Biomol Struct* 34: 173–199.
12. Chandler D. (2005). *Nature* 437: 640–647.
13. Dominy B.N., Perl D., Schmid F.X., Brooks C.L. (2002). *J Mol Biol* 319: 541–554.
14. Lin M.S., Fawzi N.L., Head-Gordon T. (2007). *Structure* 15: 727–740.
15. Berg O.G., von Hippel P.H. (1985). *Annu Rev Biophys Biophys Chem* 14: 131–160.
16. Stahlberg H., Walz T. (2008). *ACS Chem Biol* 3: 268–281.
17. Bourgeois D. , Royant A. (2005). *Curr Opin Struct Biol* 15: 538–547.
18. Bourgeois D., Schotte F., Brunori M., Vallone B. (2007). *Photochem Photobiol Sci* 6: 1,047–1,056.
19. Milani M., Nardini M., Pesce A., Mastrangelo E., Bolognesi M. (2008). *IUBMB Life* 60: 154–158.
20. Anselmi M., Nola A.D., Amadei A. (2008). *Biophys J* 94: 4,277–4,281.
21. Onuma K., Watanabe A., Kanzaki N., Kubota T. (2006). *J Phys Chem B* 110: 24,876–24,883.
22. Clore G.M., Tang C., Iwahara J. (2007). *Curr Opin Struct Biol* 17: 603–616.
23. Tang C., Ghirlando R., Clore G.M. (2008). *J Am Chem Soc* 130: 4,048–4,056.
24. Axelrod D., Koppel D., Schlessinger J., Elson E., Webb W. (1976). *Biophys J* 16: 1,055–1,069.
25. Lippincott-Schwartz J., Altan-Bonnet N., Patterson G.H. (2003). *Nat Cell Biol* Suppl: S7–14.
26. Sprague B.L., Pego R.L., Stavreva D.A., McNally J.G. (2004). *Biophys J* 86: 3,473–3,495.
27. Sprague B.L., McNally J.G. (2005). *Trends Cell Biol* 15: 84–91.
28. Braga J., McNally J.G., Carmo-Fonseca M. (2007). *Biophys J.* 92: 2,694–2,703.
29. Sprague B.L., Müller F., Pego R.L., Bungay P.M., Stavreva D.A., McNally J.G. (2006). *Biophys J* 91: 1,169–1,191.
30. Brown D.T., Izard T., Misteli T. (2006). *Nat Struct Mol Biol* 13: 250–255.
31. Förster T. (1948). *Ann. Physik* 437: 55–75.
32. Joo C., Balci H., Ishitsuka Y., Buranachai C., Ha T. (2008). *Annu Rev Biochem* 77: 1–26.

33. Toth K., Brun N., Langowski J. (2001). *Biochemistry* 40: 6,921–6,928.
34. Bussiek M., Toth K., Schwarz N., Langowski J. (2006). *Biochemistry* 45: 10,838–10, 846.
35. Sugimura S., Crothers D.M. (2006). *Proc Natl Acad Sci USA* 103: 18,510–18,514.
36. Kuznetsov S.V., Sugimura S., Vivas P., Crothers D.M., Ansari A. (2006). *Proc Natl Acad Sci USA* 103: 18,515–18,520.
37. Meulen K.A.V., Saecker R.M., Record M.T. (2008). *J Mol Biol* 377: 9–27.
38. Magde D., Elson E.L., Webb W.W. (1972). *Phys Rev Lett* 29: 705–708.
39. Elson E.L., Magde D. (1974). *Biopolymers* 13: 1–27.
40. Langowski J. (2008). *Methods Cell Biol* 85: 471–484.
41. Haustein E., Schwille P. (2007). *Annu Rev Biophys Biomol Struct* 36: 151–169.
42. Ashkin A. (1970). *Phys Rev Lett* 24: 156–159.
43. Moutt J.R., Chemla Y.R., Smith S.B., Bustamante C. (2008). *Annu Rev Biochem* 77: 205–228.
44. Brower-Toland B., Wacker D.A., Fulbright R.M., Lis J.T., Kraus W.L., Wang M.D. (2005). *J Mol Biol* 346: 135–146.
45. Brower-Toland B.D., Smith C.L., Yeh R.C., Lis J.T., Peterson C.L., Wang M.D. (2002). *Proc Natl Acad Sci USA* 99: 1,960–1,965.
46. Salomo M., Keyser U.F., Kegler K., Gutsche C., Struhalla M., Immisch C., Hahn U., Kremer F. (2007). *Microsc Res Tech* 70: 938–943.
47. Binnig, Quate, Gerber (1986). *Phys Rev Lett* 56: 930–933.
48. Cao Y., Balamurali M.M., Sharma D., Li H. (2007). *Proc Natl Acad Sci USA* 104: 15,677–15,681.
49. Hornblower B., Coombs A., Whitaker R.D., Kolomeisky A., Picone S.J., Meller A., Akeson M. (2007). *Nat Methods* 4: 315–317.
50. Krylov S.N. (2007). *Electrophoresis* 28: 69–88.
51. Mazurkiewicz J., Kepert J.F., Rippe K. (2006). *J Biol Chem* 281: 16,462–16,472.
52. Garcin E.D., Hosfield D.J., Desai S.A., Haas B.J., Björas M., Cunningham R.P., Tainer J.A. (2008). *Nat Struct Mol Biol* 15: 515–522.
53. Homola J. (2008). *Chem Rev* 108: 462–493.
54. Wilson W.D. (2002). *Science* 295: 2,103–2,105.
55. Boozer C., Kim G., Cong S., Guan H., Londergan T. (2006). *Curr Opin Biotechnol* 17: 400–405.
56. Lee H.J., Wark A.W., Goodrich T.T., Fang S., Corn R.M. (2005). *Langmuir* 21: 4,050–4,057.
57. Besenicar M., Macek P., Lakey J.H., Anderluh G. (2006). *Chem Phys Lipids* 141: 169–178.
58. Schlosshauer M., Baker D. (2004). *Protein Sci* 13: 1,660–1,669.
59. Northrup S.H., Allison S.A., McCammon J.A. (1984). *J Chem Phys* 80: 1,517–1,524.
60. Allison S.A., Ganti G., McCammon J.A. (1985). *Biopolymers* 24: 1,323–1,336.
61. Antosiewicz J., Briggs J.M., McCammon J.A. (1996). *Eur Biophys J* 24: 137–141.

62. Stroppolo M.E., Pesce A., Falconi M., O'Neill P., Bolognesi M., Desideri A. (2000). *FEBS Lett* 483: 17–20.
63. Wade R.C., Gabdoulline R.R., Lüdemann S.K., Lounnas V. (1998). *Proc Natl Acad Sci USA* 95: 5,942–5,949.
64. Wade R.C. (1996). *Biochem Soc Trans* 24: 254–259.
65. Gabdoulline R.R., Wade R.C. (1997). *Biophys J* 72: 1,917–1,929.
66. Elcock A.H., Gabdoulline R.R., Wade R.C., McCammon J.A. (1999). *J Mol Biol* 291: 149–162.
67. Gabdoulline R.R., Wade R.C. (2001). *J Mol Biol* 306: 1,139–1,155.
68. Ouporov I.V., Knull H.R., Lowe S.L., Thomasson K.A. (2001). *J Mol Recognit* 14: 29–41.
69. Sept D., McCammon J.A. (2001). *Biophys J* 81: 667–674.
70. Altobelli G., Subramaniam S. (2000). *Biophys J* 79: 2,954–2,965.
71. Fogolari F., Ugolini R., Molinari H., Viglino P., Esposito G. (2000). *Eur J Biochem* 267: 4,861–4,869.
72. Northrup S.H., Boles J.O., Reynolds J.C. (1988). *Science* 241: 67–70.
73. Rienzo F.D., Gabdoulline R.R., Menziani M.C., Benedetti P.G.D., Wade R.C. (2001). *Biophys J* 81: 3,090–3,104.
74. Haddadian E.J., Gross E.L. (2006). *Biophys J* 91: 2,589–2,600.
75. Huber G.A., Kim S. (1996). *Biophys J* 70: 97–110.
76. Rojnuckarin A., Livesay D.R., Subramaniam S. (2000). *Biophys J* 79: 686–693.
77. Lee S., Karplus M. (1987). *J Chem. Phys* 86: 1,883–1,903.
78. Zhou H.X., Szabo A. (1996). *Biophys J* 71: 2,440–2,457.
79. Song Y., Zhang Y., Bajaj C.L., Baker N.A. (2004). *Biophys J* 87: 1,558–1,566.
80. Cheng Y., Suen J.K., Zhang D., Bond S.D., Zhang Y., Song Y., Baker N.A., Bajaj C.L., Holst M.J., McCammon J.A. (2007). *Biophys J* 92: 3,397–3,406.
81. Zhou H.X. (1997). *Biophys J* 73: 2,441–2,445.
82. Selzer T., Schreiber G. (1999). *J Mol Biol* 287: 409–419.
83. Schreiber G., Shaul Y., Gottschalk K.E. (2006). *Methods Mol Biol* 340: 235–249.
84. Zhou H.X., Wong K.Y., Vijayakumar M. (1997). *Proc Natl Acad Sci USA* 94: 12,372–12,377.
85. Gabdoulline R.R., Wade R.C. (2002). *Curr Opin Struct Biol* 12: 204–213.
86. Alsallaq R., Zhou H.X. (2008). *Proteins* 71: 320–335.
87. Alsallaq R., Zhou H.X. (2007). *Structure* 15: 215–224.
88. Wang T., Tomic S., Gabdoulline R.R., Wade R.C. (2004). *Biophys J* 87: 1,618–1,630.
89. Shaul Y., Schreiber G. (2005). *Proteins* 60: 341–352.
90. Selzer T., Albeck S., Schreiber G. (2000). *Nat Struct Biol* 7: 537–541.
91. Sohn J., Buhrman G., Rudolph J. (2007). *Biochemistry* 46: 807–818.
92. Kiel C., Selzer T., Shaul Y., Schreiber G., Herrmann C. (2004). *Proc Natl Acad Sci USA* 101: 9,223–9,228.
93. Stewart R.C., Bruggen R.V. (2004). *J Mol Biol* 336: 287–301.

94. Stock A.M., Robinson V.L., Goudreau P.N. (2000). *Annu Rev Biochem* 69, 183–215.
95. Gabdoulline R.R., Wade R.C. (1999). *J Mol Recognit* 12: 226–234.
96. Frisch C., Fersht A.R., Schreiber G. (2001). *J Mol Biol* 308: 69–77.
97. Harel M., Cohen M., Schreiber G. (2007). *J Mol Biol* 371: 180–196.
98. Miyashita O., Onuchic J.N., Okamura M.Y. (2004). *Proc Natl Acad Sci USA* 101: 16,174–16,179.
99. Spaar A., Dammer C., Gabdoulline R.R., Wade R.C., Helms V. (2006). *Biophys J* 90: 1,913–1,924.
100. Tang C., Iwahara J., Clore G.M. (2006). *Nature* 444: 383–386.
101. Volkov A.N., Worrall J.A.R., Holtzmann E., Ubbink M. (2006). *Proc Natl Acad Sci USA* 103: 18,945–18,950.
102. Grünberg R., Leckner J., Nilges M. (2004). *Structure* 12: 2,125–2,136.
103. Kourentzi K., Srinivasan M., Smith-Gill S.J., Willson R.C. (2008). *J Mol Recognit* 21: 114–121.
104. Levy Y., Cho S.S., Onuchic J.N., Wolynes P.G. (2005). *J Mol Biol* 346: 1,121–1,145.
105. Levy Y., Onuchic J.N., Wolynes P.G. (2007). *J Am Chem Soc* 129: 738–739.
106. Bui J.M., McCammon J.A. (2006). *Proc Natl Acad Sci USA* 103: 15,451–15,456.
107. Bui J.M., Radic Z., Taylor P., McCammon J.A. (2006). *Biophys J* 90: 3,280–3,287.
108. Kozer N., Kuttner Y.Y., Haran G., Schreiber G. (2007). *Biophys J* 92: 2,139–2,149.
109. Kozer N., Schreiber G. (2004). *J Mol Biol* 336: 763–774.
110. Zhou H.X., Rivas G., Minton A.P. (2008). *Annu Rev Biophys* 37: 375–397.
111. Qin S., Zhou H.X. (2008). *J Phys Chem B* 112: 5,955–5,960.
112. Alsallaq R., Zhou H.X. (2008). *J Chem Phys* 128: 115108.
113. Bachut-Okrasinska E., Antosiewicz J.M. (2007). *J Phys Chem B* 111: 13,107–13,115.
114. Djordjevic M., Bundschuh R. (2008). *Biophys J* 94: 4,233–4,248.
115. Slutsky M., Mirny L.A. (2004). *Biophys J* 87: 4,021–4,035.
116. Rezania V., Tuszynski J., Hendzel M. (2007). *Phys Biol* 4: 256–267.
117. Hu L., Grosberg A.Y., Bruinsma R. (2008). *Biophys J* 95: 1,151–1,156.
118. Hu T., Shklovskii B.I. (2007). *Phys Rev E Stat Nonlin Soft Matter Phys* 76: 051909.
119. Wedemeier A., Zhang T., Merlitz H., Wu C.X., Langowski J. (2008). *J Chem Phys* 128: 155101.
120. Murugan R. (2007). *Phys Rev E Stat Nonlin Soft Matter Phys* 76: 011901.
121. Langowski J., Heermann D.W. (2007). *Semin Cell Dev Biol* 18: 659–667.
122. Merlitz H., Klenin K.V., Wu C.X., Langowski J. (2006). *J Chem Phys* 125: 014906.
123. Arya G., Zhang Q., Schlick T. (2006). *Biophys J* 91: 133–150.
124. Wong H., Victor J.M., Mozziconacci J. (2007). *PLoS ONE* 2: e877.
125. Fan L., Roberts V.A. (2006). *Proc Natl Acad Sci USA* 103: 8,384–8,389.
126. Kampmann M. (2005). *Mol Microbiol* 57: 889–899.

CHAPTER 5

Evolutionary Trace of Protein Functional Determinants

Olivier Lichtarge

Department of Molecular and Human Genetics, Baylor College of Medicine,
One Baylor Plaza, Houston, TX 77030, USA
E-mail: lichtarge@bcm.tmc.edu

Protein–protein interactions are the elementary units from which molecular pathways and cellular networks are built. A complete description of the functional surfaces that determine protein binding still eludes us, however. The Evolutionary Trace (ET) approach to this problem is to analyse jointly sequences, evolutionary trees and structures to reveal the key amino acid determinants of protein function. I will show that these amino acids cluster spatially in the structure and match functional sites. The activity of many proteins may then be traced to narrow sets of relevant amino acids that form 'elementary units of function and of interaction'. Their discovery allows experimentalists to rationally design activity through targeted mutagenesis, for example along the G protein-signalling pathway, or to build three-dimensional templates that predict the likely function of new protein structures. The scalability and generality of ET further suggest that widespread functional site annotation and engineering are within reach, leading to proteome-wide manipulation of the molecular basis of protein function.

5.1 Introduction

The work described in this chapter is rooted in a very concrete question. How to control G protein signalling? This pathway is universally found in eukaryotes. It contains the single largest gene family in humans: G protein coupled receptors. And its basic mechanism is to change the intracellular concentration of second messengers (such as Ca++, K+, cAMP, cGMP, DAG, IP3) in response to the extracellular binding of a ligand to its specific G protein coupled receptor. This is accomplished in a series of steps: first, these seven transmembrane helices receptors change conformation and activating intracellular G proteins. G proteins, which are a complex of three alpha, beta and gamma subunits, in turn become activated when the alpha subunit releases GDP and exchanges it for GTP. It then separates from the receptor and from the beta-gamma subunits to diffuse along the membrane and activate a membrane-bound channels or enzymes. So pervasive is this series of protein–ligand and protein–protein interactions that it mediates fundamental human sensing mechanisms such as vision, taste, smell, and about 70% of all our hormonal signalling. It is often quoted that 50% of all drugs target this signalling pathway. For this reason, many resources are dedicated to understand and gain control over the basic molecular events that mediate ligand binding, conformational induced allosteric signal transduction, and the formation and dissolution of an orchestrated series of protein–protein complexes in G protein signalling. Not only would this give us new insights into the basic mechanisms of protein function, but also it would yield new inroads into the prevention or control of human ailments including stroke, schizophrenia, pain, hypertension, asthma and migraines among many others.

More abstractly, however, the problems of how proteins sense inputs, respond with precise outputs, and allosterically carry information between input and output sites (or perhaps more precisely compute which inputs should be linked to which outputs) are at the heart of protein function and they go far beyond the special context of G protein signalling. These questions are representative of a much broader general search for the principles that govern protein interactions and the formation of complexes, pathways and networks. The hope in finding solutions is on the one hand to decipher the molecular basis of the relationship between protein-structure-function, and increase our fundamental knowledge of biology.

Table 5.1. Top-selling GPCR drugs (2005) (1). P2Y: purinergic; H: histamine; D: dopamine 5HT: serotonin; AT: angiotensin; ADR: adrenergic. Adapted from Ref. 4.

Ailment	Target GPCR	Drug	Sales $M
Stroke	P2Y12 antagonist	clopidrogel	5,277
Schizophrenia	5HT2/D1/D2	olanzapine	4,905
Pain	GABAB agonist	GABApentin	2,480
Hypertension	AT1 antagonist	valsartan	2,214
Allergies	H1 antagonist	fexofenadine	1,792
Migraine	5HT1D agonist	sumatriptan	1,454
Cancer	LH-RH agonist	leuporelin	904
Asthma	1ADR agonist	salmeterol	679
Gastric ulcer	H2 antagonist	famotidine	656
Schizophrenia	5HT2/D2 antagonist	risperidone	371

But on the other hand, the possibility that we may identify which molecular determinants mediate interactions also opens two important practical goals for biological engineering and ultimately therapeutic intervention, namely the possibility that we could modify and predict function rationally.

At first, a reductive approach to these problems may seem daunting. The complexity of modelling a protein, the surrounding solvent, multiple interacting partners, and diverse extracellular, intra-membranous and intracellular environments is clearly vast, and for the most part beyond current means. Even though the basic forces at play are quite well understood in terms of their physics, such as for example electrostatic, hydrophobic, van der Waals, osmotic or hydrogen bonding forces, yet they cannot be modelled completely accurately. For this, one would need a complete quantum mechanical description. While well understood in theory, the computational reality of every particle influencing every other particle of the system leads to a combinatorial computational explosion far beyond current and foreseeable computers. This therefore forces us to use semi-classical force field approximations. But these inherently contain errors or empirical biases, causing simulations to typically accumulate inaccuracies and drift away from observables unless one adds empirical constraints. A conundrum then is that it becomes difficult to discover new biological states that lie far from those for which the equilibrium is already known. For example, describing the active state of the G protein coupled receptor when the only known structures described the conformationally distinct inactive state.

Given these difficulties, we turn to a different type of reductive question and ask: given a protein of known structure, what are its functionally important amino acids? Although this query may seem initially vague, it does simplify modelling considerably as it seeks to filter out the parts of a biological system that contribute to noise from those that directly impact function. For example as Fig. 5.1 illustrates, the answer to this question would allow us to pinpoint where to direct mutational studies efficiently and successfully.

FROM FUNCTIONAL DETERMINANTS TO RATIONAL ENGINEERING

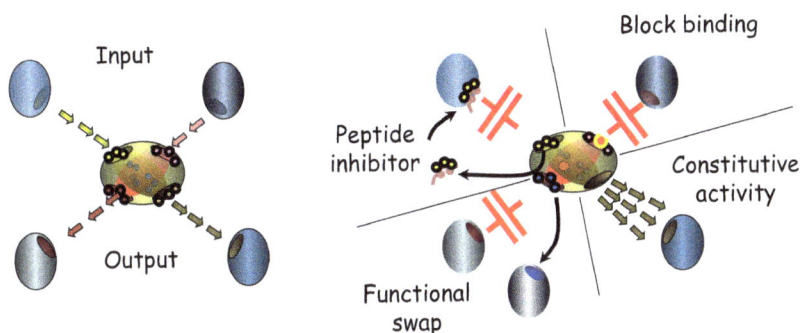

Fig. 5.1. Proteins as machines. Although simplistic, the view of a protein as having input sensing, output mediating and information transfer residues (left), lends itself to simple schemes for rational protein (right, see text).

To be precise, consider the interaction between a protein and its partners as being mediated by precise residues at 'input sites', precise residues at 'output sites', and a third type of residues in the body of the protein that shuttle information between the input and the output sites.

Then it becomes obvious that knowledge of the amino acid determinants of the input, the transfer and the output sites would allow us to generate rational, precise hypotheses on how they can be modified.

First, a single non-conservative mutation at one of these important amino acids could destroy the site. This would block that specific function – but not necessarily impair any of the other functional sites in the protein, thereby creating a separation of function mutation. A second possibility would be to transplant together the few amino acids identified as the key mediators of an interaction into another protein or peptide scaffold. This new molecule could then mimic the original site and by itself bind the interacting partner and thereby either inhibit the original interaction or bypass its need for further signalling. Yet a third possibility would be to rewire function by swapping in and out those key functional determinants between related family members and thereby transferring function of one onto the other and redirecting an interaction accordingly. Fourth, it may even be possible to target for mutation the internal amino acids that transfer information between an input and an output and thus hardwire a protein's response, which normally would depend on the presence of an interaction. For example, this could result in the constitutive activity of a receptor or of an enzyme.

Finally, if one could identify important amino acids in any protein structure, then it would be possible to compare the important functional sites that they define pairwise across the entire structural proteome. While most comparison would show large differences, there may be on occasion nearly perfect matches between previously unrelated structures. Such significant local structural similarity precisely over important functional regions might logically suggest that the underlying proteins carry out similar functions.

The case studies presented below support the approach suggested in Fig. 5.1: rational protein engineering and function prediction did follow from systematical identification of key amino acid mediators of function. And this was achieved by computing which residues are the most important in a protein.

To complete this preamble, a few more words are necessary on methods, which focus on the computational analysis of large-scale data sets produced by genomics techniques. Whether through sequencing, structural genomics or a variety of high-throughput experimental techniques such as expression microarrays, 'omics' biology yields vast quantities of sequences, increasingly large numbers of protein structures and massive bits of data on various aspects of protein function.

The problem with such data is that the process of high-throughput data extraction inherently erases the details of their biological relationships so that their biological meaning is lost. The challenge is then is to recover, or at least to formulate reasonable and testable hypothesis, on the true biological context of the sequences the structures and their functions. Ideally, we would like to be able do this on an equally large scale as that on which sequences and structures are produced; we would also like to produce results or predictions that are quantitative, or at least attached to known statistical significance values. Finally, we would hope to discover new, emergent rules that bring new insights into the sequence-structure-function relationship.

In this chapter, I show that it is possible to approach these questions from an evolutionary perspective and as a result integrate the analyses of sequence structure and function on a very large-scale, and identify both emerging new proteomic rules but also individual determinants of interaction among proteins leading, in practice, to general guidelines for the rational re-design of protein behaviour and of the molecular pathways they define.

5.2 Evolutionary Trace Basics: Which Amino Acids are Important in a Protein?

In the laboratory, it is relatively simple to identify functionally important protein residues. One simply targets mutations to amino acids and then tests which ones change the read-out on relevant assays. Together, mutations and assays can therefore pinpoint all the amino acids that play important functional roles. But there are problems with this approach. It is near impossible to systematically probe every possible amino acid with every possible substitution. And for most new proteins (and possibly for many proteins that are already well studied) the most relevant assays may be unknown, unavailable or both. It is therefore impractical to identify important amino acids experimentally on a proteomic scale. This makes it critical that we develop computational alternatives to experimental studies.

The evolutionary trace model (ET)[2] aims to perform the computational equivalent of mutational analysis. It postulates that during evolution an amino acid variation is equivalent to a laboratory mutation and that an evolutionary divergence is equivalent to a functional essay. If one treats these hypotheses as statements in mathematical logic, then it follows that just as one identifies important residues in the laboratory by correlating mutations with changes in an assay read-out, one could identify important residues *in silico* by correlating sequence variations with evolutionary divergences.

For example, consider a multiple sequence alignment of homologous proteins and their evolutionary divergence tree. If one considers all the sequences to be part of a single common branch, then the relevant sequence variation pattern that correlates with just a single branch would be absolute invariance for some position across the alignment. This defines Rank 1. Next, if one considers the first two branches of the tree as defining two distinct groups, then the relevant sequence variation pattern that correlates with now two branches would be a position in the alignment that displayed absolute invariance in one branch and also absolute invariance in the other branch, but variation between the two. This defines Rank 2. Likewise, if then one further considers the first n branches of the evolutionary tree and the n subgroups of proteins they define, then the relevant sequence variation pattern that correlates with them would be a position that are invariant within each of the first n branches, but variable among some of them. This defines Rank n.

In this way it is possible to iteratively split the aligned protein family into successively more and more branches that contain fewer and fewer sequences and ask each time whether a position in the alignment is perfectly correlated with the branches, meaning that for n branches of the position is invariant in each of the first n subgroups but variable among some of them. In this manner we can define the ET rank of a residue as the first node of the evolutionary tree after which it varies no further in any descendent branches. By definition then, top ranked residues (such as one, two, three, etc.) have important properties. They become fixed earlier in evolution; and their variations are linked to more profound evolutionary differences, suggesting also that they are linked to more profound functional differences. This is illustrated in Fig. 5.2.

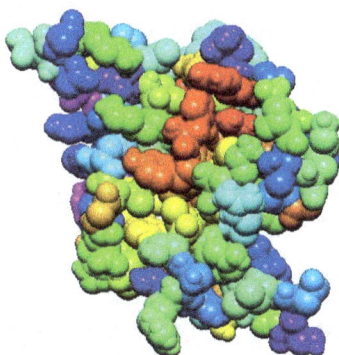

Fig. 5.2. Evolutionary Trace of SH2. The importance rank of every residue in the structure of the SH2 domain (PDB code 1skj) is colour-coded from most important (red) to least important (blue). Dropping the code 1skj into the Evolutionary Trace Viewer input box at URL http://mammoth.bcm.tmc.edu/traceview/ will query a precomputed database of ET analyses and produce a clickable link to download the results. This link opens a JAVA ET Viewer molecular display application that can show the molecule viewed from any angle, in bond or spacefill mode, in backbone or full view mode, and in colour by cluster or in ('gobstopper') colour rainbow of importance mode (as shown here). The user controls the importance threshold so as to colour only residues in the top 5% of importance, or the top 10% of importance, or any other percent coverage of importance.

This example makes it apparent that a rank of evolutionary importance can be assigned to every amino acid in the structure: that the branching hierarchy of the evolutionary tree translates into difference ranks for different residues; and that top-ranked residues, shown above in the red-orange, are precisely the binding site of the phosphorylated tyrosine peptide that a SH2 domains recognise. This site is then enlarged into a yellow cluster of slightly less important residues, and so on with green then cyan then dark blue amino acids. In the original publication, a review of available mutations suggested that substitutions at the top-ranked residues were as is fitted with complete loss of function. Substitutions at slightly lesser-ranked residues modulated function. And substitutions at unimportant residues even very close to the ligand had no impact on function. Such an implied correlation between ET rank and the functional impacts of an amino acid variations is consistent with the seminal observation by Wells and colleagues that only a fraction of interface residues contribute significantly to binding interactions, while other residues also at the interface can be substituted much more freely.[3]

At this point it is important to note that the algorithm for assigning ET ranks is well defined, simple and that it makes no assumptions about which substitutions may or may not be conservative. Finally, we note that tree-based ET analysis has some unique advantages:[4]

1. The tree naturally accounts for the over-representation of sequences from nodes that are highly populated relative to others. This is because ET assigns ranks based on the distance from the root of the evolutionary tree rather than based on the number of leaves, or sequences, that are in any one branch. This is an important deviation from simply measuring 'invariance'.

2. The tree filters out much of the noise inherent to sequence analysis. This is because it defines *a priori* which patterns of evolutionary variations are important. This bias allows us to classify residues as important, or not, in a straightforward manner that can then be tested experimentally. Other algorithms by contrast tend to gauge residue importance through measures of side chain conservation. But this is inherently problematic because the ways in which a side chain varies is context-dependent,[5] and hence a poor measure of functional significance among homologs.[6]

3. The tree confers onto ET a strategy akin to experimental mutational analysis. In the laboratory, mutational analysis builds causal links between residues and function by assaying the function of specific mutants. Similarly, ET links specific sequence variations with

functional differences, using tree branch-points as 'virtual functional assays'. This is also profoundly different from other types of sequence analyses that reason by analogy (i.e., if proteins A and Z share a motif, they likely share some function). In that light, ET simply sorts the mutations and functional assays that already occurred during evolution, and interprets them as one would in the laboratory.

4. This approach provides far more mutations and assays than are achievable in the laboratory. Specifically, pairwise sequence comparisons yield many more functionally competent variations than can be constructed in the laboratory. Additionally, and crucially, a tree with N proteins has N-1 branch-points, each of which is equivalent to a virtual functional assay. Even if only a third of these 'assays' are useful, this is many more than the one or two assays typically available in the laboratory for a specific protein family.

Thus the tree allows us to explicitly take evolution into account and thereby link a vast number of evolutionary mutations with natural selection assays. We now show that the richness of these evolutionary data translates into a large number of biologically relevant conclusions. In the next section I shall focus on a series of experiments that aimed to validate or test the limitations of ET, in order to identify functional sites and their key determinants.

5.3 Validation Through Prospective Case Studies

5.3.1 *Separation of Function*

The first and simplest test of the productive power of ET is to predict novel functional sites and disable it through targeted mutations. A recent example is the efficient separation of function experiments that enabled to identify the distinct structural regions of the Ku70/80 protein complex that were responsible for telomere maintenance as well as double strand DNA break repair.[7] Evolutionary traces were performed separately on the Ku80 and the Ku70 components of the complex. Both identified novel functional sites that were targeted for single point non-conservative substitutions in the hope of disrupting the function of these putative sites without otherwise altering other activities of this multifunctional complex which normally binds DNA.

Of 17 mutations at or near the fifth alpha-helix of Ku80, nine were associated with significant telomeric defects, but none caused any double strand break repair defect. In sharp contrast, of 13 mutations at or near the fifth a helix of the ancestor of the related Ku70 domain, five caused significant double strand break repair defects while none of the mutations were associated with any telomeric defects. This ET guided mutational study thus explained the paradoxical presence of DNA on a telomeric maintenance protein end joining activity (since the joining of telomeric ends from different chromosomes would have catastrophic results). The two functions are in effect completely segregated on opposite ends of the complex. Both are centred on the divergently related Alpha five helices so that the double strand break repair activity of Ku70 is oriented towards the centromere, while the telomere maintenance activity is oriented towards the telomere.

It is important to note that this study produced in six months 14 separation of function mutations, compared to a prior functional screen and a yeast system that had produced three such mutations over a period of two years. This illustrates the efficiency associated with leading evolutionary experiments guide laboratory investigations.

5.3.2 *Rewiring Functions*

This type of application follows from two complementary features of ET. First, functional areas can be outlined, as a whole, from the surface clusters of top-ranked residues. Second, within such a cluster every residue is tagged by its degree of importance, namely its ET rank. Logically, one may then try to swap cognate residues in descending order of importance (starting from the single most important) in order to switch functions between ancestrally related proteins. In essence, the hypothesis is that top-ranked residues define a function specificity code for that protein family. Exchanging top-ranked residues should then be necessary and sufficient to rewire function.

Regulators of G protein signalling: past studies in the family of regulators of G protein signalling (RGS) proteins provide the most thorough demonstration yet of evolution-directed discovery of allostery, Specificity and 4° Structure.[8,9] These studies also (a) anticipated mutational and crystallographic analyses, and (b) demonstrated ET-based recoding of function specificity. We review this work in some detail since it illustrates best the interplay between computation and experiments.

RGS proteins bind onto activated G proteins, which are bound to GTP, and they accelerate the rate at which these G proteins hydrolyse GTP back to GDP. This stops G protein signalling. Thus RGS play a fundamental role in regulating the strength of signalling.

In G protein signalling, the activated G protein is a Gα•GTP complex that activates effectors until it reverts to its inactive Gα•GDP state. Regulators of G protein signalling (RGS) proteins help limit G protein signalling by increasing Gα's rate of GTP hydrolysis.[10] However, not all RGS proteins act alike. For example, PDEγ (the γsubunit of the visual effector cGMP phosphodiesterase) enhances the inactivation of Gα-transducin by RGS9, but inhibits the GTP-ase enhancing effect of RGS4, RGS16, GAIP, RGS6 and RGS7. To understand the basis for this difference, we traced all 42 known RGS proteins and mapped top-ranked residues onto the available RGS4 structure,[10] to discover the novel functional surface R2, shown in red and blue in Fig. 5.3.

Two observations suggest that R2 was an interface whereby the effector influences RGS domain activity. First, side chain variations in R2 correlate with the specific activity of each RGS in the presence of the PDEγ. For example, going from proteins inhibited by PDEγ to those enhanced by it, the residues cognate to position 387 in RGS7 vary from acidic to basic, and those at position 394 vary from polar (or hydrophobic) to basic. Second, in the Gα-RGS complex, site R2 is contiguous to a part of trace cluster A2 in Gα (Fig. 5.2) that: (a) does not interact with Gα, and (b) contains residues linked to PDEγ interaction. This led us to predict that the effector binds the RGS-Gα complex by straddling both A2 and R2, and that R2 residues such as 387 and 394 are key mediators of the effector's modulation of the RGS effect on GTP-ase activity.[9]

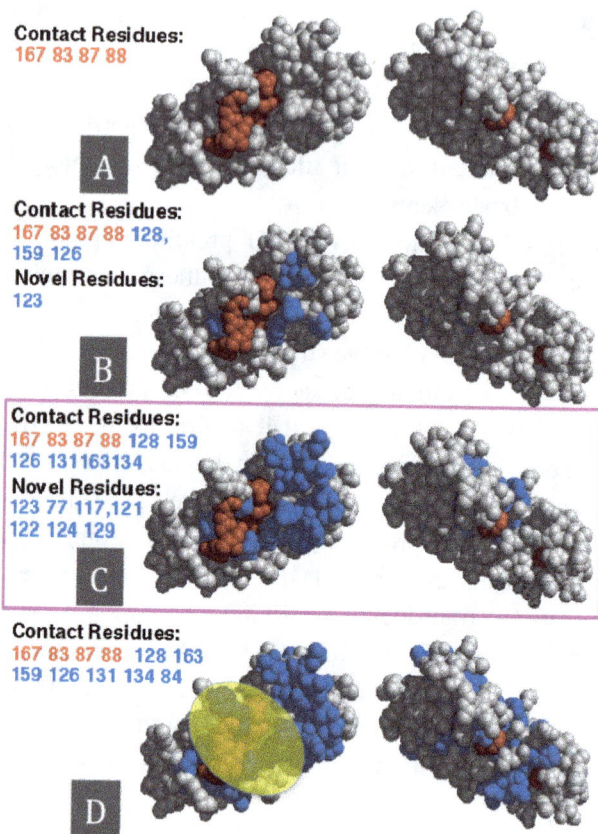

Contact Residues:
167 83 87 88

A

Contact Residues:
167 83 87 88 128,
159 126
Novel Residues:
123

B

Contact Residues:
167 83 87 88 128 159
126 131 163 134
Novel Residues:
123 77 117, 121
122 124 129

C

Contact Residues:
167 83 87 88 128 163
159 126 131 134 84

D

Fig. 5.3. RGS Trace. Successive rows from A to D show RGS traces at increasing evolutionary rank. Invariant residues are red and other trace residues are blue. A large cluster of 17 trace residues emerges on one side of the surface (left), while the opposite face (right) remains essentially free of signal. Part of the cluster (in yellow, D) includes 10 of the 11 RGS residues at the Gα interface. The remaining 7 trace residues form a novel cluster, R2, that extends beyond the Gα-binding site and whose unknown function was surmised to be an effector binding site through which Gα activity is modulated.[9]

We next tested these predictions by mutating R2 trace residues 348, 387 and 394 in RGS7, normally inhibited by PDEγ,[11] into their cognate RGS9 residues that are closely related to RGS7 but are normally enhanced by PDEγ. The first finding was that the (E387L/P394R) double mutant reduced the GTP-ase basal activity of the Gα-RGS7/9 complex to slightly less than when the wild-type complex is in the presence of PDEγ. Adding PDEγ produces no further additional inhibition. Hence these mutations constitutively inhibited RGS7, as if in the presence of PDEγ. (The additional mutation of residue 348 had little effect by itself, and the triple mutant behaves nearly the same as the E387L/P394R mutant). Remarkably, residues 387 and 394 are not in direct contact with Gα since R2 is not part of the interface to Gα. Thus their mutations not only mimic the PDEγ̃-inhibited form of the wild-type protein but they affect GTP hydrolysis by Gα at a distance. Thus (387,394) behaves as an allosteric switch of GTP-ase activity, which upon mutation recapitulates the effect of PDEγ on RGS-Gα.

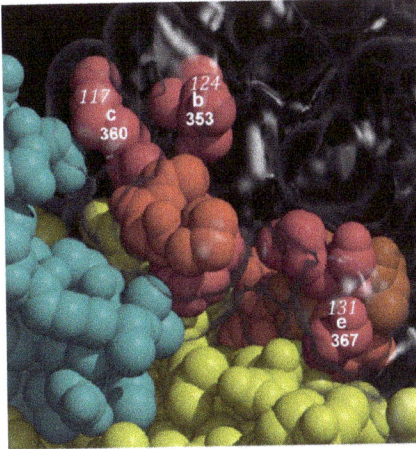

Fig. 5.4. The PDEγ-RGS-Gα structure supports trace predictions. The structure of the catalytic core domain of RGS9 in complex with both Gt/i1α•GDP•AlF4- and the C-terminal 38 amino acids of PDEγ reveals that PDEγV66 contacts R2 at class specific residue RGS9-W362. A second R2 residue RGS9-R360 is within 5 Å from PDEγ. The other residues in R2 form a network of contacts extending from 387/394 to the Gα contact residue 401. Thus residues 387 and 394 may exert their influence by communicating through this loop to the catalytic interface, with specificity determined by the amino acids that comprise both the α5/α6-connecting loop and the RGS/Gα interaction surface. (RGS 7 numbering is on top, RGS 9 numbering is below, PDEγ is blue, Gα is yellow, and R2 residues are red and orange, data from Slep et al.[12]

To further characterise the molecular basis of PDEγ's variable effect on different RGS, we targeted for mutations at three other class specific residues from the RGS/Gα interface (adjacent to R2) that also differed markedly between RGS7 and RGS9 (RGS7-A396, S401 and Y404). The mutations at A396 and Y404 had no significant effects, but the (L348Q/E387L/P394R/S401G) RGS7-Gα had basal GTP-ase activity nearly equal to that of the wild type RGS9-Gα-PDEγ complex. Furthermore, adding PDEγ causes a slight enhancement to match exactly the activity of the RGS9-Gα-PDEγ complex. Thus, residue S401 in RGS7 is a critical determinant of the direction of the PDEγ effect on Gα (enhancement or reduction). Moreover, S401 requires the assistance of 387 and 394 since when S401 is mutated alone, the protein remains inhibited by PDEγ, further indicating an allosteric relationship between 387/394 and 401.[9] These three critical RGS residues are shown in Fig. 5.4, where it is clear that the remainder of R2 residues form a direct pathway between the (387/394) allosteric switch and the effector residue (401).

The final test of our predictions came in the form of the actual structure of the PDEγ-RGS-Gα complex shown in Fig. 5.4, solved by Slep *et al.*[12] It clearly shows that PDEγ and RGS7 share an interface precisely along R2. These studies illustrated ET's ability to identify active sites, predict 4° structure interactions, and guide targeted mutagenesis to reveal the elements of an inter-protein allosteric pathway and the molecular basis of function. To our knowledge, this work was the first instance of a published computational prediction of a functional interface then followed by dual mutational and crystallographic validation.

5.3.3 *Redirecting Protein Binding Specificity to DNA*

The RGS study was the first experimental evidence in support of an evolutionary-based protein function specificity key. The concept however had first been raised in the context of the DNA binding domain (DBD) of nuclear hormone receptors in which it was noticed that the top-ranked trace residues of the DBD at the DNA binding site made direct structural contact with the most invariant nucleic acid bases of the consensus response element to which it binds.[13] This suggested a model of interfaces in which functional determinants on

O. Lichtarge

either side were in direct contact so that evolutionary pressure would be mirrored structurally across the interface. This would explain that swapping key residues on one side would directly impact interaction on the other side. The RGS study although at a protein–protein interface was a partial validation of this model, which showed at least that function to load the exchange of key residues.

In order to test this model more thoroughly, a later study analysed 22 different families of transcription factors, such as nuclear receptors, basic helix loop helix, homeodomains and others.[14] In each one, the evolutionary importance of the DNA bases in their respective response elements was measured using information entropy. It was then correlated to the average ET evolutionary importance rank among all of the contact amino acids of each base, as observed in crystal structure of protein–DNA complexes between the transcription factor and their target response elements. These correlations ranged from .94 for the USF 1 bHLHL factor to −.27 for the GCN4 b-Zip factor. In fact, all four b-Zip transcription factors had poor correlations, thought to be due to the fact that other interactions determine their transcription specificity aside from their small DNA binding domain. However, all the remainder of the transcription factors had correlations above 0.5, and they yielded, overall, an average Pearson correlation of 0.73, and a nonparametric Spearman rank-order correlation of 0.75, both with p-values less than 10^{-5}. Thus, there is correlated evolution between protein residues in their contact DNA bases at least among most transcription factors, the b-Zip family being the exception.

This result suggests that the key components on either side of an interface can be identified based on their distinctive evolutionary signatures but also that they are in close contact structurally, in essence mirroring each other structurally. One may then test correlated evolution experimentally by swapping key protein residues and observing whether this redirects a transcription factor's DNA binding specificity. For this, we engineered the orphan receptor LRH-1, which is a monomeric C4 zinc finger protein. Two mutations were introduced: G462V and E458G in order to change these top-ranked DNA interface residues to their cognate side chains in steroid receptors. The hope was thus to also swap DNA binding specificity from TCAAGGTCA, the natural consensus element to the mutant TCAAGAACT sequence recognised by steroid responsive transcription factors. Indeed, an in vitro binding competition assays showed that the mutation dramatically decreased the mutants' ability to compete for binding to the wild type LRH-1 response element, but that it now had much higher affinity for the steroid response element. These experiments show that we generated a mutant LRH-1 with ortholog DNA binding specificity by targeting mutations to only two top-ranked DNA interacting amino acid residues that are different in the LRH and the steroid receptors. Finally, we note that a similar functional swap experiment was also successfully carried out, but *in vivo*.

5.3.3 *Other Case Studies*

In fact, there are now many examples of ET-directed discovery and manipulation of protein function, as summarised in Table 5.2, Refs: 7–9, 15–23).

Table 5.2. Evolutionary trace directed discovery and manipulation of protein function

PROTEIN	BINDING SITE PREDICTION	FUNCTION SWAP	FUNCTION BLOCK	FUNCTION SEPARATION	FUNCTION INHIBITION
Gα transducin	√		√		
Steroid receptors	√	√	√		
Nuclear Transport	√		√		
RGS	√	√	√		
Proneural bHLH	√	√	√		
GPCRs	√	√	√	√	
Cohesin	√		√		√
Ku70/80	√		√	√	√
GRK	√		√	√	√
RecA	√		√	√	
LexA	√		√	√	

The key points of these studies is that they repeatedly illustrate in different proteins, and in collaboration with varied laboratories, how clusters of top-ranked ET residues on protein surfaces anticipate the location of functional sites, that can then be validated through subsequent targeted mutations. In turn, in multifunctional proteins these mutations can efficiently create separation of function, rewire function, or when key residues are transplanted in appropriately designed peptides, create molecular inhibitors of protein–protein interactions (data for the latter studies are preliminary, but consistent in three different protein–protein interaction systems).

The weakness of these data, however, is that they are merely case studies. Although they show that ET has been successfully applied in a wide range of both eukaryotic and prokaryotic, and even though there are now many such cases, they do not themselves guarantee that ET will be equally successful for any other given protein. In other words, these case studies provide proof of principle, but there remains to gather proteome-wide evidence to suggest that ET will in fact identify functional sites for any protein. The evidence for this is presented next.

5.4 Proteomics Properties of Evolutionary Important Residues

A number of technical studies have sought to demonstrate in a variety of retrospective control proteins including usually both enzymes and non-enzymes, and both eukaryotic as well as prokaryotic species that as shown above in the SH2 example, one may generally expect that (a) any protein can have its residues ranked by evolutionary importance; (b) any protein will have identifiable clusters of top-ranked ET residues; (c) these clusters will match known sites (or predict others yet to be discovered); and (d) the ET can be automated so that the series of steps a-c can is amenable to high-throughput and optimization so that accuracy is maintained even during large scale use.

First, it is easy to compute top-ranked residues as described in Fig. 5.1 for any protein. Unless a protein of interest has no other known homolog, or too few to sample its evolution, both alignments and tree can be obtained, and hence a trace rank as well for every residue. Whether these rankings are meaningful, however, is a relevant and non-trivial question. Since, ET's goal has been to identify functional sites in the protein structure, the question of relevance has been approached by asking whether top-ranked residues were distributed in the structure in a random fashion or if in fact they aggregated into statistically significant clusters.

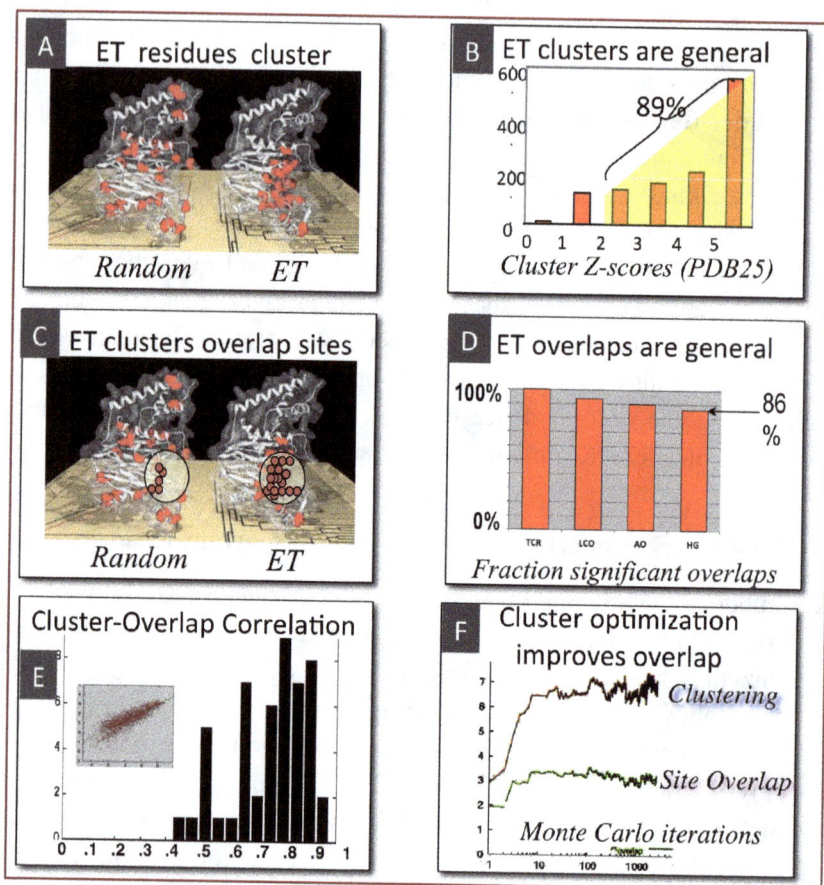

Fig. 5.5. Proteomic Properties of Evolutionarily Important Residues. Trace residues cluster together in structures much more than expected from random chance (A), and this is a phenomena observed in most proteins (B).[24] Moreover, ET clusters overlap functional sites much more than expected by chance (C), and this again is a general phenomena (d).[25] In fact, the quality of clustering is directly related the the quality of the overlap (or prediction) (E),[26] and this in turn guides the automated optimization of ET (f).[27]

For this it is easy to compare the number of clusters, or the size of the largest cluster and that one may obtain by picking n residues randomly to the number or size observed by taking the first n top ranked residues. As shown in Fig. 5.5a, the former yields very many small clusters of one or two residues scattered randomly over the entire structure whereas the latter yields a very large dominance cluster that contains nearly all the residues by itself and located in a precise region of the protein. When this comparison is repeated over the non-redundant PDB25 (meaning a set of all known protein structures that share no more than 25% sequence pairwise identity) one can see in Fig. 5.5b that about 90% of all proteins have nonrandom trace clusters for which the statistical Z-score is greater than two (meaning that the random chance of generating a set of residues that cluster as well or better is at least two standard deviations above the expected average). These data are therefore demonstrating that in the vast majority of proteins top-ranked residues can be reliably generated and they will non-randomly cluster in the structure.

Next, in order to establish the functional significance of ET clusters, we also need to demonstrate that they overlap with functional sites more so then if the residues were picked at random. This may be done much as above by comparing the election on functional sites of a randomly picked residues versus the same number of top-ranked ET residues, as in Fig. 5.5c. A key result is that by any number of statistics the match between ET clusters and functional sites is seen to be statistically significant. It reaches 100% of the proteins tested by the least stringent statistical measure, and 86% by the most stringent one, as shown in Fig. 5.5d. Thus, ET clusters will reliably identify functional sites.

It is important to note however that in these studies of overlap, the traces were obtained manually. Therefore the alignments of homologous proteins were curated to remove gaps, fragments and obvious errors, perhaps even entire branches if they were so deeply divergent that they seemed unrelated to the function of interest. This highlights the fact that the statistical significance of the output of the ET algorithm is a direct reflection of the quality of the input, meaning that the extent to which sequences are error-free, well-aligned, and properly partitioned into a functionally relevant treaty (as approximated by the sequence identity dendrogram) determined whether trace residues will cluster. In that light, the possibility arises to select input sequences under the constraints that they should maximise the clustering quality of ET residues, since this is an observable feature, and in the hope to maximize the overlap with functional sites, which is the unknown feature we wish to predict.

To carry out this program, we first note that the quality of ET residues structural clustering is directly correlated with the quality of the overlap between an ET cluster and a functional site, as shown in the inset in Fig. 5.5e. Moreover, this correlation is repeatedly observed in a set of over 50 diverse proteins (Fig. 5.5e). In practice, this suggests that optimizing the choice of input sequences identified the most statistically significant cluster of top-ranked residues will also maximize the overlap of that cluster with a functional site.

At a more theoretical level, this establishes a number of universal features of evolutionary important residues: protein sequence residues may be ranked by evolutionary importance; top-ranked residues cluster non-randomly in the protein structure; these clusters mark and predict functional sites; the better the quality of the cluster, the better the overlap, or prediction, of the functional site. Other retrospective control studies have also shown that there is structural symmetry of evolutionary importance across the protein-DNA interface suggesting that the molecular determinants of function, and specificity, on one molecule are directly in contact with those of the binding partner. This is consistent with the case studies in RGS and in steroid receptors that suggest that binding specificity resides with the variations among top-ranked interfacial ET residues, and therefore accompanies them as they are swapped among homologs to rewire function.

In practice, these data first yields a recipe by which the evolutionary trace may be optimized for high-throughput by selecting input sequences to maximize clustering and as a consequence maximize functional site overlap/prediction as shown in Fig. 5.5f. More generally, however, they suggest that the case studies previously discussed are representative of results one may expect in any protein family, provided enough homologs are available to generate significant trace clusters. This may be readily ascertained at http://mammoth.bcm.tmc.edu/ETserver.html which is the site of the

ET Server and its associated functional site prediction tools: the
ET_Viewer and the ET_Report_Maker.[28,29]

Having established the general validity of ET analysis to identify
functional sites and their key residues, we now turn to its application in
order to decipher the molecular determinants of GPCR function, as an
example of an in depth study on a specific system, and to the
prediction of protein function for Structural Genomics, as an example
of its high-throughput capabilities.

5.5 Molecular Determinants of GPCR Signal Transduction

There are a large number of different ligands that bind an equally large
number of GPCRs. The differences in size and molecular type among
ligands are linked to different sites of ligand action such as the
extracellular domain, the intracellular domain or the transition between
both. This variety of binding locations suggests that comparative
analysis is unlikely to review a universal binding pocket. However, all
receptors upon activation couple to G proteins or become
phosphorylated by dedicated kinases. Since all of these tend to be well
conserved during evolution, a reasonable hypothesis is that GPCRs
will share significance commonalities in their conformational switch
and G protein coupling mechanisms. If so, a joint evolutionary trace of
diverse Class A (rhodopsin-like) receptors should identify be
associated residues.

A study of 343 Class A receptors transmembrane spanning helices
including bioamine, chemokine, visual and olfactory GPCRs identified
top-ranked amino acids that mapped into a tight, statistically
significant cluster located towards the cytoplasmic end of the
transmembrane section of the rhodopsin structure.[17] Support for this
being a universal signal transduction switch into GPCRs came from a
large number of prior experimental studies that had direct mutations at
these amino acids. Nearly all (88%) had a documented impact on
function in at least one receptor, and as many as 73% had documented
impacts in three or more receptors. By contrast, residues predicted to
be of the least importance were reported to have a mutational impact
on the function of three or more receptors in only 19% of cases. The

same studies also suggest that fact this ET-identified switch could be divided into three structural subsites: one linked mostly to light and sensitivity; one linked to ligand coupling; and a site linked conformational switching and structural stability. As illustrated in Fig. 5.6 these sites are respectively closer to the ligand, closer to the G protein and in between.

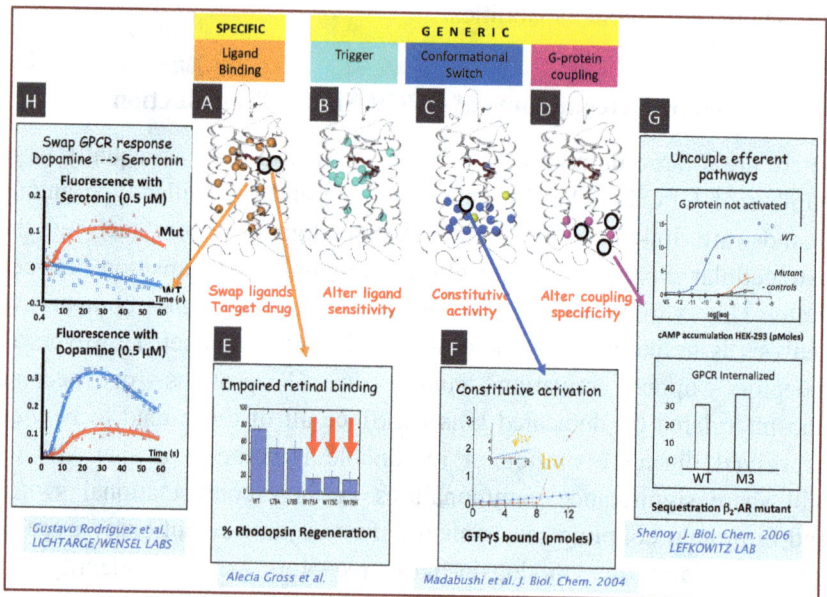

Fig. 5.6. An ET model of the molecular determinants of GPCR signalling. ET analysis suggests a model of the key functional residues that mediate signal transduction in GPCRs. Comparison with the literature then further divides these residues into subdomains, A through D (see text).[17] Subsequent experiments confirm the model based on diminished retinal binding and constitutive activity,[17] or uncoupling of G protein activation from internalization[22] (E through F). Preliminary data further suggests that a dopamine receptor may be mutationally rewired to respond to serotonin by swapping appropriate top-ranked ET residues (unpublished, with Ted Wensel).

To complete this picture it is also important to identify the ligand binding pocket. This requires that ET analysis be constrained to receptors that all share the same again, for example, the visual receptors which all bind retinal. Such rhodopsin specific ET analysis identifies another set of key functional residues, presumably important to vision. Since some of these will be important to vision because they are generally important in all GPCRs, they can be subtracted to reveal the amino acids that are uniquely important to vision. This completes a full evolutionary model of single construction in rhodopsin and Class A receptors.

Subsequent experimental studies with the Wensel laboratory were consistent with this model. Some mutations directed at the retinal binding site diminished binding of this ligand. Other mutations directed at the conformational switch produced constitutive activity in rhodopsin. Yet a third set of mutations (performed in the Lefkowitz laboratory) directed at the G protein coupling site created a mutant of the beta-adrenergic receptor that could bind its ligand but could not activate the G protein although it could still be phosphorylated, and then be internalised after binding to the scaffold protein beta-arrestin. This last experiment showed that the two different signalling branches of an activated GPCR could be functionally separated.

Finally, in order to test predictions of the key determinants of ligand binding and sensitivity, we have begun to swap into the dopamine receptor (DR) top-ranked, cognate and bioamine specific amino acids from the serotonin receptor (SR), that cluster near the predicted ligand binding pocket and that are variable between DR and SR. A number of these single point mutational swaps prove sufficient to either diminish dopamine response, increase serotonin response and occasionally both. Controls however show that the same mutations have no impact on responsiveness to norepinephrine, suggesting that the effect observed is specific to dopamine and serotonin. Moreover, similar mutations targeted to poorly-ranked residues in the vicinity of the putative ligand binding site do not change either dopamine or serotonin or norepinephrine responsiveness. These preliminary data, gathered by Gustavo Rodriguez in collaboration with the laboratory of Ted Wensel, support the ability to identify the key molecular determinants of the allosteric pathway, linking a given input (here either dopamine or serotonin) to a given output (signalling or none, respectively) and then to rewire the pathway by swapping residues. While in programming terms this is equivalent to recoding the transformation between an input and output; a more concrete analogy is that these mutations adapt the GPCR lock to a different ligand key by altering the lock's tumblers. More generally, the hypothesis and supporting data from Fig. 5.6 suggests that one can manipulate the input, output and connecting pathway sites in GPCR based on comparative analyses such as ET.

5.6 Protein Function Prediction

At the other end of the spectrum, the possibility of identifying molecular determinants of function on a large scale suggests that they may be compared across all available protein structures in order to identify proteins with identical functions. This idea is a generalization of the observation that all proteases share the same three residues in the same geometry – referred to as the catalytic triad. If similar structural motifs of just a few residues, that are the hallmark of various functions, can be identified and then recognised in structures, then their matches may reveal which proteins perform which functions. This would be particularly useful in the context of structural genomics, since many of the novel protein structures solved are chosen because they have little or no homology to previously known proteins and therefore cannot be assigned a function through simple sequence comparison.

To carry out this programme requires the availability of functionally relevant small structural motifs thereafter referred to as 3D templates. Unfortunately, there are very few proteins for which the key functional residues are known from experiments, and approximate templates based on proximity to ligands or catalytic sites tend by themselves to have many non-specific geometric matches across the set of all known protein structures.

To address this problem, an evolutionary trace functional annotation (ETA) pipeline has been built based on the following operations.[30] First, 3D templates are built without any prior knowledge of the function or mechanism involved, simply by identifying the most prominent surface cluster of top-ranked residues, and selecting among these the six closest to one another. The six Ca atoms of these amino acids, their relative geometry and their side chain type define a 3D template.[31]

Next, these of 3D templates are geometrically searched and matched across the PDB up to a certain threshold in least root mean square deviation. Unfortunately, most of those matches are nonspecific. As a result, it is necessary to introduce further filters to eliminate random geometric matches. One such filter exploits the fact that the random matches are likely to fall in the areas of proteins that lack any evolutionary importance. By computing an evolutionary trace on the matched protein and ranking the importance of the matched residues it is thus possible to train a support vector machine to reject nearly 90% of the matches.[31] The next filter demands that functionally relevant matches be supported by other matches, two distinct structures that bear the same functional information. In other words, functionally relevant matches should achieve a vote plurality.[32] Finally, the last filter reasons that a functionally relevant match from protein A to a protein B, should be reciprocated by a match from protein B back to protein A.[33]

A set of retrospective studies in enzymes that were solved by Structural Genomics and already annotated (and more recently including non-enzymes as well) show that when these filters are used all together to maximise specificity it is possible to suggest functions for 55% of the cases (coverage), and to reach a positive predictive value (PPV) above 95%. Should sensitivity be a major goal, it is possible to drop some of the filters mentioned above. If so, one may raise coverage to 88% of enzymes but the PPV drops to 90% overall. (In non-enzymes, specificity may be even better but coverage is lower, about 40%; this can rise to nearly 70% coverage at the expense of PPV that drops to 90% overall – unpublished data.)

In practice, these studies represent an approach to functional annotation that is complementary and orthogonal to the leading method, namely lateral transfer of annotation based on homology recognised by BLAST, or PSI-BLAST, or hidden Markov Models (HMM). The high specificity that is reached when all filters are used is especially significant since the concern is that as sequence identity falls, annotations become increasingly unreliable and fill databases with errors that will then propagate and mushroom.

More broadly, these studies further support the universal value of identifying top-ranked residues across the proteome and interpreting them when they cluster on surfaces as the key determinants of function and specificity. In that light, ETA is the high-throughput equivalent of the case studies mentioned previously.

References

1. Jacoby E., Bouhelal R., Gerspacher M., Seuwen K. (2006). *Chem Med Chem* 1: 761.
2. Lichtarge O., Bourne H.R., Cohen F.E. (1996). *J Mol Biol* 257: 342.
3. Cunningham B.C., Wells J.A. (1993). *J Mol Biol* 234: 554.
4. Lichtarge O., Sowa M.E. (2002). *Curr Opin Struct Biol* 12: 21.
5. Lipscomb L.A., *et al.* (1998). *Protein Sci* 7: 765.
6. Wilson C.A., Kreychman J., Gerstein M. (2000). *J Mol Biol* 297: 233.
7. Ribes-Zamora A., Mihalek I., Lichtarge O., Bertuch A.A. (2007). *Nat Struct Mol Biol* 14: 301.
8. Sowa M.E., *et al.* (2001). *Nat Struct Biol* 8: 234.
9. Sowa M.E., He W., Wensel T.G., Lichtarge O. (2000). *Proc Natl Acad Sci USA* 97: 1483.
10. Tesmer J.J., Berman D.M., Gilman A.G., Sprang S.R. (1997). *Cell* 89: 251.
11. Natochin M., Artemyev N.O. (1998). *Biochemistry* 37: 13,776.
12. Slep K.C., *et al.* (2001). *Nature* 409: 1,071.
13. Lichtarge O., Yamamoto K.R., Cohen F.E. (1997). *J Mol Biol* 274: 325.
14. Raviscioni M., Gu P., Sattar M., Cooney A.J., Lichtarge O. (2005). *J Mol Biol* 350: 402.
15. Cushman I., *et al.* (2004). *J Mol Biol* 344: 303.
16. Lichtarge O., Bourne H.R., Cohen F.E. (1996). *Proc Natl Acad Sci USA* 93: 7,507.
17. Madabushi S., *et al.* (2004). *J Biol Chem* 279 : 8,126.
18. Onrust R., *et al.* (1997). *Science* 275 : 381.
19. Quan et X.J., *et al.* (2004). *Development* 131: 1,679.
20. Rajagopalan L., *et al.* (2006). *J Neurosci* 26: 12,727.
21. Raviscioni M., He Q., Salicru E.M., Smith C.L., Lichtarge O. (2006). *Proteins* 64: 1,046.
22. Shenoy S.K., *et al.* (2006). *J Biol Chem* 281: 1,261.
23. Yang M., *et al.* (2002). *Mol Endocrinol* 16: 814.
24. Madabushi S., *et al.* (2002). *J Mol Biol* 316: 139.

25. Yao H., *et al.* (2003). *J Mol Biol* 326: 255.
26. Mihalek I., Res I., Lichtarge O. (2006). *Proteins* 63: 87.
27. Mihalek I., Res I., Lichtarge O. (2006). *Bioinformatics* 22: 149.
28. *Ibid.*, 1,656.
29. Morgan D.H., Kristensen D.M., Mittelman D., Lichtarge O. (2006). *Bioinformatics* 22: 2,049.
30. Ward R.M., *et al.* (2009). *Bioinformatics* 35: 1,267.
31. Kristensen D.M., *et al.* (2006). *Protein Sci* 15: 1,530.
32. Kristensen D.M., *et al.* (2008). *BMC Bioinformatics* 9: 17.
33. Ward R.M., *et al.* (2008). *PLoS ONE* 3: e2136.

CHAPTER 6

Protein–Protein Docking

Adrien Saladin and Chantal Prevost

Laboratoire de Biochimie Théorique, CNRS UPR 9080,
Institut de Biologie Physico-Chimique et Université Paris 7,
13 rue Pierre et Marie Curie,
F-75005 Paris, France
E-mail: chantal.prevost@ibpc.fr

Methods for predicting the three-dimensional structure of protein–protein complexes are necessary tools in the post-genomics context. They are based on three ingredients: an appropriate representation of the macromolecule, an efficient search algorithm and a discriminatory scoring function. The present chapter describes the main algorithms underlying protein–protein docking methods, together with the strategies that have been developed to efficiently search the interaction space within limited calculation times. Assessment of the current protein–protein docking programmes through the community wide CAPRI (Critical Assessment of PRrotein Interactions) experience pictures a rapidly progressing field, where diverse successful solutions are available for complex predictions provided that the protein partners present little structural differences between the free and the bound form. Challenges that are presently tackled by the developers include the development of robust scoring functions adapted to low-resolution models, multicomponent docking and flexibility account.

6.1 Introduction

Assembling macromolecules has been a major goal of theoretical modelling ever since sufficient structural information has been known on the main building blocks, protein or nucleic acids. While individual dynamic properties of macromolecules play an essential role in catalytic activity or in general biological processes, it is noteworthy that most biomolecular entities exert their function within complexes, ranging from binary complexes to huge macromolecular machineries like the ribosome. Being able to predict the three-dimensional structure of macromolecular assemblages therefore appears as a crucial step to gain knowledge on the mechanisms that rules cell life.[1-3] Due to the size of the systems to assemble, the development of the docking branch of macromolecular modelling has long been conditioned by the progress of computer processing and storage capacities. Nevertheless, the very first protein–protein docking simulations, performed in 1978 by Wodak and Janin, already set the main principles that guide docking methods.[4] The study used a simplified protein representation for the association partners and a discrete, systematic search strategy (Fig. 6.1), followed by refinement of the best solutions. Scoring was based on the degree of surface complementarity and the size of the interface. This precursor study permitted to delineate the problems specific to docking methods and the basis of their resolution: quick assemblage of macromolecular systems requires simplifying as much as possible the system representation, the search process and the scoring criteria[5] provided that these simplifications (a) do not impede the generation of a geometry close to the native one and (b) are compatible with successful ranking of docking predictions.

Section 6.2 of this chapter comes back to the different choices that developers are faced with regarding the level of complexity of the system representation, the algorithm used to explore the possible geometries of association and the degree of precision of the scoring function. Docking strategies can be viewed as efficient combinations of such choices. Section 6.3 describes several representative docking methods, either systematic or guided. We will insist on the underlying strategy used in each method. In order for the docking programmes to increase their performance, they must be evaluated on a common basis (see Section 6.4). The protein–protein docking field presents the particularity to be structured around the CAPRI experience (http://www.ebi.ac.uk/msd-srv/capri),[6-8] a blind prediction docking test that aims at assessing the progress of the field and accompanying its development. This programme was initiated in 2001 as a result of the first conference covering protein–protein docking.[9] Groups developing docking methods benefit from a comparative evaluation of diverse docking strategies.

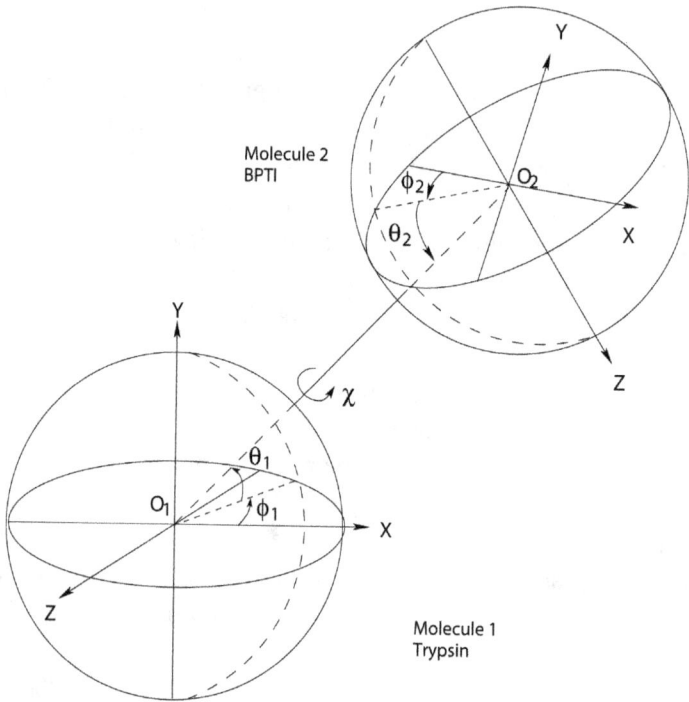

Fig. 6.1. Definition of the degrees of freedom that are sampled during the docking search process developed by Wodak and Janin[4] for extensive docking of trypsin with BPTI. The six degrees of freedom comprise the (O1, O2) distance, where O1 and O2 are, respectively, the mass centres of Trypsin and BPTI, the spherical angles $\Phi 1$ and $\theta 1$ for trypsin, $\Phi 2$ and $\theta 2$ for BPTI and the rotational angle χ around axis (O1,O2, after Wodak and Janin[4]).

Section 6.4.1 summarises the main conclusions of the sixteen CAPRI rounds that have taken place until now.[6–8,10–13] CAPRI also helps identifying the current limitations of docking methods and the bottlenecks hindering further development. We will present the

challenges being tackled by the community in applying the docking
programmes to any family of proteins (Section 6.4.3).

6.2 Definition and Goals of Macromolecular Docking

6.2.1 Protein–Protein Docking Terminology

With a wide acceptance of the term, macromolecular docking refers to
any theoretical procedure capable of predicting the three-dimensional
structure of a binary protein complex starting from the structures of its
individual components. This covers a large spectrum of situations,
ranging from high precision docking, where the main goal is to predict
precise interactions between amino acid functional groups of two
protein partners, to extensive docking, precursor of high-throughput
screening of protein databases. High precision docking is generally
performed when information is available on the location of at least one
partner binding site, permitting to concentrate the sampling efforts on
internal degrees of freedom.[14,15] In those cases, sampling relies on
methods like molecular dynamics (MD) simulation or the Monte Carlo
(MC) search. This type of docking does not consider the search time as
a limiting factor, as long as precise interactions are finally predicted.

At the other end of the spectrum, what can be called 'extensive
docking' consists of the wide exploration of relative positions and
orientations of the partners in order to generate all possible geometries
of association, which are scored and ranked in a predictive purpose.[5,16–20]
This necessitated the development of specific methods and strategies
that must obey the stringent requirements in terms of speed and
conformational space exploration. It can be noted that prediction of
precise interactions between the partners is not necessarily required as
long as correct scoring is achieved. This family of methods makes up
the protein–protein docking methods that constitute the object of this
chapter.

A specific terminology has developed concomitantly to the
protein–protein docking methods. It is useful to define here the
different terms that will be used in this chapter. The proteins to be

docked, or in other words the association partners, are generally referred to as the receptor (generally the bigger of the two partners) and the ligand. When they are in a bound form, their structure is directly taken from the known three-dimensional structure of the complex. The bound problem constitutes a preliminary test for docking methods. It consists of separating out the elements of a macromolecular complex with known 3D-structure and verifying whether the method is capable of reconstructing the crystal complex and attributing a top score to that geometry of association.

The unbound problem is the 'real life' docking problem, corresponding to situations where the result is unknown. In those cases, the protein structures are taken from the Protein Data Bank (PDB)[21] in a free form or in complex form with another molecule. This structure is referred to as the unbound form and it can present from slight (mostly, side chain torsional changes)[22] to huge (loop remodelling, change in secondary structure or domain motion) differences with respect to the bound form.[13,23] Slight differences can be accounted for using soft docking methods.[24–29] In that case, the surface resolution or the sensitivity of the scoring function are purposely decreased, allowing a certain degree of interpenetration between the two partners. When a partner structure has not been solved by crystallography or NMR, it needs to be reconstructed by homology modelling from the structure of other proteins. Clearly, correct prediction is much more difficult to achieve when starting from such modelled structures, where loop insertions or modifications in secondary structures are frequent (see for example Targets T20 and T24 of CAPRI[13]). Most often their resolution, like that of systems with important surface remodelling between the unbound and the bound form, pertains to the category of flexible docking presented in Chapters 7–9 of this volume.

6.2.2 Goals and Strategies

Protein–protein docking methods combine three fundamental ingredients:[5] (a) an appropriate representation of the protein partners together with the definition of the degrees of freedom that will be searched, (b) an algorithm to explore the conformational space as completely as possible and (c) a scoring function to classify the predictions (Fig. 6.2).

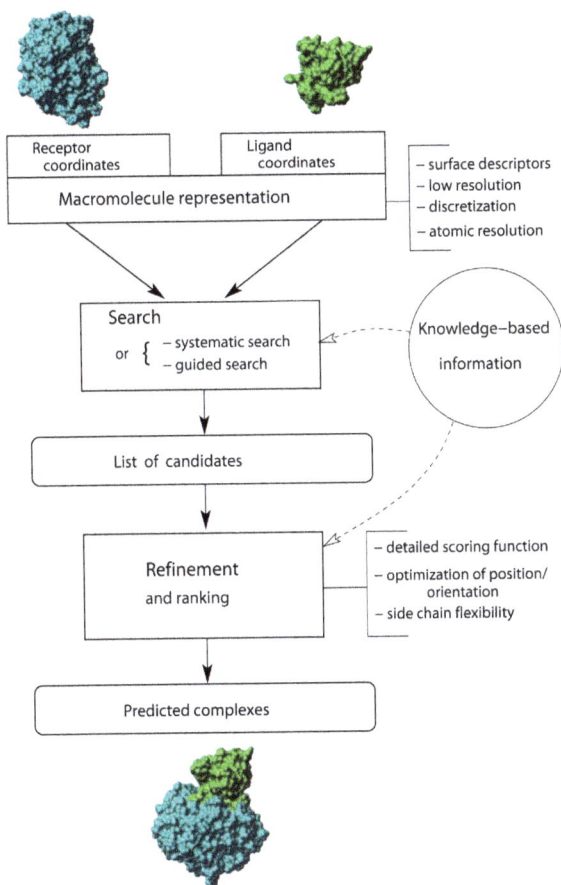

Fig. 6.2. Schematic organisation of typical docking methods. Docking strategies require the choice of a protein representation, a search method and a degree of resolution for final refinement and scoring.

Several levels of complexity for protein representation and scoring are encountered in docking programmes. Protein models vary from simplified surface representation to detailed, all atom representation via coarse grain models (Section 6.3.2.2). The partners can be considered as rigid bodies, in which case only six degrees of freedom (generally three translations and three rotations) are searched.[25,28,30] Alternatively, conformational changes of surface side chain can be explicitly taken into account by introducing additional degrees of freedom.[31,32] In the same way, scoring functions range from statistical learning or knowledge-based functions[33,34] to evaluations of the system's free energy,[35] via measures of surface fit.[36] Utilization of simplified models and simple scoring functions permits to search the whole conformational space within a reasonable calculation time, when using detailed representation and precise scoring would make such extensive search a totally prohibitive task. However, these simplifications are bound to induce biases in the scoring, producing what is called false positive solutions, i.e. predictions far from the correct complex structure but characterised by a high score. It must also be noted that in order to be efficient, a scoring function needs to be adapted to the level of resolution of the protein representation. For example, it is useless to rely on scoring functions based on the calculation of atomic potentials, which are sensitive to atomic deviations of tenths of angstroms, in procedures where the relative protein positions and the side chain conformations have not been adjusted with high precision.

Given these constraints, many docking programmes include two phases, a search phase performed at low resolution where the conformational space is extensively explored and a refinement phase where the best ranking geometries resulting from the search are re-evaluated at higher resolution (Fig. 6.2). During refinement, the search concentrates in the vicinity of the candidate geometry and the scoring uses more computer demanding criteria.

Search algorithms developed for the first phase can be divided into two categories, systematic search and guided search algorithms. The first category starts by generating all possible relative positions/orientations of the two partners before identifying and classifying potentially favourable geometries. The second category directly uses the scoring function to guide the search towards favourable geometries.

In addition to developing new representations and search algorithms, the skill of the groups creating docking programmes consists of choosing balanced combinations of these solutions that allow high computational speed and efficiency. As will be seen in Section 6.3.3.4, some methods, particularly among the most recent ones, include several levels of complexity in a unique docking simulation, in what can be called a hierarchical or a multi-scale approach.[32,37,38]

6.3 Protein–Protein Docking Methods

We organise the description of the docking methods around the two main families of search strategiess: systematic search and guided search methods. This division approximately follows the chronological appearance of the two approaches. Both continue to be developed since none of them has proven its superiority.[13] We will illustrate the presentation by detailing some representative docking programmes, bearing in mind that these are just examples among a number of valuable methods (Table 6.1).

6.3.1 Systematic Search Methods

6.3.1.1 Discrete Sampling: The Correlation Methods

In order to sample all possible relative positions of the partners, it is necessary to discretize the conformational space that is searched. In a rigid body context, this makes six degrees of freedom (three translations and three rotations) to be sampled in a combinatorial way. The step size for discretization should be smaller than the typical width of potential energy wells in order for the correct geometry not to be overlooked. This makes the number of generated positions, and therefore the number of scoring evaluations, to rapidly become impressive. For example, systematic sampling of a ligand position within a 120 Å cube centred at the receptor, using translation steps of 1 Å and rotation steps of 10 to 15 degrees, necessitates the generation of about 10^{10} configurations. As a result, the docking field really started its expansion when methods capable of accelerating both the search and the score calculation appeared.

In 1992, the group of Vakser opened the path[25] by proposing the use of fast Fourier transformation (FFT) for the calculation of a scoring function c, defined as a correlation between two discrete functions a and b respectively associated to the receptor and the ligand.

$$a_{l,m,n} = \quad \begin{array}{l} 1 \text{ on the surface,} \\ \rho \text{ inside the molecule,} \\ 0 \text{ outside} \end{array} \qquad (6.1)$$

$$b_{l,m,n} = \quad \begin{array}{l} 1 \text{ on the surface,} \\ \delta \text{ inside the molecule,} \\ 0 \text{ outside} \end{array} \qquad (6.2)$$

$$c_{\alpha,\beta,\gamma} = \sum_{l=1}^{N} \sum_{m=1}^{N} \sum_{n=1}^{N} a_{l,m,n} \cdot b_{l+\alpha,m+\beta,n+\gamma} \qquad (6.3)$$

In this representation, each protein partner has been digitalised on a three-dimensional grid indexed by l, m, n. It is partitioned into interior, exterior and surface regions. Interior parameters ρ and δ are used for discriminating overlapping regions, with ρ being a large negative value and δ a small positive value ($0 < \delta < 1$) (see Fig. 6.3 left). The correlation value for each displacement (α, β, γ) corresponds to a positive score for surface contact, accumulated on overlapping surface points, corrected by a penalty for interpenetration accumulated on overlapping interior points. Surface regions of each partner are defined using a thickness greater than a simple layer of grid points to implicitly allow for small imperfection in surface matching resulting from discretisation or from small surface readjustment (typical values are 1.5 to 2.5 Å). Surface thickness is a sensitive parameter, since increasing its value augments the chance of producing false positive predictions. The grid size appears to be another sensitive parameter, with optimal value around 0.7–0.8 Å.[25] The introduction of a correlation function as docking score permits to take advantage of well designed algorithms devised to spare calculation time. The discrete Fourier transform $C_{o,p,q}$ of c,

$$C_{o,p,q} = \sum_{l=1}^{N} \sum_{m=1}^{N} \sum_{n=1}^{N} e^{-2\pi i \frac{ol+pm+qn}{N}} \cdot C_{l,m,n} \tag{6.4}$$

can also be written as the product of the complex conjugate $A^{*}_{o,p,q}$ of the discrete Fourier transform of a and the Fourier transform $B_{o,p,q}$ of b.

$$C_{o,p,q} = A^{*}_{o,p,q} \cdot B_{o,p,q} \tag{6.5}$$

Therefore, it is possible to evaluate $C_{o,p,q}$ without calculating the product $a_{l,m;n} \cdot b_{l',m',n'}$ at each of the grid positions (l, m, n) and (l',m',n'). The Fourier transform A is directly inferred from a, and B from b, using the same formula as Eq. 6.4. $c_{l,m,n}$ is then calculated from $C_{o,p,q}$ using an inverse Fourier transform. The use of the fast Fourier transform algorithm[39] moreover permits to limit the transformation of a 3D function of N^6 values to an $N^3 \ln N$ calculation order. The correlation values can be mapped as correlation peaks as shown on Fig. 6.3 (right panel). The highest peaks correspond to regions of extensive matching of the partner surface shapes and are conserved for further discrimination using finer grid steps (Section 6.3.3). The process needs to be completed by sampling the relative orientations of the two protein partners. For each generated orientation of the ligand with respect to the receptor, a new correlation function is calculated (Fig. 6.3 right).

Following this first implementation, the FFT class of method was explored for further improvement.[40] The group of Sternberg used an electrostatic criterion instead of a shape criterion in defining the functions a and b describing the two partners,[41]

$$a_{l,m,n} = \begin{cases} 0 \text{ inside the molecule} \\ U(\,l,m,n\,) \text{ outside the molecule} \end{cases} \tag{6.6}$$

$$b_{l,m,n} = q'(\,l,m,n\,) \tag{6.7}$$

$$U(\,l,m,n\,) = \sum_{j} \frac{q_j}{\varepsilon(\,r_j\,)} \cdot r_j \tag{6.8}$$

with $U(l, m, n)$ the electrostatic potential created outside the receptor, j being a receptor atom, q_j the partial charge on j, r_j the position of atom j, $\varepsilon(r_j)$ a screening dielectric function and $q'(l, m, n)$ the ligand charge discretised on the grid.

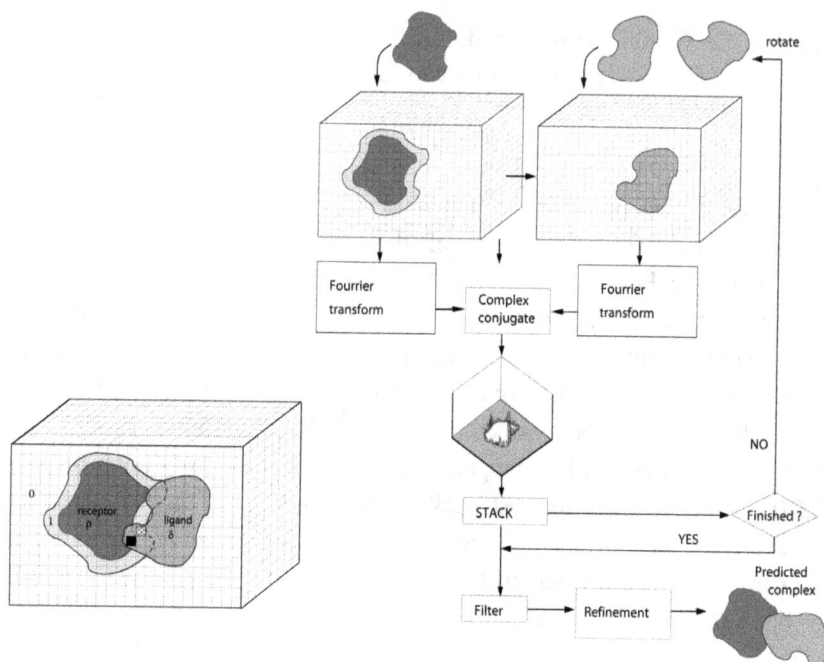

Fig. 6.3. Scoring function and search strategy in correlation methods (after Gabb *et al.*[40]). Left panel: the receptor and the ligand are discretized on three-dimensional grids and are partitioned into inside (gray), outside (white) and surface regions (light gray). In the Sternberg implementation represented here, only the receptor is attributed a surface region. Surface matching between receptor and ligand is measured by adding positive contributions due to surface/inside region overlap (hatched cube) and penalties due to inside/inside region overlap (black cube). δ and ρ are defined in the text. Right panel: for each sampled ligand orientation, the correlation function is calculated via fast Fourier transformation. The highest peaks identify the translation vectors producing favourable surface complementarity.

In this work, the electrostatics-based correlation function was used as a filter of the shape-based selected geometries. This allowed about 50% reduction of the number of geometries to be further evaluated. Other groups devised strategies to combine both steric and electrostatic approaches. Mandell *et al.*[42] found that the use of composite convolution functions accounting for both steric and electrostatic properties increased the number of correct solutions for five tested protein–protein complexes. The binding geometry could then be identified as a cluster of predictions with favourable free energies. Chen *et al.*[43] reached similar conclusions when examining an ensemble of 27 systems with the programme ZDOCK, using composite functions made of steric, desolvation and electrostatic components. The Eisenstein team[44] used grids of complex numbers with a real part related to shape complementarity while the imaginary part contained low resolution information on electrostatics. Evaluating both the steric and electrostatic complementarity improved the docking results for all tested enzyme/inhibitor systems and improved the score of near-native antibody/antigen systems. In the last case however, inclusion of electrostatics produced many false positive solutions, thus decreasing the ranking of the near native solutions. The same group later improved the algorithm by introducing composite geometric–hydrophobic scoring, particularly adapted to systems presenting large interfaces.[45]

Another point of concern with the FFT methods is the production of many false positive solutions, i.e. predictions far from the correct complex structure but characterised by a high score. These solutions appear as a result of surface softening that permits getting free of localised artificial steric clashes arising from discretization, but at the cost of accepting real steric clashes. Accordingly, FFT methods were incremented by refinement phases where precision often increased up to the atomic level and where the discrete grid search was sometimes replaced by continuous approaches (Section 6.3.3.1). As already noted, production of–positive solutions is shared by most docking methods. They appear when working at low resolution or when implicitly taking into account slight conformational changes.

A third point concerns the necessity to repeat the FFT procedure for each possible relative orientation of the association partners. The programme Hex proposes an elegant solution to this problem, by introducing polar Fourier expansions of protein shape and electrostatic properties expressed in spherical coordinates.[30,46] In this coordinate system, the six position degrees of freedom become five rotations and one inter-protein separation. For each separation value, the correlation between the functions associated to each protein partner can be calculated as overlap integrals. Hex is therefore a remarkably fast correlation method (in the order of minutes) that can also be extended to take limited amplitude conformational changes into account.[47]

Finally, a drawback of FFT-based methods, inherent to this class of methods, is that the investigated volume has to be sampled in its totality. The programme BIGGER[28] is a surface-matching programme based on a real space grid searching algorithm. By making efficient use of Boolean operators and heuristic rules, the complete search is performed in the order $O(N^{2.8})$. Instead of using Fourier transformations, the algorithm stores the geometries with highest scores in a limited size memory stack (typically, 1,000 candidate geometries). The stored geometries are then re-evaluated using four criteria, surface matching, probability of occurrence of every observed contact between pairs of amino acids across the interface, electrostatics and desolvation.

6.3.1.2 *Geometric Surface Matching*

An alternative to the systematic positioning of partitioned proteins is to focus on surface matching and to exclusively generate those geometries that present localised surface match. In this case, the protein representation is a surface descriptor capturing the essential features of the surface in terms of concave and convex regions, their size and depth, and their relative locations on the surface. The first approach to this problem was proposed by Connolly, who represented the protein surface by 'critical points', describing 'holes' (maxima of the shape function) and 'knobs' (minima).[48,49] Further adaptations were found necessary for this popular surface representation to be efficiently used for protein–protein docking[50] and to reduce the combinatorial complexity. Norel et al.[51] showed that using pairs of Connolly critical points together with their surface normals permits to successfully tackle the combinatorial problem. Another solution consists in using the critical points to partition the protein surface into concave, convex and flat parts.[52,53] The search for pattern matching is performed by geometric hashing: surface descriptors are stored in hash tables and checked for correspondence.[54]

These methodologies developed by the groups of Wolfson and Nussinov are among the quickest docking programmes so far, permitting possible geometries of partner association to be scanned within minutes. They have a strong potential of evolution towards the resolution of more sophisticated docking problems. Indeed, a flexible docking version FlexDock, capable of docking proteins presenting large amplitude internal domain movements is already available at little additional computer time cost[53,55] and a version dedicated to multi-component docking has also been released.[56] The representation is not directly compatible with carrying out finer evaluation, which requires higher resolution representation and searching approaches.[57] Nevertheless, the methodology is a highly powerful screening tool to identify possible binding geometries.

6.3.2 Guided Search Methods

The second class of methods builds on exploration algorithms that are commonly used in molecular modelling to explore internal fluctuations of macromolecules, like energy minimizations (EM),[58] molecular dynamics (MD)[15,37] and Monte Carlo (MC)[32,59] simulations, or genetic algorithms (GA).[60,62] The partner representation is accordingly more detailed than for systematic search approaches. Typically, the guided methods use atomic representations together with a force field where each atom is attributed van der Waals parameters and a partial charge. The interaction energy is defined as a sum of pairwise van der Waals and electrostatic interactions to which are often added desolvation terms, as desolvation is an essential component of the free energy of protein association. The search is performed with respect to at least the six positional degrees of freedom. Regions of the potential energy space that present minimum energy values correspond to favourable geometries of association. It is useful to keep in mind that the relative importance of the terms describing the interaction energy varies according to the distance separating the partners. At distances higher than 5 Å, the electrostatic term dominates.[63] For some complexes like barnase-bastar, this term is an important component of the driving force of association and indeed, Camacho *et al.*[64,65] have related the kinetics of association to the degree of electrostatic steering. Desolvation terms become significant when the partner proteins get closer, while van der Waals interaction terms are a determining part of the calculated interaction energy within closely packed complexes. In that case, slight misalignment of two atoms can result in a tremendous increase of the interaction energy. It has been necessary to specifically address that point when docking unbound proteins.

This section first describes a typical example of guided docking method, based on MC search. As this family of docking methods is more time-consuming than systematic search methods, algorithms had to be specifically developed to spare computer time during the search. They are discussed in Section 6.3.2.2. In Section 6.3.2.3, we also discuss a particular aspect of guided search driven by experimentally-derived data.

6.3.2.1 *Example of a Guided Search Programme: ICM-DISCO*

Figure 6.4 schematically describes the principal steps of an ICM-DISCO docking simulation. Programme ICM-DISCO, developed by the group of Abagyan, consists of a two-step procedure for exploring the potential energy space of the system.[59] The first step is based on a pseudo-Brownian Monte Carlo search of the six degrees of freedom characterising the position of the ligand with respect to the receptor. Both protein partners are modelled in atomic representation and are considered as rigid bodies. The potential energy of the system is reduced to an interaction energy E, composed of the following terms,

$$E = E_{vw} + E^{el}/_{solv} + E_{hb} + E_{hp} \qquad (6.9)$$

where E_{vw} is the van der Waals interaction term, the electrostatic term $E^{el}/_{solv}$ is a modified Coulomb term with distance dependent dielectric $\varepsilon(r) = 4r$ corrected by the solvent accessible surface, E_{hb} is a hydrogen bonding potential taken as a Gaussian centred at hydrogen bonding sites and E_{hp} is an hydrophobic potential dependent on the buried hydrophobic surface area. The repulsive part of the van der Waals term is truncated in order for the E_{vw} term to remain below a maximum value. Each of these terms is pre-calculated on a grid surrounding the receptor, which makes the calculation of Eq. 6.9 much faster[31] (Section 6.3.2.2). Starting from a given ligand position characterised by an interaction energy E_0, a possible new position is generated at random, followed by energy minimization. The energy E of the resulting geometry is compared to E_0 and accepted according to the Metropolis criterion.[66] In the description by Fernandez-Recio *et al.*[59] 120 simulations of 20,000 steps each are performed and all accepted conformations are merged and clustered according to root mean square deviation (RMSD) values, before being ranked by energy. The best energy cluster representatives are submitted to the second (refinement) step of the procedure, where the ligand side chains are now considered flexible. The values of their torsion angles are sampled along with the six positional variables, again using a Monte Carlo algorithm (see Section 6.3.3.2). In this docking programme, time reduction bears on the energy calculation, which is simplified by the use of grid calculation. Both the search and the refinement docking steps are based on the same precise protein representation, compatible with the final use of free energy estimate as a scoring function. Differences between the search stage and the refinement stage mainly reside in the definition of useful degrees of freedom and in the magnitude of MC steps.

6.3.2.2 *Speeding up the Calculation*

Grid calculation: Grid calculation is used in ICM-DISCO to decrease the computer time without sacrificing the precision of protein representation or scoring function. Typically, the potential induced by the receptor is pre-calculated at each point of a grid.[31,67] This is done for each of the energy terms described in Eq. 6.9. The van der Waals grid potential is calculated as the superposition of two potential grids, using either a hydrogen atom-sized probe or a carbon atom-sized probe at each grid point. As for any simplifying method, the use of grid potentials induces a loss in resolution.

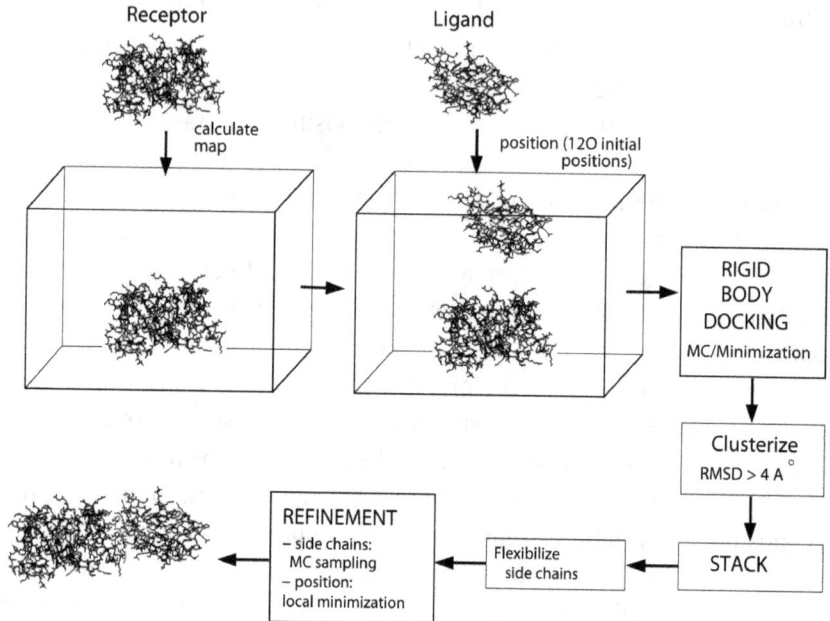

Fig. 6.4. Schematic representation of the ICM-DISCO docking strategy, after Fernandez-Recio *et al.*[31] Contributions to the scoring function are calculated at each node of a grid potential. The rigid body docking is based on a Monte Carlo/minimization sampling of the ligand position. The best candidate predictions are refined by including flexible interface side chains and sampling possible conformations (Monte Carlo) while optimising the ligand position.

Particularly, calculation errors increase when the two partner proteins get closer. The resolution loss may in fact present an advantage since it smoothes the potential energy surface, whereas the consequent ranking errors or the generation of false–positive predictions can be corrected at the refinement stage. However, the methodology remains dependent on possible conformational changes of the receptor that may modify the potential at the grid points.

Coarse graining: An alternative to discretizing the energy calculation consists in simplifying the protein representation. Coarse grain (CG) models group several heavy atoms into larger beads, thus cutting the number of particles to be considered in pairwise electrostatic or van der Waals calculations.[68] In addition to speeding up the docking calculations, reduced protein representations are coupled to simplified force fields that smooth the conformational energy landscape. The first coarse grain model ever used in docking calculations was the Levitt model,[69,70] used in the Wodak and Janin pioneering study.[4] In this model, each amino acid is represented by a bead equivalent to its time-averaged structure. A simplified protein representation is also used by the group of Baker in the initial rigid-body search step of RosettaDock.[32] In this case, each amino acid side chain is represented by a centroid pseudoatom placed at an average position determined from a PDB survey.

The Zacharias model developed in 2003[58] emphasises on obtaining faithful reproduction of the volume occupied by the protein, in a perspective of surface matching. Accordingly, in addition to beads centred on each Cα atom, amino acid side chains are represented by one to two beads with van der Waals radius that reflect the size occupied by the component atoms. The reduction factor is one bead per four to five heavy atoms (Fig. 6.5). Electrostatics representation is limited to full charges placed on the terminal bead of charged amino acids and the scoring function is taken as an interaction potential between the two partner proteins, summing a Coulomb term for electrostatics and a smoothed van der Waals term. This representation permits multi-minimization docking search, starting from tens of thousands initial configurations of the ligand, to be performed in several hours with the ATTRACT programme. Note that the search method is particularly well adapted to additional time cuts since the minimization calculations can be independently distributed on several processors with a scaling close to linear.[71]

The approach has recently been extended to treating protein–nucleic acid complexes.[72] Coarse-grained models offer a direct correspondence with the atomic model they are based upon. They are therefore compatible with an easy passage to atomic scale, which however necessitates correct sampling of side chain adjustments. Flexibility can be introduced in CG models, notably by representing flexible parts with conformational ensembles.[73] There is presently a renewed interest for such low resolution representations, considered as promising new generation models for dealing with huge molecular assemblies.[68] A large effort is being devoted to developing accurate force fields[74,75] and physical degrees of freedom accounting for protein or nucleic acids internal flexibility.[76]

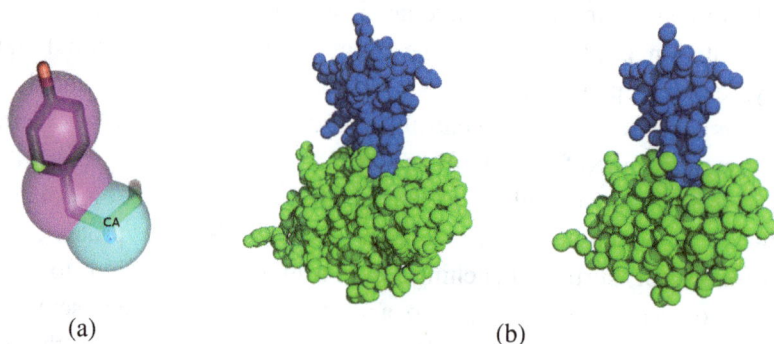

(a)   (b)

Fig. 6.5. Reduced protein representation used in the ATTRACT docking programme.[58] (a) Example of a tyrosine residue represented by three beads, one centred at the Cα carbon and two beads for the side chain. (b) Van der Waals representation of the trypsin-BPTI complex respectively in atomic and in reduced representation. The main features of the accessible surface are conserved.

6.3.2.3 *Data-driven Methods*

When available, information on the residues pertaining to the binding interface can be a potent driving force for guided methods. The programme HADDOCK[37] is worth mentioning for its particular use of guided methods, particularly those developed for Nuclear Magnetic Resonance (NMR) structure refinement. HADDOCK was initially designed for the cases where information permits to identify residues pertaining to the interface, without precisely knowing which residues of their partner they interact with. NMR provides such information when comparing chemical displacements in the bound and free forms of a same protein.[77] But the information can come from biochemical or biophysical results as well.[78] It is translated into a set of ambiguous interaction restraints (AIR), used to guide the ligand approach during a first rigid body MD search phase. From that point on, the search is limited in space in the vicinity of the contact interface of the best solutions (typically, 200 best solutions). This restriction allows addition of internal degrees of freedom to the six initial search variables, during subsequent internal coordinate simulated annealing MD processes where side chains are free to adapt their conformation and main chain deformations are possible. In order to cope with the possibility that some of the data may not be pertinent for interface identification, random subsets of AIRs are sequentially used. Note that HADDOCK's emphasis on flexibility places it within the flexible docking class of methods. More information can be found in Chapter 7 specifically devoted to that question.

More generally, taking advantage of even limited information on interface residues appears as one promising route to increase the accuracy of docking methods and, further, to extend them towards flexible docking.[15] Data-based information has already been incorporated into docking strategies to bias the search[79-81] or to filter out false positive predictions.[12,40,62,82] It can also be a direct component of the scoring function, at initial[83] or final[33] stages of docking. Availability of robust and reliable prediction of interface regions should permit to focus the search on regions of interest, in a strategy reminding popular methods for small ligand docking that require preliminary identification of putative binding pockets.[84,85] A number of investigations are under way to predict interface residues, based on Evolutionary Trace analysis,[63,86-88] interface property analysis[89,90] or knowledge-based methods.[91-93]

6.3.3 *Refinement*

In most docking programmes, the search phase is followed by a refinement step (see Fig. 6.6). The main reason is that scoring functions are not presently capable of correctly ranking the predictions, or even distinguishing false–positive predictions from correct ones, at the level of resolution necessary to perform a complete search (see Section 6.3.4). Typically, only the best predictions are concerned by the refinement step, hundreds out of tens of thousands predictions, generally selected among representatives of clusters of solutions. Refinement can be performed independently of docking procedures[57,94] and this is specifically addressed in Chapter 9 of this book. However, as it constitutes an intrinsic part of several docking strategies presented here, we rapidly discuss how it is incorporated into these methods.

6.3.3.1 *Increasing the Resolution*

A first possibility for refining a prediction is to reproduce the search process at higher search resolution while using a more precise scoring function. For discrete search processes, this means reducing the search volume and the search step. For example, FFT docking can be performed in a reduced volume around the geometry to be refined, using finer translation and rotation increments.[25] This process allows elimination of a great part of the false–positive predictions. Further refinement may require changing for a continuous search process like local energy minimization.[95,96]

6.3.3.2 *Accounting for Side Chain Conformational Change*

These steps may be sufficient to solve docking problems of the bound type. In the unbound case however, soft representations are generally unable to account for conformational change of long side chains, occurring preferentially at protein-protein interfaces.[22] A notable exception is found in the programme BIGGER, where such induced deformations are implicitly accounted for in a way compatible with the simplified representation used in the search phase.[28] BIGGER partitions the proteins between inside, outside and surface regions and penalises the superposition of inside regions. To take long surface side chains into account, the regions they occupy are simply given an 'outside' status, thus avoiding that steric clashes due to side chain rotation may eliminate a correct geometry. In the general case, the passage from a soft docking approach to a representation where the details of the protein surface are restored requires to explicitly account for side chain conformational changes induced by complex formation.[96]

Searching the space of side chain conformations is often performed in internal coordinate representation, where the degrees of freedom are side chain torsional angles. MC simulations are well adapted to such search and have been used in various refinement procedures.[59,83] An alternative method consists in searching the side chain conformations among a panoply of amino acid rotamers, where rotamers are side chain conformations that are found with a high propensity in the Protein Data Bank. Rotamers are searched either sequentially,[58] or they can be taken into account as an ensemble and treated by the mean field theory[95] or using a linear programming formulation.[94] In any case, side chain sampling needs to be coupled with adjustments of the protein relative position.

The combination between position search and exploration of side chain conformation is critical for the success of a docking programme[83] since incomplete exploration can lead to incorrectly ranking the docking predictions (Fig. 6.6). In RosettaDock, the Baker group proposes a perturbation-based protocol to fully relax a given complex geometry.[83] It consists in generating numerous starting points around the ligand for launching a series of minimizations with periodic MC sampling of the side chains. Introduction of this final refinement process in the RosettaDock protocol dramatically improved the ranking of putative solutions and therefore the overall docking efficiency.[32,97]

6.3.3.3 *Explicit Solvation*

Finally, another element that plays an essential role for the stability of protein–protein complexes is the hydrated environment.[98] Water molecules play a role not only as a dielectric medium and a component of hydrophobic interactions, but also individually via hydrogen bond bridging interactions at the interface.[98–100] Few groups have proposed solutions to explicitly account for solvation in docking methods. The group of Sternberg[95] implemented a method based on dipole representation interacting with the association partners during the refinement phase. Van Dijk *et al.*[101] mimics desolvation within the encounter complex by allowing water molecules initially present in the first water shell to be expelled during optimization of the complex, using a biased Monte Carlo procedure.

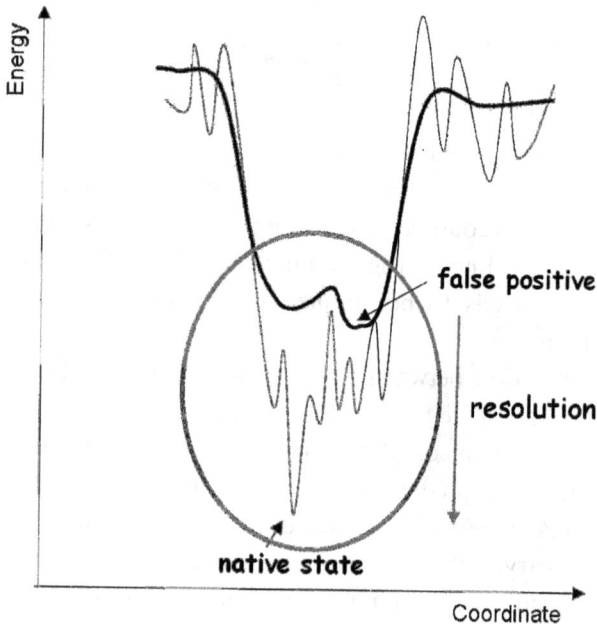

Fig. 6.6. Schematic representation of the refinement procedure. The thick line represents the potential energy surface along a given coordinate, obtained at low resolution using a soft scoring function. Increasing the resolution in terms of protein representation, step size of the search or scoring function reveals the details of the potential energy surface (thin line) and allows distinguishing the native state of the complex from false positive solutions.

6.3.3.4 *Hierarchical Approaches*

In order to optimise both the efficiency of exploration and the computation time, the two-step approach comprising a search step followed by refinement can be extended to hierarchical multi-scale strategies. RosettaDock is a representative example of such a strategy.[83] The initial phase of extensive rigid body MC search is performed at low resolution (see Section 6.3.2.2), using a rough amino-acid based scoring function comprising a statistical residue environment and residue–residue interaction potential, with a penalty for overlapping residues. This score is used to eliminate geometries below a given threshold. The geometries that pass the low resolution filter are treated at finer resolution using more precise scoring function. The resolution is increased by adding flexibility to the side chains and submitting the structures to series of relaxations (small variations in displacement and orientation followed by MC cycles of side chain repacking and rigid body minimization, see Section 6.3.3.2), whereas the scoring gets closer to a free energy function. Each new step towards precise resolution permits to 'dig' the potential energy surface, thus revealing free energy minima that can finally be ranked by energy values (Fig. 6.6).

6.3.4 *Scoring the Predictions*

Scoring is an essential component of docking strategies. While a necessary condition for successful docking is that extensive search should generate complex geometries close to the native one, the scoring function must be able to identify those structures among the thousands generated structures.

In theory, the ideal scoring function permitting to distinguish correct geometries of association from incorrect ones is the free energy, directly related to the dissociation constant of the predicted complex. In practice, this quantity is very difficult to access theoretically and its calculation requires large computer resources.[102,103] Consequently, the scoring functions used in docking programmes are at best free energy approximations, based on robust formulations that have been trained on a number of systems. Eq. 6.10 lists the terms entering the composition of a free energy-like function in ICM,[31]

$$E = \omega_0 E_{int} + \omega_1 E_{vw} + \omega_2 E^{solv}/_{el} + \omega_3 E_{hb} + \omega_4 E_{hp} + \omega_5 E^{solv}/_{SA} + \omega_6 E_{config}$$

$$(6.9)$$

where E_{int} is a sum of van der Waals, electrostatics (Coulomb term with distance dependent dielectric function $\varepsilon(r) = 4r$) and torsional potential energy terms, E_{vw}, E_{hb}, $E^{solv}/_{el}$ and E_{hp} are the terms defined in Eq. 6.9 with the difference that E_{vw} is no more truncated, $E^{solv}/_{SA}$ is a surface based solvation energy and E_{config} is a term for side chain configurational entropy. The weights ω_0 to ω_6 can be adjusted to optimise the ranking of docking predictions.[104]

As mentioned earlier, using atomic free energy scoring functions like that presented in Eq. 6.10 necessitates the predicted structures to reach X-ray levels of resolution, which can be obtained only via thorough exploration of the side chain conformations, coupled to protein fine positioning. While this strategy has led to successful results,[32,59] the search for robust scoring functions capable of correctly ranking low resolution predictions remains an important challenge for the docking community, so as to avoid the time-consuming phase of structure optimization.[34] A strong effort is being devoted to this challenge. This issue is reviewed in Chapter 9 of this volume.

6.4 Evaluation of the Docking Methods

From the beginning of its development, the protein–protein docking community has decided to organise a common evaluation process.[105] This is one important key for the strength of this discipline. Initiated in 2001 and managed since then by leading docking developers (John Moult, Joel Janin, Ilya Vakser, Kim Henrick, Lynn Ten Eyck, Michael Sternberg, Sandor Vajda and Shoshana Wodak), CAPRI is a community wide experience aiming at evaluating the docking methods on the basis of blind predictions trials (see http://www.ebi.ac.uk/msd-srv/capri/index.html). In addition to the CAPRI experience, benchmarks for protein–protein docking have been proposed to the community for training their docking programmes on a common basis. Docking methods that actively participate in CAPRI rounds are listed in Table 6.1.

6.4.1 *The CAPRI Experience*

Sixteen rounds of CAPRI have been run between 2001 and 2008, summing up to 37 targets. The targets are protein–protein complexes with structure recently determined but not yet released. Rounds are launched each time structural biologists bring a target to the CAPRI management team, i.e. they willingly accept to wait for the round to be closed before releasing a newly-solved structure by crystallography or NMR. CAPRI has recently opened its field of applications to protein–nucleic acid complexes. As far as possible, unbound structures of the partner proteins are provided to the predictor groups. This was not always possible in the first CAPRI rounds and 12 out of 16 targets of CAPRI rounds 1 to 5 had one of their partners supplied in the bound form.[6,7,11,12]

More recently, target structures had to be modelled from homologous proteins by the participant groups (targets T11, T20 and T24[11,106]). Predictors are given between one and four weeks to propose ten ranked predictions. They can use any means they judge necessary in addition to their docking programme, like for example experimental information on the location of the docking interface, or the use of specifically human skills like visual appreciation. Note that in one occasion, the complex between pancreatic-amylase and camelid antibody, (targets T04 and T05[7]), the complementarity determining regions (CDRs) that were expected to be part of the binding region were not involved in the interaction, leading all predictor groups to select a wrong geometry of association. Some groups have developed servers that perform calculations and select the ten predictions without any human interference: ClusPro[107,108] and more recently SKE-DOCK,[109] PatchDock[110] and GRAMM-X[111] have participated in CAPRI rounds. RosettaDock,[81] recently released as a server,[112] has been used as a third-party programme constitutive of docking strategies developed by different groups[82,97,113] (see Table 6.1).

Successes and failures: The submitted predictions are evaluated after the consensus criteria described in Table 6.2, once predictions presenting too many steric clashes have been eliminated.[7] They are classified as high, medium, acceptable or incorrect solutions depending on the fraction of correctly predicted native contacts f_{nat}, on the ligand backbone root mean square displacement L_rms between the predicted and target conformations, after superposition of the receptors, and on the RMSD limited to the interface residues I_rms.

Table 6.1. Characteristics of the main docking programmes in CAPRI.

Programme	Search method	Protein representation	Website	Ref.
3D Dock	correlation: FFT	discrete	http://www.bmm.icnet.uk/docking/	40
ATTRACT (PTOOLS)	guided: mult-minimization	coarse grain	http://www.ibpc.fr/chantal/www/ptools/	58
BIGGER	systematic	discrete	no web site supplied	114
CLUSPRO[a]	correlation: FFT[b]	discrete	http://nrc.bu.edu/cluster/	107
DOT	correlation: FFT	discrete	http://www.sdsc.edu/CCMS/DOT	42
GRAMM-X	correlation: FFT	discrete	http://vakser.bioinformatics.ku.edu /resources/gramm/grammx	111
HADDOCK	guided: data-driven, MD	atomic	http://www.nmr.chem.uu.nl/haddock	37
Hex	polar FFT expansion	discrete	http://www.csd.abdn.ac.uk/hex/	30
ICM-Disco	guided: MC minimization	atomic	http://www.molsoft.com/icm_pro.html	59
MolFit	correlation: FFT	discrete	http://www.weizmann.ac.il/Chemical_ Research_Support/molfit/	45
PatchDock	geometric	surface	http://bioinfo3d.cs.tau.ac.il/PatchDock	110
RosettaDock	guided: MC minimization	coarse grain to atomic	http://graylab.jhu.edu/docking/rosetta	32
SKE-DOCK	geometric	surface	http://www.pharm.kitasato-u.ac.jp/bmd/files/SKE_DOCK.html	109
ZDOCK[c]	correlation: FFT	discrete	http://zdock.bu.edu/software.php	43

Bold entries in the first column correspond to programmes that can be run on a web server. [a] Refined with SMOOTHDOCK. [b] Uses DOT or ZDOCK as search methods; [c] Refined with RDOCK.

Additional descriptors of the prediction accuracy can be found in the CAPRI website (http://www.ebi.ac.uk/msd-srv/capri/): fraction of non-native contacts $f_{non-nat}$, fraction of native interface residues f(IR) for the ligand and for the receptor, fraction of non-native interface residues f(OP), interface surface area IA, distance between the geometric centres of the predicted and the target ligand conformation d(L), and angle mismatch theta(L) between the predicted and target ligand orientation.

If one excludes one cancelled round (due to accidental structure disclosure), 24 targets that have been submitted for prediction have given way to published evaluation up to now. The overall progression of the field cannot be easily depicted as the target complexity has notably increased along the rounds and most recent targets present some type of flexibility. It can nevertheless be observed that from target T06 on, all bound/unbound targets have given way to at least 1 and up to 21 high quality prediction, with the exception of target T18 showing a flexible interface loop. Alternatively, the results for unbound/unbound targets continue to be highly dependent on the degree of surface remodelling between unbound and bound forms of the partners. Encouragingly target T11, where dockerin was modelled by homology and cohesin was an unbound structure, gave way to medium quality predictions. This indicates that the docking methods are sufficiently robust to predict low resolution models starting from approximate structures of the compounds when the amplitude of backbone conformational change at the interface is low (which is the case for T11).

Table 6.2. Quality criteria for CAPRI predictions.

Rank	f_{nat}	L_rms	or	I_rms
High	> 0.5	< 1.0	or	< 1.0
Medium	> 0.3	1.0 < x < 5.0	or	1.0 < x < 2.0
Acceptable	> 0.1	5.0 < x < 10.0	or	2.0 < x < 4.0
Incorrect	< 0.1			

Reproduced from Mendez *et al.*[7] fnat, L_rms and I_rms are defined in the text. In addition to the f_{nat} criterion, either the L_rms or the I_rms criteria must be fulfilled to determine the rank.

Scoring in CAPRI is a new branch of the CAPRI experience, initiated in 2005 starting from round 8, exclusively focusing on scoring. The idea came from the observation that good quality predictions generated by a docking method are sometimes overlooked at the scoring level. The principle of the scoring experience is to take advantage of the thousands of predictions generated by all predictor groups during the search and refinement phases to evaluate the scoring procedures. Predictor groups are invited to share their generated structures with the rest of the community and the scoring participants have one week to extract ten correct solutions from the merged generated structures. Up to now, the scoring contest has given rise to only one evaluation.[13] Globally, the scoring predictions were enriched in acceptable or medium grade solutions with respect to docking predictions, however high grade solutions that were included in the merged ensemble were lost.

6.4.2 *Docking Benchmarks*

Independently of the CAPRI experience, protein–protein docking benchmarks have been published,[115–117] recently incremented by a protein–DNA docking benchmark.[118] These benchmarks contain all non-redundant complex structures found in the PDB for which the unbound structure of independent protein partners is available. There are presently 124 four such structures in the PDB for protein–protein complexes, which have been classified according to the biological function and the degree of docking difficulty. Benchmarks are useful to extensively test docking programmes, methodological developments or scoring functions against a variety of systems and complexity levels, from small rigid to huge flexible complexes.[23,83]

6.4.3 *Challenges*

CAPRI evaluation has allowed identification of three challenges to be faced by developers. The first one concerns the capacity of correctly scoring the generated predictions. This issue has already been discussed in Sections 6.3.4 and 6.4.1 and is specifically addressed in Chapter 9. The second one is related to multi-component docking, i.e. predicting the 3D structure of assemblages containing several proteins or nucleic acids. For the moment, CAPRI has only addressed one particular class of multi-component docking, which is oligomeric assemblage. For target T10 consisting of the trimeric form of the coat glycoprotein E of the encephalitis virus,[119] predictors used symmetry restraints to reduce the search complexity.[11] The groups of Wolfson and Nussinov have proposed an efficient general treatment of multi-component docking using geometric hashing[56] (Section 6.3.1.2). The programme independently docks pairs of proteins, before assembling only the two-protein solutions that present spatial compatibility.

In addition to the computational performance for dealing with such high combinatorial complexity, an interesting observation is that the two-protein solutions that finally lead to the correct geometry of the total system are often poorly ranked in the preliminary binary docking. Similar results were found by Saladin *et al.* using a multi-component version of ATTRACT and a coarse grain representation.[71]

Finally, the third challenge facing docking developers is flexibility. About a quarter of the complexes found in the protein–protein benchmarks or among the CAPRI targets present large amplitude conformational changes induced by the complex formation. Whether they concern loop refolding, domain movement or unfolding–folding events, these conformational changes generally result in remodelling the protein interface. As a consequence, the correct geometry is bound to be rejected during the initial search phase by most docking programmes, notably those that concentrate the search on detecting surface complementarity. Novel approaches are therefore developed to tackle this problem.

6.5 Conclusions and Outlook

We presently witness an outburst of the protein–protein docking field, due to a convergence of the post-genomic era that unravelled a huge quantity of genomic sequences with unknown associated protein function, and to the progress of computer technology and algorithms.

The first purpose of this field is to fill the gap between the huge number of three-dimensional (3D) structures of isolated proteins found in the Protein Data Bank (about 50,000 in 2008) and the small number of 3D structures of complexes (thousands). Novel algorithmic strategies have been developed to face the complexity of extensively searching the possible position/orientation of the protein partners. Several types of low resolution protein representations have been investigated and present interesting possibilities, inclusive for flexibility issues. Such representations open promising perspective for the prediction of huge macromolecular assemblies.

In the present state of the art, assessed by the CAPRI experience, good prediction results are obtained for systems that do not present high amplitude backbone conformational change between the free and bound forms. They are borne upon scoring functions that are efficient at atomic level provided that the side chain packing has been optimised. Promising lines of investigation bear on the development of scoring functions able to discriminate the correct complex geometries even at low resolution, and on methods accounting for possible interface remodelling, i.e. that can explore internal conformational change simultaneously to the exploration of relative position/orientation. Preliminary detection of putative interface regions on each association partner should favour the generalization of flexible docking strategies in docking methods.

Acknowledgements

This work was supported by the French CNRS, MESRT and Paris 7 University. A. Saladin benefited from a UFA fellowship. The authors would like to thank K. Bastard for stimulating docking discussions.

References

1. Russell R.B., Alber F., Aloy P., Davis F.P., Korkin D., Pichaud M., Topf M., Sali A. (2004). *Curr Opin Struct Biol* 14: 313–324.
2. Wolfson H.J., Shatsky M., Schneidman-Duhovny D., Dror O., Shulman-Peleg A., Ma B., Nussinov R. (2005). *Curr Prot Pept Sci* 6: 171–183.

3. Bravo J, Aloy P. (2006). *Curr Opin Struct Biol* 16: 385–392.
4. Wodak S.J., Janin J. (1978). *J Mol Biol* 124: 323–342.
5. Halperin I., Ma B., Wolfson H., Nussinov R. (2002). *Proteins* 47: 409–443.
6. Janin J., Henrick K., Moult J., Eyck L.T., Sternberg M.J.E., Vajda S., Vakser I., Wodak S. J. (2003). *Proteins* 52: 2–9.
7. Mendez R., Leplae R., Maria L.D., Wodak S.J. (2003). *Proteins* 52: 51–67.
8. Janin J. (2005). *Protein Sci* 14: 278–283.
9. Vajda S., Vakser I.A., Sternberg M.J.E., Janin J. (2002). *Proteins* 47: 444–446.
10. Wodak S.J., Mendez R. (2004). *Curr Opin Struct Biol* 14: 242–249.
11. Janin J. (2005). *Proteins* 60: 170–175.
12. Mendez R., Leplae R., Lensink M.F., Wodak S.J. (2005). *Proteins* 60: 150–169.
13. Lensink M.F., Mendez R., Wodak S.J. (2007). *Proteins* 69: 704–718.
14. Bastard K., Thureau A., Lavery R., Prevost C. (2003). *J Comput Chem* 24: 1,910–1,920.
15. Qin S., Zhou H.X. (2007). *Proteins* 69: 743–749.
16. Sternberg M.J.E., Gabb H.A., Jackson R.M. (1998). *Curr Opin Struct Biol* 8: 250–256.
17. Smith G.R., Sternberg M.J.E. (2002). *Curr Opin Struct Biol* 12: 28–35.
18. Schneidman-Duhovny D., Nussinov R., Wolfson H.J. (2004). *Curr Med Chem* 11: 91–107.
19. Vakser I.A., Kundrotas P. (2008). *Curr Pharm Biotechnol* 9: 57–66.
20. Ritchie D.W. (2008). *Curr Protein Pept Sci* 9: 1–15.
21. Berman H.M., Battistuz T., Bhat T.N., Bluhm W.F., Bourne P.E., Burkhardt K., Feng Z., Gilliland G.L., Iype L., Jain S., Fagan P., Marvin J., Padilla D., Ravichandran V., Schneider B., Thanki N., Weissig H., Westbrook J.D., Zardecki C. (2002). *Acta Crystallogr D Biol Crystallogr* 58: 899–907.
22. Betts M.J., Sternberg M.J.E. (1999). *Protein Eng* 12: 271-283.
23. Vajda S. (2005). *Proteins* 60: 176–180.
24. Jiang F., Kim S.H. (1991). *J Mol Biol* 219: 79–102.
25. Katchalski-Katzir E., Shariv I., Eisenstein M., Friesem A.A., Aalo C., Vakser I.A. (1992). *Proc Natl Acad Sci USA* 89: 2,195–2,199.
26. Walls P.H., Sternberg M.J.E. (1992). *J Mol Biol* 228: 277–297.
27. Robert C.H., Janin J. (1998). *J Mol Biol* 283: 1,037–1,047.
28. Palma P.N., Krippahl L., Wampler J.E., Moura J.J. (2000). *Proteins* 39: 372–384.
29. Li N., Sun Z., Jiang F. (2007). *Proteins* 69: 801–808.
30. Ritchie D.W., Kemp G.J. (2000). *Proteins* 39: 178–194.
31. Fernandez-Recio J., Totrov M., Abagyan R. (2002). *Protein Sci* 11: 280–291.
32. Schueler-Furman O., Wang C., Baker D. (2005). *Proteins* 60: 187–194.
33. Bernauer J., Aze J., Janin J., Poupon A. (2007). *Bioinformatics* 23: 555–562.
34. Huang S.Y., Zou X. (2008). *Proteins* 72: 557–579.
35. Camacho C.J., Ma H., Champ P.C. (2006). *Proteins* 63: 868–877.
36. Norel R., Petrey D., Wolfson H.J., Nussinov R. (1999). *Proteins* 36: 307–317.

37. Dominguez C., Boelens R., Bonvin A.M.J.J. (2003). *J Am Chem Soc* 125: 1,731–1,737.
38. de Vries S.J., van Dijk A.D.J., Krzeminski M., van Dijk M., Thureau A., Hsu V., Wassenaar T., Bonvin A.M.J.J. (2007). *Proteins* 69: 726–733.
39. Press W.H., Teukolsky S.A., Vetterling W.T., Flannery B.P. (2005). *Numerical Recipes: The Art of Scientific Computing*, Cambridge University Press, Second Edition.
40. Eisenstein M., Katchalski-Katzir E. (2004). *C R Biol* 327: 409–420.
41. Gabb H.A., Jackson R.M., Sternberg M.J.E. (1997). *J Mol Biol* 272: 106–120.
42. Mandell J.G., Roberts V.A., Pique M.E., Kotlovyi V., Mitchell J.C., Nelson E., Tsigelny I., van Eyck L.F.T. (2001). *Protein Eng* 14: 105–113.
43. Chen R., Weng Z. (2002). *Proteins* 47: 281–294.
44. Heifetz A., Katchalski-Katzir E., Eisenstein M. (2002). *Protein Sci* 11: 571–587.
45. Berchanski A., Shapira B., Eisenstein M. (2004). *Proteins* 56: 130–142.
46. Ritchie D.W. (2003). *Proteins* 52: 98–106.
47. Mustard D., Ritchie D.W. (2005). *Proteins* 60: 269–274.
48. Connolly M.L. (1983). *Science* 221: 709–713.
49. Connolly M.L. (1986). *Biopolymers* 25: 1,229–1,247.
50. Wang H. (1991). *J Comput Chem* 12: 746–750.
51. Norel R., Lin S.L., Wolfson H.J., Nussinov R. (1994). *Biopolymers* 34: 933–940.
52. Duhovny D., Nussinov R., Wolfson H.J. (2002). Efficient Unbound Docking of Rigid Molecules. In: *Algorithms in Bioinformatics*, Springer.
53. Schneidman-Duhovny D., Inbar Y., Nussinov R., Wolfson H.J. (2005). *Proteins* 60: 224–231.
54. Wolfson H., Rigoutsos I. (1997). *Computational Science & Engineering, IEEE* 4: 10–21.
55. Schneidman-Duhovny D., Nussinov R., Wolfson H.J. (2007). *Proteins* 69: 764–773.
56. Inbar Y., Benyamini H., Nussinov R., Wolfson H.J. (2005). *J Mol Biol* 349: 435–447.
57. Mashiach E., Schneidman-Duhovny D., Andrusier N., Nussinov R., Wolfson H.J. (2009). *Nucleic Acids Res* 36: W229–W232.
58. Zacharias M. (2003). *Protein Sci* 12: 1,271–1,282.
59. Fernandez-Recio J., Totrov M., Abagyan R. (2003). *Proteins* 52: 113–117.
60. Hou T., Wang J., Chen L., Xu X. (1999). *Protein Eng* 12: 639–648.
61. Taylor J.S., Burnett R.M. (2000). *Proteins* 41: 173–191.
62. Kanamori E., Murakami Y., Tsuchiya Y., Standley D.M., Nakamura H., Kinoshita K. (2007). *Proteins* 69: 832–838.
63. Fitzjohn P.W., Bates P.A. (2003). *Proteins* 52:28-32.
64. Camacho C.J., Weng Z., Vajda S., DeLisi C. (1999). *Biophys J* 76: 1,166–1,178.
65. Camacho C.J., Vajda S. (2002). *Curr Opin Struct Biol* 12: 36–40.

66. Metropolis N., Rosenbluth A., Rosenbluth M., Teller A., Teller E. (1953). *J Chem Phys* 3: 266–278.

67. Totrov M., Abagyan R. (1997). *Proteins Suppl* 1: 215–220.

68. Clementi C. (2008). *Curr Opin Struct Biol* 18: 10–15.

69. Levitt M., Warshel A. (1975). *Nature* 253: 694–698.

70. Levitt M. (1976). *J Mol Biol* 104: 59–107.

71. Saladin A., Fiorucci S., Poulain P., Prevost C., Zacharias M. (2009). *BMC Struct Biol* 9: 27–35.

72. Poulain P., Saladin A., Hartmann B., Prevost C. (2008). *J Comput Chem* 29: 2,582–2,592.

73. Bastard K., Prevost C., Zacharias M. (2006). *Proteins* 62: 956–969.

74. Izvekov S., Voth G.A. (2005). *J Phys Chem B* 109: 2,469–2,473.

75. Basdevant N., Borgis D., Ha-Duong T. (2007). *J Phys Chem B* 111: 9,390–9,399.

76. Voth G.A. (ed.) (2009). Coarse-Graining of Condensed Phase and Biomolecular Systems, CRC Press.

77. Bonvin A.M.J.J., Boelens R., Kaptein R. (2005). *Curr Opin Chem Biol* 9: 501–508.

78. van Dijk A.D.J., Boelens R., Bonvin A.M.J.J. (2005). *FEBS J* 272: 293–312.

79. Ben-Zeev E., Eisenstein M. (2003). *Proteins* 52: 24–27.

80. Ma X.H., Li C.H., Shen L.Z., Gong X.Q., Chen W.Z., Wang C.X. (2005). *Proteins* 60: 319–323.

81. Chaudhury S., Sircar A., Sivasubramanian A., Berrondo M., Gray J.J. (2007). *Proteins* 69: 793–800.

82. Gong X.Q., Chang S., Zhang Q.H., Li C.H., Shen L.Z., Ma X.H., Wang M.H., Liu B., He H.Q., Chen W.Z., Wang C.X. (2007). *Proteins* 69: 859–865.

83. Gray J.J., Moughon S., Wang C., Schueler-Furman O., Kuhlman B., Rohl C., Baker D. (2003). *J Mol Biol* 331: 281–299.

84. Kuntz I., Blaney J., Oatley S., Langridge R., Ferrin T. (1982). *J Mol Biol* 161: 269–288.

85. Brooijmans N., Kuntz I. (2003). *Annu Rev Biophys Biomol Struct* 32: 335–373.

86. Yao H., Kristensen D.M., Mihalek I., Sowa M.E., Shaw C., Kimmel M., Kavraki L., Lichtarge O. (2003). *J Mol Biol* 326: 255–261.

87. Kufareva I., Budagyan L., Raush E., Totrov M., Abagyan R. (207). *Proteins* 67: 400–417.

88. Engelen S., Trojan L.A., Sacquin-Mora S., Lavery R., Carbone A. (2009). *PLoS Comput Biol* 5, e1,000,267.

89. Mintz S., Shulman-Peleg A., Wolfson H.J., Nussinov R. (2005). *Proteins* 61: 6–20.

90. Bahadur R.P., Zacharias M. (2008). *Cell Mol Life Sci* 65: 1,059–1,072.

91. Fernandez-Recio J., Totrov M., Skorodumov C., Abagyan R. (2005). *Proteins* 58: 134–143.

92. Bordner A.J., Abagyan R. (2005). *Proteins* 60: 353–366.

93. de Vries S.J., Bonvin A.M.J.J. (2006). *Bioinformatics* 22: 2,094–2,098.
94. Andrusier N., Nussinov R., Wolfson H.J. (2007). *Proteins* 69: 139–159.
95. Jackson R.M., Gabb H.A., Sternberg M.J.E. (1998). *J Mol Biol* 276: 265–285.
96. Gray J.J. (2006). *Curr Opin Struct Biol* 16: 183–193.
97. Pierce B., Weng Z. (2008). *Proteins* 72: 270–279.
98. Janin J. (199). *Structure* 7: R277–R279.
99. Li Z., Lazaridis T. (2007). *Phys Chem Chem Phys* 9: 573–581.
100. Bueno M., Camacho C.J. (2007). *Proteins* 69: 786–792.
101. van Dijk A.D.J., Bonvin A.M.J.J. (2006). *Bioinformatics* 22: 2,340–2,347.
102. Beveridge D.L., DiCapua F.M. (1989). *Annu Rev Biophys Biophys Chem* 18: 431–492.
103. Meirovitch H. (2007). *Curr Opin Struct Biol* 17: 181–186.
104. Fernandez-Recio J., Abagyan R., Totrov M. (2005). *Proteins* 60: 308–313.
105. Strynadka N., Eisenstein M., Katchalski-Katzir E., Shoichet B., Kuntz I., Abagyan R., Totrov M., Janin J., Cherfls J., Zimmerman F., Olson A., Duncan B., Rao M., Jackson R., Sternberg M., James M. (1996). *Nat Struct Biol* 3: 209–210.
106. Janin J. (2007). *Proteins* 69: 699–703.
107. Comeau S.R., Gatchell D.W., Vajda S., Camacho C.J. (2004). *Nucleic Acids Res* 32: W96–W99.
108. Comeau S.R., Kozakov D., Brenke R., Shen Y., Beglov D., Vajda S. (2007). *Proteins* 69: 781–785.
109. Terashi G., Takeda-Shitaka M., Kanou K., Iwadate M., Takaya D., Umeyama H. (2007). *Proteins* 69: 866–872.
110. Schneidman-Duhovny D., Inbar Y., Nussinov R., Wolfson H.J. (2005). *Nucleic Acids Res* 33: W363–W367.
111. Tovchigrechko A., Vakser I.A. (2006). *Nucleic Acids Res* 34: W310–W314.
112. Lyskov S., Gray J.J. (2008). *Nucleic Acids Res* 36: W233–W238.
113. Heifetz A., Pal S., Smith G.R. (2007). *Proteins* 69: 816–822.
114. Krippahl L., Moura J.J., Palma P.N. (2003). *Proteins* 52:19–23.
115. Chen R., Mintseris J., Janin J., Weng Z. (2003). *Proteins* 52: 88–91.
116. Mintseris J., Wiehe K., Pierce B., Anderson R., Chen R., Janin J., Weng Z. (2005). *Proteins* 60: 214–216.
117. Hwang H., Pierce B., Mintseris J., Janin J., Weng Z. (2008). *Proteins* 73:705-709.
118. van Dijk M., Bonvin A.M.J.J. (2008). *Nucleic Acids Res* 36:e88.
119. Berchanski A., Segal D., Eisenstein M. (2005). *Proteins* 60:202–206.

CHAPTER 7

Data-driven Docking: Using External Information to Spark the Biomolecular Rendez-vous

Adrien S. J. Melquiond and Alexandre M. J. J. Bonvin

NMR Research Group, Bijvoet Center for Biomolecular Research, Utrecht University, Padualaan 8, 3584 CH Utrecht, The Netherlands
E-mail: a.m.j.j.bonvin@uu.nl

Advances in biophysics and biochemistry have pushed back the limits of structural characterization of biomolecular assemblies. Mixing even a limited amount of experimental and/or bioinformatic data with modelling methods such as macromolecular docking represents a valuable strategy to predict 3D structures of complexes. In this chapter, we discuss and illustrate the various sources of data that can be used for this purpose, with emphasis on their combination with docking methods. Finally, we discuss the place of data-driven docking in the modelling of the 3D structures of biomolecular assemblies.

7.1 Introduction

Considering the wide efforts during the last decades in genomics, proteomics, metabolomics and other related fields, the science of the molecular biology of the cell is in a state of explosive growth. Coping with this large amount of information is the present-day challenge of the molecular biology of the 21st century.[1] As a consequence, structural biology is drifting towards systems biology, linking molecules to systems and trying to understand how the biomolecular units work together to fulfil their task.[2-4] Systematic, high-throughput analysis of proteomes has led to the characterization of many complexes.[5-8] These complexes are however often only known in the sense that their constituents have been determined, but their three-dimensional (3D) structures are most of the time unknown and it will probably take years before high-resolution structures of most of them are available.[9] To overcome experimental difficulties surrounding the structure determination of complexes, computational approaches such as molecular docking became a paramount alternative to gain atomic insight into their structure.[10] Docking methods generate a model of a complex based on the known 3D structures of its own components.[11,12] To produce the best models, emerging methods aim at combining the atomic structures of the individual components with a mixed bag of experimental data.[13,14] For instance, biochemical and biophysical experiments are widely used to derive structural information on biomolecular interactions. Combining these data with docking seems an obvious path to follow considering the large number of known 3D structures of single proteins. By combining external information with modelling techniques such as docking, the hope is to 'spark' the biomolecular rendez-vous and somewhat decrease the gap between the number of known structures of complexes and of their constituents.

Conventional crystallography and NMR structural biology techniques, which have achieved in the past great successes with membrane associated proteins[15-18] and large macromolecular complexes,[19-21] are often confronted with challenges when dealing with weak or short-lived complexes. These are often of the utmost biological importance. X-ray crystallography is the most common technique for the structural analysis of proteins and protein complexes, and remains the 'gold standard' in terms of accuracy and resolution.[13] Crystallography, however, requires milligram quantities of a pure and monodisperse sample and its crystallization, which is far from trivial, especially for complexes. NMR spectroscopy is also widely used for the determination of atomic structures of single proteins or domains thereof [22] (currently ~14% of the Protein Data Bank[23]). It has proven particularly useful in the identification of residues involved in biomolecular interactions.[24] The main limitation for NMR is that large complexes cause severe line broadening which narrows its applicability to ~80 kDa.[25,26] Moreover, collecting structural restraints as NOEs for large systems as well as transient complexes requires a considerable effort.

In this chapter, we focus on docking approaches which rely on the use of biochemical and/or biophysical data to guide their prediction. We will not discuss the methods which use experimental information only as an a posteriori filter to validate or reject models. Our discussion will further be limited to macromolecular complexes, omitting protein-small ligand complexes; however, much of what is presented here is also valid for that class of complexes. For a review on 'guided docking' for studying protein–ligand complexes, see Ref. 27.

The chapter is organised as follows. For each critical issue specific to the problem of biomolecular complex structure prediction, we list the different sources of experimental and/or bioinformatic data which can provide help to tackle the problem. Then, we explain how they are implemented in our data-driven docking program HADDOCK (High Ambiguity Driven DOCKing)[28–30] to support the prediction procedure. We also discuss the position of data-driven docking in the field of computational methods for modelling structures of protein complexes, with references to the Critical Assessment of PRedicted Interactions (CAPRI) experiment.[31,32] CAPRI is a 'blind' docking experiment in which participants have a limited time to predict the structure of a complex given only the structures of the constituents. Finally, we conclude with an outlook on what could be the future potential of this technique.

7.2 Stoichiometry and Composition

To determine the structure of a biomolecular complex, one first needs to know which components are interacting and in which proportion (Fig. 7.1). This is also the first requirement to run a docking simulation: who is interacting with whom and in how many copies? From this only knowledge, a molecular *ab initio* docking simulation can be performed. Information about the composition and stoichiometry of a complex is non-trivial to obtain and should be interpreted with care in order to avoid high false–positive rates.

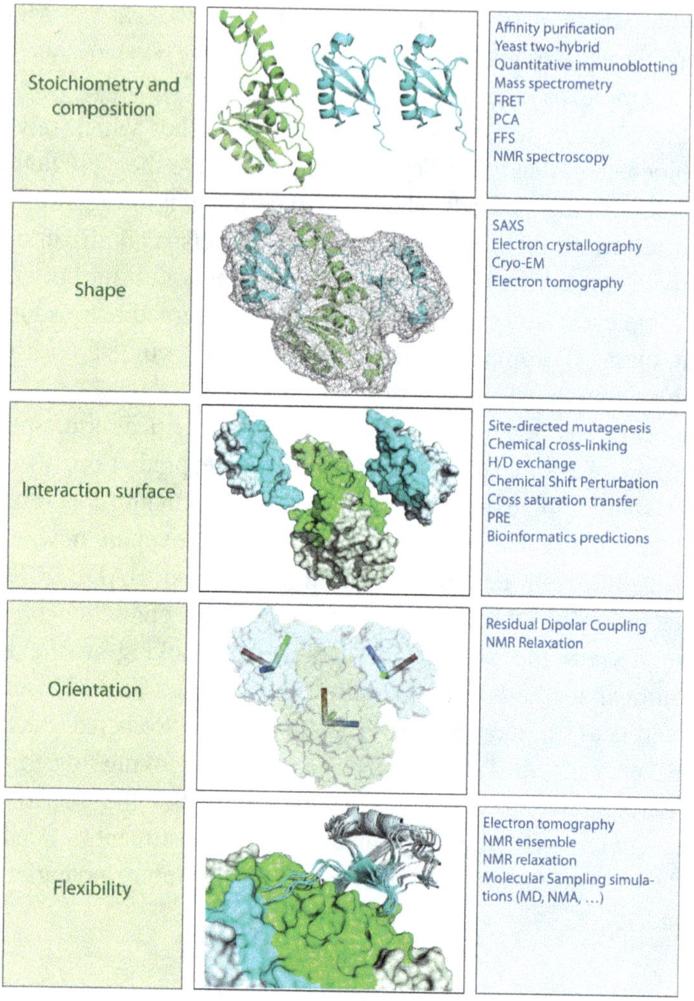

Fig. 7.1. Overview chart of the main issues addressed to the field of biomolecular complex structure prediction. The last column summarises the experimental methods commonly used to come up to each specific problem. The biomolecular complex used to illustrate this table is the solution structure of the Josephin domain of Ataxin-3 in complex with ubiquitin molecule (PDB code: 2JRI). This complex was generated with HADDOCK using CSP data.[156] Pictures generated with Pymol.[157]

The stoichiometry and composition of protein subunits in an assembly can be determined by several methods such as quantitative immunoblotting and mass spectrometry. For a stable complex, measuring the molecular weights of the native assembly and the constituent chains on gel electrophoresis under denaturing conditions represents an accurate and easy reproducible technique to determine efficiently these parameters. Affinity purification[33,34] which combines the purification of a protein complex with the detection of its individual components by mass spectrometry, was a breakthrough in the high-throughput analysis of the subunit stoichiometry of biomolecular complexes.[35-37] More problematic is the study of transient interactions because of their dynamic properties. Indeed, the equilibrium between bound and unbound forms hampers their purification and makes them hard to keep intact for mass spectroscopy detection. To overcome such limitations, it is usually possible determine first the interacting partners by the yeast two-hybrid system,[38,39] Fluorescence Resonance Energy Transfer (FRET)[40,41] or Protein Fragment Complementation Assay (PCA): these techniques consist of fusing two halves of a split reporter to two proteins of interest; if a signal is detected, it means that the reporter properly folded as a consequence of the interaction of the fused proteins).[42] Once the composition of the complex is known, it is then time to tackle the question of its stoichiometry. One method consists of strengthening the interaction by introducing chemical cross-links between the components and detecting the complex by electrospray ionization mass spectrometry.[43,44] Fluorescence Fluctuation Spectroscopy (FFS), in which proteins are labelled with fluorescent tags, allows real time monitoring of the interactions: by brightness analysis, clues can be obtained to quantify the stoichiometry of protein homo- or hetero-complexes in living cells.[45] The stoichiometry of a complex can also be determined from titration experiments in conventional NMR spectroscopy; this method however provides much more information on the protein interfaces as will be discussed later.

Thinking now in terms of docking, knowledge of the stoichiometry is obviously the first prerequisite to start a prediction. While the nature of the components involved in a complex formation is usually well known, their exact stoichiometry is not always reliable, which can cause serious problems in the modelling process. This is an important issue that should be sorted out before starting any docking studies, irrespective of the modelling method chosen.

7.3 Shape of a Biomolecular Complex

'Powerful as single methodologies may be, they are enhanced when applied in judiciously integrated combinations.'[46] This statement, which could refer to the data-driven docking methodology, highlights the recent successes in the integration of low-resolution technique data to refine the comprehensive analysis of large and dynamic macromolecular machines studied by primary X-ray diffraction and NMR experimental techniques. Therefore, while the techniques detailed below are usually described as a low-resolution structural techniques, it is perhaps more appropriate to describe them as high precision technique for determining large distances or molecular envelopes (see Fig. 7.1), that can act as very tight restraints in the modelling of bio-macromolecules 3D structure.[47,48] The modelling or refinement procedures using both 'low' and 'high' resolution techniques for structural determination of a macromolecular assembly are commonly referred to 'hybrid methods'.[49]

A first technique to investigate the organisation of a complex is Small-angle X-ray Scattering (SAXS). This technique yields information on the structure of a multi-body system in terms of average component sizes and shapes. It is usually performed in aqueous solution leading to an averaging of the scattering pattern. SAXS, which is applicable to macromolecules of nanometer size dimensions, is accurate, mostly non-destructive and usually requires only a minimum of sample preparation. It provides a low-resolution model of a protein and can be used to fit separate domains solved at atomic resolution into a 'SAXS envelope'. Even though small-angle scattering is not applicable to determine the atomic positions within the molecule, the current size limits of this technique (50–250 kDa) makes it suitable for the study of biomolecular complexes, and complementary to cryo-electron microscopy (cryo-EM) (lower size limit ~110 kDa[50]) or NMR spectroscopy (upper limit ~80 kDa[25]). In addition, it is quite fast (a few days at most for acquisition and interpretation of the data) compared to other structural methods. Recently, a few methods have been proposed to assemble high resolution

NMR structures of domains based on a 'SAXS filter'.[51-53] In another study, SAXS data were combined with NMR, X-ray and electron microscopy data to reconstruct the quaternary structure of a multi-domain protein.[54]

Electron crystallography[55] and single-particle electron microscopy (cryo-EM/TEM) can also be used to reveal the shape and symmetry of an assembly, sometimes at near-atomic resolution,[56] but more frequently at an intermediate resolution. When the object of interest is a complex for which a cryo-EM density map has been obtained at moderate resolution (~ 10–25 Å) and high resolution structures for individual subunits are available, the resolution of the complex may be leveraged by fitting the components into the density map with an accuracy approaching one-tenth the resolution of the EM reconstruction.[57,58] Finally, Electron Tomography (ET), based on multiple tilted views of the same object and its reconstruction, can also be used to study the structure of macromolecular assemblies despite its relatively low resolution (in the range of 20–40 Å).[59] This technique has the advantage of being truly single-particle: the structural information it delivers is derived from a single instance of the molecule, and offers important benefits in dynamic applications such as the study of flexible proteins and conformational changes that occur during binding interactions between proteins.[60]

These experimental techniques are very promising in combination with docking as they provide spatial information on the organisation of the biomolecular complex.[61] Most studies have treated the fitted components as rigid bodies,[62-64] but discrepancies can be encountered that require adjustments in the components ('flexible docking').[65,66]

7.4 Nature of the Interface: Which Residues are Engaged in a Date?

While the techniques described above can provide essential information on the nature of the interacting subunits and the overall shape of their assembly, knowledge about which specific residues are involved in the interaction is much more valuable to drive the docking prediction. Several experimental techniques can give insights into the residues located at the interface of a complex. Critical issues in this case are the level of detail (e.g. are the data residue-specific or not?) and the reliability of the information (e.g. are we observing a direct or an indirect effect?). Notwithstanding these limitations, we discuss here the different techniques commonly used to investigate the interface of a complex (Fig. 7.1).

A well-known approach to obtain information on residues at an interface is site-directed mutagenesis.[67] Single-mutagenesis studies measure the change in binding affinity after the mutation of a particular residue (see Fig. 7.2). If the mutation affects the interaction, it suggests, but does not guarantee, that the mutated residue mediates the interaction. In contrast, a 'silent' change on the binding affinity does not necessarily mean that the residue is not located at the interface, as it may simply not contribute much to the binding energy or be substituted by water or surrounding side chains.[68] Another possibility is that the mutation causes a conformational change and, as a consequence, the binding affinity change could thus reflect an indirect effect and potentially be misinterpreted. Assuming that no such problems occur, mutagenesis can depict a very sensitive map of the interface of a biomolecular complex.[69] Target residues for mutagenesis can be selected based on evolutionary knowledge or systematically scanned as in alanine scanning mutagenesis studies.[70] Many experimental methods can be used to find out whether a given mutation influence complex formation, such as various binding assays, surface plasmon resonance, mass spectrometry, yeast two-hybrid systems and phage display libraries. More detailed information can be obtained using so-called 'double mutant cycles'.[71] After the creation of series of mutants for both proteins, one can assess whether the influence on complex formation of mutation X in protein A depends on mutation Y in protein B. This is done by measuring Kd values for combinations of mutants. If a dependence is identified, the mutations are said to be coupled and one infers that the residues are close in space, i.e. that they are in contact or at close proximity across the interface.

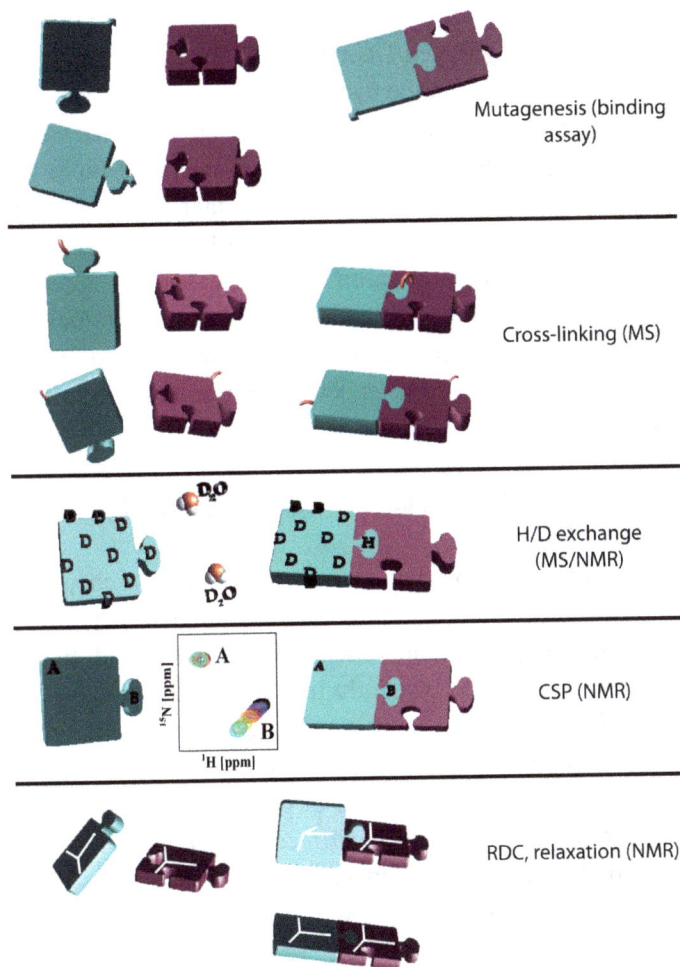

Fig. 7.2. Illustration of the various experimental data that can be combined with docking to restrain the interaction surface definition. The puzzle pieces are representing the two components of the complex. *Mutagenesis:* the arrow on the cyan puzzle piece indicates the mutated residue; *Cross-linking:* the red line indicates the cross-link; *H/D exchange:* D and H are respectively standing for deuterium and hydrogen; *CSP:* the HSQC spectrum shows the chemical shifts of two residues A and B, one of which is shifting during the titration (addition of the interacting protein to the sample solution) and the other not; *RDCs, relaxation:* this figure illustrates the orientational ambiguity which can be unravelled by these techniques; the axis system in white indicates the tensor which provides orientational information.

Besides mutagenesis, other techniques are also used to unravel protein interaction surfaces. Hydrogen/Deuterium (H/D) exchange, for example, consists in exchanging the labile hydrogen atoms, usually amides protons in the backbone, by a deuterium atom. In an aqueous solution, the accessible amide protons on the surface will exchange rapidly with the deuterated solvent, whereas the ones at the interface will be protected (Fig. 7.2). Repeating the same experiment on both the isolated proteins and the complex gives information about the solvent accessibility of various parts of the molecule, and thus the residues located at the interface. The H/D exchange can be followed by NMR spectroscopy, by recording a series of ^{15}N HSQC spectra at different time points.[72] Mass spectrometry is also very efficient in detecting the resulting change in mass caused by the deuteration; it is possible to achieve residue resolution by using fragmentation of the peptides by tandem mass spectrometry.[73,74]

Another possibility to characterise interfaces is chemical cross-linking, followed by identification of the cross-linked peptides by mass peptide fingerprinting. The approach relies on bi- and tri-functional cross-linking reagents that covalently link proteins interacting with each other (see Fig. 7.2). Proteolytic cleavage and subsequent mass spectroscopic identification of the cross-linked species reveal their composition.[75,76] *In vivo* cross-linking of protein complexes using photo-reactive amino acid analogs has been recently reported.[77] Although this method is promising, it has been reported that the cross-linkers may disrupt the structure of the protein complex, and that care should therefore be taken to interpret the results.[78] A new experimental approach, known as Radical Probe Mass Spectrometry (RP-MS), has been proposed to probe the structure of proteins and their interaction with other macromolecules.[79] The principle of this method relies on a limited oxidation of the side chains of the residues exposed to the solvent by fluxes of hydroxyl radicals on millisecond timescales. The small size of the probe and the short exposure timescale ensure the integrity of the protein complex.[80] A new docking algorithm (Proximo) was recently described using RP-MS data to map the interacting surfaces.[81]

Finally, we will focus on the various NMR spectroscopy techniques that can be used to map the interacting surfaces in a biomolecular complex. In its classical use, NMR spectroscopy allows determination of atomic-resolution structures of large subunits and even weak complexes in solution under near-native conditions.[82–84] Conventional restraints obtained from NMR include distance bounds between pairs of atoms (from NOEs experiments, typically < 5–6Å) and dihedral angles. They are used in the structure determination of biomolecules and their complexes. If no significant conformational changes takes place upon complexation, intermolecular NOE restraints, possibly reinforced by RDCs (see below), can be sufficient to determine the structure of the biomolecular complex by simple rigid-body minimization.[85] The structure determination of the complex between Nck-2 SH3 domain and PINCH-1 LIM4 domain by J. Vaynberg *et al.* provides a good illustration of the power of *de novo* NMR spectroscopy when it comes to studying weak interactions.[86]

However, in the case of most transient (or 'short-lived') biomolecular complexes, obtaining enough NOE restraints can be very challenging or even impossible. Therefore, other NMR experiments have been developed that are very suited to map interfaces. First, the NMR chemical shift is very sensitive to the environment: the chemical shift of residues at an interface will change as a result of the altered environment caused by the presence of the interacting partner.[87] The easiest way to observe these changes by recording Heteronuclear Single Quantum Coherence (HSQC) spectra of one ^{15}N (or ^{13}C) labelled partner in the complex. These HSQC spectra are recorded in the absence and presence of increasing amounts of the partner protein ('titration experiment'). This allows following the chemical shift changes of the labelled partner in the complex (*cf.* Fig. 7.2). One then repeats this procedure with the second molecule labelled. Under the assumption that the perturbed residues correspond to the interacting residues, a detailed map of the interface is thus obtained and can be used in molecular docking approaches,[88] or complemented by orientational restraints such as RDCs.[89,90] A potential problem here is that chemical shift changes can also be caused by indirect allosteric effects (e.g. remote conformational changes) (false positives). In addition, not all residues within an interface show chemical shift changes upon binding (false negatives). The data should therefore be interpreted with caution.

Two other NMR techniques are also used to map the interaction surfaces in protein–protein complexes. In cross-saturation experiments, an unlabelled protein, 'the target', is complexed with a perdeuterated ^{15}N-enriched protein, 'the reporter'.[91] Saturation of the unlabelled protein results in an attenuation of the magnetization of the reporter H_N signals due to spin diffusion.[92,93] This is measured in the form of a decrease of the peak intensities in a ^{15}N-HSQC spectrum. The largest effects are observed for residues at the interface since these are the ones affected by spin diffusion coming from the 'target' protein. By reversing the labelling scheme, one can map the other interface. Generally, cross-saturation experiments are more accurate to depict the interface than chemical shift perturbation data, as the latter are very sensitive to conformational changes. Cross-saturation is also applicable to weak and transient complexes since it involves steady-state NOE type experiments.[94]

The last NMR parameters that we would like to introduce to define an interaction surface are paramagnetic effects arising from magnetic dipolar interactions between a nucleus and the unpaired electrons of a paramagnetic center. The use of paramagnetic tags attached to a protein can induce this phenomenon. Two types of effects can be measured: Paramagnetic Relaxation Enhancement (PRE), which corresponds to line-broadening due to enhanced relaxation, and Pseudocontact Shifts (PS), which correspond to a change in the chemical shift itself. Because the magnetic moment of the unpaired electron is large, both PRE and PS effects are large and can provide long-range distance information (up to ~40 Å); they can therefore efficiently complement NOE restraints or CSP data. It is also possible to add paramagnetic ions to the sample solution, causing a line-broadening effect for the solvent-exposed residues while the residues located at the interface in a complex will be protected allowing their identification.[95,96] Finally, paramagnetic restraints can also be used to break the symmetry in symmetrical complexes[97] or shed light on the orientation of the subunits,[98] as it will be explained below. This method has its limits too: due to the strength of the effect, intermolecular PREs do not only report on the specific complex but also on pre-equilibrium encounter complexes that can be sampled in the early stages of the protein–protein association[99,100] and care should therefore be taken in their proper use as restraints.

7.5 Orientation and Symmetry Problems

Whereas the ambiguities on composition, stoichiometry, shape and interface of a biomolecular complex can be unravelled by the different experimental data discussed so far, one major geometrical unknown still remains for the community of molecular 'dockers': indeed, orientational and symmetrical problems are widely encountered when performing a molecular docking on a homomeric system (symmetrical) and more generally when the complex implies a rather non-specific or 'flat' interface.[101–103] Paramagnetic restraints were previously mentioned as one possibility to break the orientational ambiguity. NMR spectroscopy provides, however, other restraints amenable to resolve these uncertainties. We will first discuss how to possibly lift the veil on orientational ambiguity of the subunits and then tackle the symmetry problem.

In the absence of unambiguous distance restraints, the relative orientation of the components of a complex remains to be defined. In some cases, shape complementarity and electrostatic considerations might be sufficient to resolve the orientational ambiguity. The use of additional experimental information, if available, can however give more confidence in the generated model. One experimental source that can provide such information is the Residual Dipolar Coupling (RDC) between two nuclei.[104] In solution, proteins can freely rotate, 'tumble', and will thus sample all the orientations with respect to the external magnetic field of the spectrometer. As a consequence, the dipolar coupling averages to zero. However, in an anisotropic medium obtained, for example by adding bicelles to the sample, the protein will sample preferred and less preferred orientations resulting in a slight anisotropy of orientations. This results in a slight alignment with respect to magnetic field, which partially reintroduces dipolar couplings, i.e. RDCs. RDCs provide information on the orientation of the bond-vector between the two observed nuclei with respect to the alignment tensor. A *priori*, the alignment tensor is unknown. However, when the structure of the protein is known, it can be determined from the experimental set of RDCs representing different (bond-)vectors in the protein. In the case of a complex, the alignment tensor can be determined separately for both components. This allows orienting the two components with respect to each other, since their alignment tensors in the complex should be co-linear (see Fig. 7.2). This orientational information can be introduced in various ways in docking.[85,98,105] Nevertheless, multiple solutions might exist due to the intrinsic degeneracy of RDCs. This ambiguity can be removed if multiple sets of RDCs are used simultaneously. In addition, since RDCs do not provide the translational component, they are often used in combination with additional restraints to refine the molecular docking prediction.

In a comparable way, NMR ^{15}N relaxation rates can be used to orient the two components of a complex with respect to each other. For ^{15}N nuclei located in secondary structure elements, i.e. rigid regions in the protein structure, the value of the relaxation rates will depend on the rotational diffusion of the protein, which can be described by a rotational diffusion tensor. More specifically, the ratio between the ^{15}N transversal (R_2) and longitudinal (R_1) relaxation rates will depend on the orientation of the $^{15}N-^1H$ bond-vector with respect to the rotational diffusion tensor. As a consequence, in the case of anisotropic rotational diffusion, the R_2/R_1 rates will provide orientational information. The diffusion tensor of two components of a biomolecular complex can, as described above for the RDCs, be determined from the experimental relaxation rates of the components within the complex. Since the rotational diffusion will be determined by the shape of the whole complex, the two components can be oriented with respect to each other in such a way that their diffusion tensors are co-linear. As RDCs, relaxation data can be used in addition to CSP data in docking approaches to guide the formation of the complex.[106]

There is a strong tendency for symmetry in oligomeric proteins, even for heteromeric complexes.[107,108] Symmetry, also affects experimental measurements. In NMR spectra of symmetric oligomers, for example, symmetry-related nuclei have degenerate chemical shifts due to their identical chemical environments. It is thus intrinsically impossible for symmetrical oligomers to distinguish which cross peaks arise from intra-

or inter-monomer interactions in Nuclear Overhauser Effect Spectroscopy (NOESY) spectra. Several experimental approaches have been proposed to solve the symmetry degeneracy problem: the basic idea is to break the symmetry by mixing isotopically labelled and unlabelled monomers within the symmetric oligomers.[109] However, for higher-order oligomers, ambiguity still remains. The order of an oligomer can be determined by a variety of experimental methods such as equilibrium ultracentrifugation, dynamic light scattering or NMR diffusion experiments. The most efficient experimental method to date is the usage of a paramagnetic probe to break the symmetry of the system by collecting long-range restraints.[97] Recently, two *ab initio* docking programmes (ZDOCK[110] and PatchDock[111]) gave birth to specific versions for the prediction of the structure of a C_n symmetric multimer. Both programmes, respectively M-ZDOCK[112] and SymmDock,[113] are performing a rotational search of order n about a symmetry axis for the best conformation of a multimer based on the structure of a monomer. In HADDOCK,[28,30] one can enforce the symmetry either within or between the molecules. The symmetry relationship is defined in the form of symmetry distance restraints as proposed by M. Nilges *et al.*:[114–116] for each restraint two distances are specified which are required to remain equal during the calculations, irrespective of the actual distance. One advantage of this method is that it is not restricted to cyclic symmetries, other symmetries (like for example D2 symmetry) can be enforced by combining multiple symmetry restraints). This was first implemented in the Molfit[117] to enforce C_n and D_n symmetries.[118,119]

7.6 Flexibility: How to Cope with Conformational Changes Occurring upon Complex Formation?

Incorporating flexibility in predicting the structure of a biomolecular complex is a major challenge for the docking algorithms.[120–122] As we are focusing on the data-driven docking, we aim to present the methods to analyse the protein flexibility and how they can be used as an *a priori* information to start the docking prediction. For a detailed overview of the treatment of flexibility in docking, with a stress on the different algorithms to simulate the flexibility during the different stages of the docking procedure, please refer to the related chapters in this book.

First, an implicit treatment of flexibility can be achieved by generating an ensemble of discrete conformations. Ensembles of conformations are widely used to mimic the induced slight conformational changes which may occur upon protein–protein association. This is commonly used in cross-docking simulations, where the (rigid-body) docking process is repeated from an ensemble of starting structures, and it is usually time-consuming considering the number of possible combinations. This relies on the assumption that the unbound monomer might sample its 'bound state' even in the absence of its binding partner, provided that no major conformational change is occurring during the binding.[123] The conformations can span various degrees of flexibility, from small mainly side chain re-arrangements to large-scale global backbone motions. The recognition step primarily involves an appropriate pairwise matching between the ensembles of conformations. These ensembles can be obtained by using different solved 3D structures of the same protein (by X-ray or more typically NMR ensembles) or by using Molecular Dynamics (MD) snapshots.[124,125] Exposed side chains in an ensemble of NMR structures usually sample various conformations; their use in data-driven docking was shown to increase both accuracy and hit rate.[88] While ensemble docking improves the performance in terms of the number of near native solutions, it also leads to an increased number of false positives (wrong solutions with high scores) and consequently renders the scoring more difficult.[126,127] In addition, no clear correlation has been found so far between RMSD from the bound form and success rate.

Another possible treatment of flexibility can be to determine a continuous protein conformation space, which can be used as a search space for refinement algorithms or to generate a set of discrete conformations (as the former strategy). In this case, the flexibility analysis methods create a set of vectors which describes the flexibility of the protein. Both Normal Mode Analysis (NMA) and Essential Dynamics end in this group. Ensembles of structures generated from collective motion-based methods have been applied to protein–protein docking: these, usually, suit very well the conformational changes between the bound and free forms, even if they sample larger conformational changes involving backbone atoms.[128–130] However, a recent study over a set of 134 proteins from a docking benchmark shows that, in the majority of interactions with substantial conformational change (RMSD > 2Å), a single low-frequency normal mode was found that describes well the direction of the observed conformational change for only one-third of the proteins.[131] This suggests that, if the normal modes can be used to predict reliably the extent of observed conformational change, the collective motions are substantially altered during the final stage of the recognition process. Besides the NMA, Essential Dynamics analysis (often named Principal Component Analysis or PCA) depicts the main flexible degrees of freedom of a protein, starting from a set of conformations.[132] An ensemble of structures calculated with CONCOORD[133] have been used in CAPRI.[31] Structures obtained from flexibility analysis were shown to improve the quality of results for protein docking with both small molecules and peptides.[134] At last, it is worth mentioning that complexes have been refined against cryo-EM maps using elastic network models, a simplified version of NMA.[135–137]

Finally, the last relevant information considering the flexibility that one would like to use *a priori* to drive the docking predictions is the definition of the flexible and rigid segments in the proteins. This can be obtained by rigidity theory[138] and hinge detection algorithms,[139] but also by experimental methods like NMR relaxation[140] or TEM.[60]

7.7 How to Implement Data-driven Docking, the HADDOCK Example

Most experimental and/or predicted data are highly ambiguous and only provide information about putative interface residues, but not about the specific contacts made. To reflect this, such data are incorporated into HADDOCK[28–30] in the form of ambiguous interaction restraints (AIRs).[114] Prior to docking, the user must supply for every molecule a list of active residues (residues that are known to make contact within the complex) and passive residues (residues that potentially make contact). For every active residue, a single AIR restraint is defined between the atoms of that residue and the atoms of all active and passive residues on the partner. An explicit AIR energy term (a classical distance restraint term) is introduced into the calculation through a flat-bottom harmonic potential (becoming linear after a given cut-off distance) that depends on an effective distance calculated as follows:

$$d_{iAB}^{eff} = \left(\sum_{m_{iA}=1}^{Natoms} \sum_{k=1}^{NresB} \sum_{n_{kB}=1}^{Natoms} \frac{1}{d_{m_{iA}n_{kB}}^{6}} \right)^{-\frac{1}{6}} \tag{7.1}$$

where N_{atoms} indicates all atoms of a given residue and N_{res} the sum of active and passive residues for a given protein. An upper limit to the effective distance (typically 2 Å) is enforced by HADDOCK. If this limit is exceeded, the AIR energy becomes positive and the active residue experiences an attractive force towards the active and passive residues of the partner molecule. If not, the restraint is satisfied. Since many atom–atom distances inversely contribute to the effective distance, an AIR restraint is typically satisfied if a residue comes within 4–5 Å of any active or passive residue of the partner molecule (the effective distance is always shorter than the shortest distance entering the sum). In this way, (putative) interface residues are enforced to make contact with (a surface region on) the partner protein, but not with any specific partner residue. These ambiguous restraints drive the docking in the same way that Nuclear Overhauser Effect (NOE) distance restraints drive the calculation of an NMR structure. In addition, in order to account for possible false positives in the identification of interface residues, typically 50% of the AIR restraints are randomly deleted for each docking trial. HADDOCK also supports the use of orientational restraints in the form of RDC[105] or diffusion anisotropy restraints.[106] In addition, it is possible to impose various symmetry restraints in the case, for example of homomeric complexes.[114] Conformational flexibility in HADDOCK can be described in different manners, e.g. by starting from an ensemble of structures (an NMR ensemble for example) and/or by defining flexible segments, either manually or in an automated manner during the calculations based on an on-the-fly contact analysis.

Many other docking methods have been used in combination with experimental data to filter the generated solutions or limit the sampling to pre-define regions. For example, the FFT programs Hex and GRAMM-X have been used with mutagenesis data,[141-144] DOT has been used with H/D exchange data, [145,146] BIGGER has often been used in combination with CSP data[147] and FTDOCK has been used with CSP and RDC data.[148,149] Next to experimental information, computational and bioinformatics approaches can also be used to predict interfaces. A large number of such methods have been developed over the last years. For recent reviews see Refs 150 and 151. Several of these prediction methods have also been used in combination with docking, for example WHISCY,[152] ProMate[153] and cons-PPISP.[154]

7.8 Conclusions and Outlook

The inclusion of experimental information in the prediction of the 3D structure of biomolecular complexes by molecular modelling is becoming more and more common. Next to the docking-related approaches discussed in this chapter, other methods are emerging. Particularly promising are the 'hybrid methods' that combine low resolution information (cryo-EM, TEM, SAXS) with high resolution structures of its components to model the large macromolecular assemblies by satisfaction of spatial restraints. Such approaches are however not applicable for transient complexes.

Another class of techniques model interactions by homology, namely by 'comparative modeling'. The idea is simple: use known protein–protein complexes for which coordinates are available to model the interactions between their homologues. The method is based on techniques that have been borrowed from protein-fold recognition (or threading); these methods typically assess how well a homologous pair of sequences 'fits' onto a previously determined structure of a complex. These approaches are however far from perfect (the target-template alignment is often not trivial), and are limited in the case of conformational changes or insertions/deletions at the interface with respect to the chosen template. They are also limited when it comes to predicting specificity in a family of related complexes in which all components are homologous (e.g. E2/E3 complexes along the ubiquitination pathway). Related to comparative modelling are the Comparative Patch Analysis methods: they derive sets of restraints from known interactions of the complex components (or their homologs) with other proteins, limiting as a consequence the sampling search to defined regions (patches) were the complex association is more probable to occur. This technique is somewhere a hybrid of comparative modelling and data-driven methods, as the docking procedure consists more in a refinement of pair wise assemblies generated on the basis of the predicted patches. This relies on evolutionary principles suggesting that protein-binding sites are conserved whatever the fold of their binding partner.[155]

We have discussed in this chapter the combination of limited amounts of experimental data with molecular docking methods and shown that they can give valuable structural insights on biomolecular interactions. While *ab initio* docking algorithms are becoming more and more performing and accurate, it is still clearly beneficial to include some kind of 'smart' bias in the interaction search in order to enrich the 'near correct' configurations before subsequent refinement stages (these are usually the most time-consuming). This will be especially important when dealing with large macromolecular systems or when rather unspecific interaction surfaces are involved. We are now entering an era in which the combination of all the existing knowledge with modelling techniques will allow to shine light on macromolecular assemblies. The time has come to 'combine and conquer'!

Acknowledgements

This work was supported by the Netherlands Organization for Scientific Research (VICI grant #700.56.442 to A.B.) and by the European Community (FP6 integrated Project SPINE2-COMPLEX contract no. 032220). We thank Dr Klaartje Houben and Mickaël Krzeminski (Utrecht University) for helpful discussions about NMR spectroscopy and help in molecular rendering, respectively.

References

1. Alberts B. (1998). *Cell* 92: 291–294.
2. Kitano H. (2002). *Science* 295: 1,662–1,664.
3. Sali A., Glaeser R., Earnest T., Baumeister W. (2003). *Nature* 422: 216–225.
4. Beltrao P., Kiel C., Serrano L. (2007). *Curr Opin Struct Biol* 17: 378–384.
5. Gavin A.C., Bosche M., Krause R., Grandi P., Marzioch M., Bauer A., Schultz J., Rick J.M., Michon A.M., Cruciat C.M., Remor M., Hofert C., Schelder M., Brajenovic M., Ruffner H., Merino A., Klein K., Hudak M., Dickson D., Rudi T., Gnau V., Bauch A., Bastuck S., Huhse B., Leutwein C., Heurtier M.A., Copley R.R., Edelmann A., Querfurth E., Rybin V., Drewes G., Raida M., Bouwmeester T., Bork P., Seraphin B., Kuster B., Neubauer G., Superti-Furga G. (2002). *Nature* 415: 141–147.
6. Ho Y., Gruhler A., Heilbut A., Bader G.D., Moore L., Adams S.L., Millar A., Taylor P., Bennett K., Boutilier K., Yang L., Wolting C., Donaldson I., Schandorff S., Shewnarane J., Vo M., Taggart J., Goudreault M., Muskat B., Alfarano C., Dewar D., Lin Z., Michalickova K., Willems A.R., Sassi H., Nielsen P.A., Rasmussen K.J., Andersen J.R., Johansen L.E., Hansen L.H., Jespersen H., Podtelejnikov A., Nielsen E., Crawford J., Poulsen V., Sorensen B.D., Matthiesen J., Hendrickson R.C., Gleeson F., Pawson T., Moran M.F., Durocher D., Mann M., Hogue C.W., Figeys D., Tyers M. (2002). *Nature* 415: 180–183.
7. Aloy P., Bottcher B., Ceulemans H., Leutwein C., Mellwig C., Fischer S., Gavin A.C., Bork P., Superti-Furga G., Serrano L., Russell R.B. (2004). *Science* 303: 2,026–2,029.
8. Gavin A.C., Aloy P., Grandi P., Krause R., Boesche M., Marzioch M., Rau C., Jensen L.J., Bastuck S., Dumpelfeld B., Edelmann A., Heurtier M.A., Hoffman V., Hoefert C., Klein K., Hudak M., Michon A.M., Schelder M., Schirle M., Remor M., Rudi T., Hooper S., Bauer A., Bouwmeester T., Casari G., Drewes G., Neubauer G., Rick J.M., Kuster B., Bork P., Russell R.B., Superti-Furga G. (2006). *Nature* 440: 631–636.
9. Aloy P., Pichaud M., Russell R.B. (2005). *Curr Opin Struct Biol* 15: 15–22.
10. Aloy P., Russell R.B. (2006). *Nat Rev Mol Cell Biol* 7: 188–197.
11. Smith G.R., Sternberg M.J. (2002). *Curr Opin Struct Biol* 12: 28–35.
12. Halperin I., Ma B., Wolfson H., Nussinov R. (2002). *Proteins* 47: 409–443.
13. Russell R.B., Alber F., Aloy P., Davis F.P., Korkin D., Pichaud M., Topf M., Sali A. (2004). *Curr Opin Struct Biol* 14: 313–324.
14. Alber F., Forster F., Korkin D., Topf M., Sali A. (2008). *Annu Rev Biochem* 77: 443–477.
15. Weiss M.S., Abele U., Weckesser J., Welte W., Schiltz E., Schulz G.E. (1991). *Science* 254: 1,627–1,630.
16. Cowan S.W., Schirmer T., Rummel G., Steiert M., Ghosh R., Pauptit R.A., Jansonius J.N., Rosenbusch J.P. (1992). *Nature* 358: 727–733.

17. Ketchem R.R., Hu W., Cross T.A. (1993). *Science* 261: 1,457–1,460.
18. Opella S.J., Marassi F.M., Gesell J.J., Valente A.P., Kim Y., Oblatt-Montal M., Montal M. (1999). *Nat Struct Biol* 6: 374–379.
19. Xu Z., Horwich A.L., Sigler P.B. (1997). *Nature* 388: 741–750.
20. Schluenzen F., Tocilj A., Zarivach R., Harms J., Gluehmann M., Janell D., Bashan A., Bartels H., Agmon I., Franceschi F., Yonath A. (2000). *Cell* 102: 615–623.
21. Griswold I.J., Dahlquist F.W. (2002). *Nat Struct Biol* 9: 567–568.
22. Wüthrich K. (1986). *NMR of Proteins and Nucleic Acids* (Series, B. L., Ed.), Wiley, John & Sons, New York, USA.
23. Berman H., Henrick K., Nakamura H. (2003). *Nat Struct Biol* 10: 980.
24. Bonvin A.M., Boelens R., Kaptein R. (2005). *Curr Opin Chem Biol* 9: 501–508.
25. Tugarinov V., Choy W.Y., Orekhov V.Y., Kay L.E. (2005). *Proc Natl Acad Sci USA* 102: 622–627.
26. Grishaev A., Tugarinov V., Kay L.E., Trewhella J., Bax A. (2008). *J Biomol NMR* 40: 95–106.
27. Fradera X., Kaur J., Mestres J. (2004). *J Comput Aided Mol Des* 18: 635-650.
28. Dominguez C., Boelens R., Bonvin A.M. (2003). *J Am Chem Soc* 125: 1,731–1,737.
29. van Dijk A.D., de Vries S.J., Dominguez C., Chen H., Zhou H.X., Bonvin A.M. (2005). *Proteins* 60: 232–238.
30. de Vries S.J., van Dijk A.D., Krzeminski M., van Dijk M., Thureau A., Hsu V., Wassenaar T., Bonvin A.M. (2007). *Proteins* 69: 726–733.
31. Janin J., Henrick K., Moult J., Eyck L.T., Sternberg M.J., Vajda S., Vakser I., Wodak S.J. (2003). *Proteins* 52: 2–9.
32. Wodak S.J., Mendez R. (2004). *Curr Opin Struct Biol* 14: 242–249.
33. Krogan N.J., Cagney G., Yu H., Zhong G., Guo X., Ignatchenko A., Li J., Pu S., Datta N., Tikuisis A.P., Punna T., Peregrin-Alvarez J.M., Shales M., Zhang X., Davey M., Robinson M.D., Paccanaro A., Bray J.E., Sheung A., Beattie B., Richards D.P., Canadien V., Lalev A., Mena F., Wong P., Starostine A., Canete M.M., Vlasblom J., Wu S., Orsi C., Collins S.R., Chandran S., Haw R., Rilstone J.J., Gandi K., Thompson N.J., Musso G., St Onge P., Ghanny S., Lam M.H., Butland G., Altaf-Ul A.M., Kanaya S., Shilatifard A., O'Shea E., Weissman J.S., Ingles C.J., Hughes T.R., Parkinson J., Gerstein M., Wodak S.J., Emili A., Greenblatt J.F. (2006). *Nature* 440: 637–643.
34. Collins S.R., Kemmeren P., Zhao X.C., Greenblatt J.F., Spencer F., Holstege F.C., Weissman J.S., Krogan N.J. (2007). *Mol Cell Proteomics* 6: 439–450.
35. Gygi S.P., Aebersold R. (2002). *Curr Opin Chem Biol* 4: 489–494.
36. Bauer A., Kuster B. (2003). *Eur J Biochem* 270: 570–578.
37. Hernandez H., Robinson C.V. (2007). *Nat Protoc* 2: 715–726.
38. Rual J.F., Venkatesan K., Hao T., Hirozane-Kishikawa T., Dricot A., Li N., Berriz G.F., Gibbons F.D., Dreze M., Ayivi-Guedehoussou N., Klitgord N., Simon C., Boxem M., Milstein S., Rosenberg J., Goldberg D.S., Zhang L.V., Wong S.L.,

Franklin G., Li S., Albala J.S., Lim J., Fraughton C., Llamosas E., Cevik S., Bex C., Lamesch P., Sikorski R.S., Vandenhaute J., Zoghbi H.Y., Smolyar A., Bosak S., Sequerra R., Doucette-Stamm L., Cusick M.E., Hill D.E., Roth F.P., Vidal M. (2005). *Nature* 437: 1,173–1,178.

39. Parrish J.R., Gulyas K.D., Finley R.L. Jr (2006). *Curr Opin Biotechnol* 17: 387–393.
40. Truong K., Ikura M. (2001). *Curr Opin Struct Biol* 11: 573–578.
41. Yan Y., Marriott G. (2003). *Curr Opin Chem Biol* 7: 635–640.
42. Michnick S.W., Ear P.H., Manderson E.N., Remy I., Stefan E. (2007). *Nat Rev Drug Discov* 6: 569–582.
43. Sharon M., Robinson C.V. (2007). *Annu Rev Biochem* 76: 167–193.
44. Gingras A.C., Gstaiger M., Raught B., Aebersold R. (2007). *Nat Rev Mol Cell Biol* 8: 645–654.
45. Chen Y., Muller J.D. (2007). *Proc Natl Acad Sci USA* 104: 3147–3152.
46. Steven A.C., Baumeister W. (2008). *J Struct Biol* 163: 186–195.
47. Topf M., Sali A. (2005). *Curr Opin Struct Biol* 15: 578–585.
48. Neylon C. (2008). *Eur Biophys J* 37: 531–541.
49. Robinson C.V., Sali A., Baumeister W. (2007). *Nature* 450: 973–982.
50. Jawhari A., Uhring M., De Carlo S., Crucifix C., Tocchini-Valentini G., Moras D., Schultz P., Poterszman A. (2006). *EMBO Rep* 7: 500–505.
51. Mattinen M.L., Paakkonen K., Ikonen T., Craven J., Drakenberg T., Serimaa R., Waltho J., Annila A. (2002). *Biophys J* 83: 1,177–1,183.
52. Grishaev A., Wu J., Trewhella J., Bax A. (2005). *J Am Chem Soc* 127: 16,621–16,628.
53. Petoukhov M.V., Monie T.P., Allain F.H., Matthews S., Curry S., Svergun D.I. (2006). *Structure* 14: 1,021–1,027.
54. Tidow H., Melero R., Mylonas E., Freund S.M., Grossmann J.G., Carazo J.M., Svergun D.I., Valle M., Fersht A.R. (2007). *Proc Natl Acad Sci USA* 104: 12,324–12,329.
55. Nogales E., Wolf S.G., Downing K.H. (1998). *Nature* 391: 199–203.
56. Ludtke S.J., Baker M.L., Chen D.H., Song J.L., Chuang D.T., Chiu W. (2008). *Structure* 16: 441–448.
57. Baker T.S., Johnson J.E. (1996). *Curr Opin Struct Biol* 6: 585–594.
58. Conway J.F., Watts N.R., Belnap D.M., Cheng N., Stahl S.J., Wingfield P.T., Steven A.C. (2003). *J Virol* 77: 6,466–6,473.
59. Nitsch M., Walz J., Typke D., Klumpp M., Essen L.O., Baumeister W. (1998). *Nat Struct Biol* 5: 855–857.
60. Bongini L., Fanelli D., Piazza F., De Los Rios P., Sandin S., Skoglund U. (2004). *Proc Natl Acad Sci USA* 101: 6,466–6,471.
61. Forster F., Webb B., Krukenberg K.A., Tsuruta H., Agard D.A., Sali A. (2008). *J Mol Biol* 382: 1,089–1,106.
62. Roseman A.M. (2000). *Acta Crystallogr D Biol Crystallogr* 56: 1,332–1,340.

63. Chacon P., Wriggers W. (2002). *J Mol Biol* 317: 375–384.
64. Volkmann N., Hanein D. (2003). *Methods Enzymol* 374: 204–225.
65. Valle M., Zavialov A., Sengupta J., Rawat U., Ehrenberg M., Frank J. (2003). *Cell* 114: 123–134.
66. Tama F., Miyashita O., Brooks C.L. 3rd (2004). *J Mol Biol* 337: 985–999.
67. Cunningham B.C., Jhurani P., Ng P., Wells J.A. (1989). *Science* 243: 1,330–1,336.
68. DeLano W.L. (2002). *Curr Opin Struct Biol* 12: 14–20.
69. Sivasubramanian A., Chao G., Pressler H.M., Wittrup K.D., Gray J.J. (2006). *Structure* 14: 401–414.
70. Clackson T., Wells J.A. (1995). *Science* 267: 383–386.
71. Carter P.J., Winter G., Wilkinson A.J., Fersht A.R. (1984). *Cell* 38: 835–840.
72. Emerson S.D., Palermo R., Liu C.M., Tilley J.W., Chen L., Danho W., Madison V.S., Greeley D.N., Ju G., Fry D.C. (2003). *Protein Sci* 12: 811–822.
73. Mandell J.G., Falick A.M., Komives E.A. (1998). *Proc Natl Acad Sci USA* 95: 14,705–14,710.
74. Lanman J., Prevelige P.E. Jr (2004). *Curr Opin Struct Biol* 14: 181–188.
75. Bouveret E., Rigaut G., Shevchenko A., Wilm M., Seraphin B. (2000). *EMBO J* 19: 1,661–1,671.
76. Back J.W., de Jong L., Muijsers A.O., de Koster C.G. (2003). *J Mol Biol* 331: 303–313.
77. Suchanek M., Radzikowska A., Thiele C. (2005). *Nat Methods* 2: 261–267.
78. Peters K., Richards F.M. (1977). *Annu Rev Biochem* 46: 523–551.
79. Maleknia S.D., Chance M.R., Downard K.M. (1999). *Rapid Commun Mass Spectrom* 13: 2,352–2,358.
80. Maleknia S.D., Downard K. (2001). *Mass Spectrom Rev* 20: 388–401.
81. Gerega S.K., Downard K.M. (2006). *Bioinformatics* 22: 1,702–1,709.
82. Fiaux J., Bertelsen E.B., Horwich A.L., Wuthrich K. (2002). *Nature* 418: 207–211.
83. Fernandez C., Wider G. (2003). *Curr Opin Struct Biol* 13: 570–580.
84. Gross J.D., Moerke N.J., von der Haar T., Lugovskoy A.A., Sachs A.B., McCarthy J.E., Wagner G. (2003). *Cell* 115: 739–750.
85. Clore G.M. (2000). *Proc Natl Acad Sci USA* 97: 9,021–9,025.
86. Vaynberg J., Fukuda T., Chen K., Vinogradova O., Velyvis A., Tu Y., Ng L., Wu C., Qin J. (2005). *Mol Cell* 17: 513–523.
87. Zuiderweg E.R. (2002). *Biochemistry* 41: 1–7.
88. Dominguez C., Bonvin A.M., Winkler G.S., van Schaik F.M., Timmers H.T., Boelens R. (2004). *Structure* 12: 633–644.
89. McCoy M.A., Wyss D.F. (2002). *J Am Chem Soc* 124: 2,104–2,105.
90. Clore G.M., Schwieters C.D. (2003). *J Am Chem Soc* 125: 2,902–2,912.
91. Takahashi H., Nakanishi T., Kami K., Arata Y., Shimada I. (2000). *Nat Struct Biol* 7: 220–223.
92. Kalk A., Berendsen H.J.C. (1976). *Journal of Magnetic Resonance (1969)* 24: 343–366.

93. Akasaka K. (1981). *Journal of Magnetic Resonance (1969)* 45: 337–343.
94. Shimada I., Ueda T., Matsumoto M., Sakakura M., Osawa M., Takeuchi K., Nishida N., Takahashi H. (2009). *Progress in Nuclear Magnetic Resonance Spectroscopy* 54: 123–140.
95. Arumugam S., Van Doren S.R. (2003). *Biochemistry* 42: 7,950–7,958.
96. Sakakura M., Noba S., Luchette P.A., Shimada I., Prosser R.S. (2005). *J Am Chem Soc* 127: 5,826–5,832.
97. Gaponenko V., Altieri A.S., Li J., Byrd R.A. (2002). *J Biomol NMR* 24: 143–148.
98. Fushman D., Varadan R., Assfalg M., Walker O. (2004). *Progress in Nuclear Magnetic Resonance Spectroscopy* 44: 189–214.
99. Iwahara J., Clore G.M. (2006). *Nature* 440: 1,227–1,230.
100. Tang C., Iwahara J., Clore G.M. (2006). *Nature* 444: 383–386.
101. Argos P. (1988). *Protein Eng* 2: 101–113.
102. Jones S., Thornton J.M. (1996). *Proc Natl Acad Sci USA* 93: 13–20.
103. Chakrabarti P., Janin J. (2002). *Proteins* 47: 334–343.
104. Bax A. (2003). *Protein Sci* 12: 1–16.
105. van Dijk A.D., Fushman D., Bonvin A.M. (2005). *Proteins* 60: 367–381.
106. van Dijk A.D., Kaptein R., Boelens R., Bonvin A.M. (2006). *J Biomol NMR* 34: 237–244.
107. Goodsell D.S., Olson A.J. (200). *Annu Rev Biophys Biomol Struct* 29: 105–153.
108. Levy E.D., Pereira-Leal J.B., Chothia C., Teichmann S.A. (2006). *PLoS Comput Biol* 2: e155.
109. Arrowsmith C., Pachter R., Altman R., Jardetzky O. (1991). *Eur J Biochem* 202: 53–66.
110. Chen R., Li L., Weng Z. (2003). *Proteins* 52: 80–87.
111. Schneidman-Duhovny D., Inbar Y., Polak V., Shatsky M., Halperin I., Benyamini H., Barzilai A., Dror O., Haspel N., Nussinov R., Wolfson H.J. (2003). *Proteins* 52: 107–112.
112. Pierce B., Tong W., Weng Z. (2005). *Bioinformatics* 21: 1,472–1,478.
113. Schneidman-Duhovny D., Inbar Y., Nussinov R., Wolfson H.J. (2005). *Nucleic Acids Res* 33: W363–367.
114. Nilges M. (1003). *Proteins* 17: 297–309.
115. O'Donoghue S.I., Junius F.K., King G.F. (1993). *Protein Eng* 6: 557–564.
116. O'Donoghue S. I., Chang X., Abseher R., Nilges M., Led J.J. (2000). *J Biomol NMR* 16: 93–108.
117. Katchalski-Katzir E., Shariv I., Eisenstein M., Friesem A.A., Aflalo C., Vakser I.A. (1992). *Proc Natl Acad Sci USA* 89: 2,195–2,199.
118. Berchanski A., Eisenstein M. (2003). *Proteins* 53: 817–829.
119. Berchanski A., Segal D., Eisenstein M. (2005). *Proteins* 60: 202–206.
120. May A., Zacharias M. (2005). *Biochim Biophys Acta* 1754: 225–231.
121. Bonvin A.M. (2006). *Curr Opin Struct Biol* 16: 194–200.

122. Andrusier N., Mashiach E., Nussinov R., Wolfson H.J. (2008). *Proteins* 73: 271–289.
123. Tsai C.J., Kumar S., Ma B., Nussinov R. (1999). *Protein Sci* 8: 1,181–1,190.
124. Li X., Keskin O., Ma B., Nussinov R., Liang J. (2004). *J Mol Biol* 344: 781–795.
125. Rajamani D., Thiel S., Vajda S., Camacho C.J. (2004). *Proc Natl Acad Sci USA* 101: 11,287–11,292.
126. Smith G.R., Sternberg M.J., Bates P.A. (2005). *J Mol Biol* 347: 1,077–1,101.
127. Grunberg R., Nilges M., Leckner J. (2006). *Structure* 14: 683–693.
128. Tama F., Sanejouand Y.H. (2001). *Protein Eng* 14: 1–6.
129. Tobi D., Bahar I. (2005). *Proc Natl Acad Sci USA* 102: 18,908–18,913.
130. Petrone P., Pande V.S. (2006). *Biophys J* 90: 1,583–1,593.
131. Dobbins S.E., Lesk V.I., Sternberg M.J. (2008). *Proc Natl Acad Sci USA* 105: 10,390–10,395.
132. Amadei A., Linssen A.B., Berendsen H.J. (1993). *Proteins* 17: 412–425.
133. de Groot B.L., van Aalten D.M., Scheek R.M., Amadei A., Vriend G., Berendsen H.J. (1997). *Proteins* 29: 240–251.
134. Zavodszky M.I., Lei M., Thorpe M.F., Day A.R., Kuhn L.A. (2004). *Proteins* 57: 243–261.
135. Ming D., Kong Y., Wakil S.J., Brink J., Ma J. (2002). *Proc Natl Acad Sci USA* 99: 7,895–7,899.
136. Chacon P., Tama F., Wriggers W. (2003). *J Mol Biol* 326: 485–492.
137. Tama F., Wriggers W., Brooks C.L. 3rd (2002). *J Mol Biol* 321: 297–305.
138. Jacobs D.J., Rader A.J., Kuhn L.A., Thorpe M.F. (2001). *Proteins* 44: 150–165.
139. Wriggers W., Schulten K. (1997). *Proteins* 29: 1–14.
140. Palmer A.G. 3rd (2004). *Chem Rev* 104: 3,623–3,640.
141. Azuma Y., Renault L., Garcia-Ranea J.A., Valencia A., Nishimoto T., Wittinghofer A. (1999). *Journal of Molecular Biology* 289: 1,119–1,130.
142. Gaboriaud C., Juanhuix J., Gruez A., Lacroix M., Darnault C., Pignol D., Verger D., Fontecilla-Camps J.C., Arlaud G.J. (2003). *Journal of Biological Chemistry* 278: 46,974–46,982.
143. Ritchie D.W., Kemp G.J.L. (2000). *Proteins-Structure Function and Genetics* 39: 178–194.
144. Vakser I.A. (1995). *Protein Eng* 8: 371–377.
145. Anand G.S., Law D., Mandell J.G., Snead A.N., Tsigelny I., Taylor S.S., Ten Eyck L.F., Komives E.A. (2003). *Proceedings of the National Academy of Sciences of the United States of America* 100: 13,264–13,269.
146. Mandell J.G., Roberts V.A., Pique M.E., Kotlovyi V., Mitchell J.C., Nelson E., Tsigelny I., Ten Eyck L.F. (2001). *Protein Engineering* 14: 105–113.
147. Palma P.N., Krippahl L., Wampler J.E., Moura J.J. (2000). *Proteins* 39: 372–384.
148. Dobrodumov A., Gronenborn A.M. (2003). *Proteins-Structure Function and Genetics* 53: 18–32.

149. Gabb H.A., Jackson R.M., Sternberg M.J.E. (1997). *Journal of Molecular Biology* 272: 106–120.
150. Qin S., Zhou H.X. (2007). *Proteins* 69: 743–749.
151. de Vries S.J., Bonvin A.M. (2008). *Curr Protein Pept Sci* 9: 394–406.
152. de Vries S.J., van Dijk A.D., Bonvin A.M. (2006). *Proteins* 63: 479–489.
153. Neuvirth H., Raz R., Schreiber G. (2004). *J Mol Biol* 338: 181–199.
154. Zhou H.X., Shan Y. (2001). *Proteins* 44: 336–343.
155. Korkin D., Davis F.P., Sali A. (2005). *Protein Sci* 14: 2,350–2,360.
156. Nicastro G., Menon R.P., Masino L., Knowles P.P., McDonald N.Q., Pastore A. (2005). *Proc Natl Acad Sci USA* 102: 10,493–10,498.
157. DeLano W.L. (2002). *The PyMOL Molecular Graphics System*. DeLano Scientific, Palo Alto, USA.

CHAPTER 8

High-resolution Protein–Protein Docking

Nir London and Ora Schueler-Furman

*Department of Microbiology and Molecular Genetics,
Institute for Medical Research Israel-Canada,
Hadassah Medical School, The Hebrew University,
P.O. Box 12272, 91120 Jerusalem, Israel
E-mail: oraf@ekmd.huji.ac.il*

Advances in the field of protein–protein docking allow us today to create models of protein complexes that approach in quality experimentally derived structures. Crucial to this is the explicit modelling of the conformational changes that the individual protein monomers undergo upon binding. A stringent energy function that allows the distinction between well-packed interfaces and wrong alternatives can then be used to select the correct model. The models can provide important insights into the biological function of the protein–protein interaction, and provide guidance to experimentalists. Thanks to their high resolution, these models are amenable to computational interface design methods that were until recently restricted to experimentally derived structures, thereby opening up the way towards docking-based redesign of interactions, or targeted inhibition. Nevertheless, proteins that undergo larger conformational changes upon binding are still difficult to model, but steady advances in this field allow for optimistic outlooks, even if challenges remain ahead.

8.1 Introduction

8.1.1 *From Molecules to Networks: Making Sense of Large-scale Data, Starting from the Atomic Details of Protein–Protein Interactions*

Proteins are major players in the cell, and often they perform their function by interacting with other proteins. Protein–protein interactions come in various flavours, they may form stable complexes in macromolecular multi-protein machineries (e.g. proteasomes, chaperones, synthesis of DNA, RNA and proteins, etc.), or associate transiently (e.g. for chemotaxis, signal transduction, enzyme catalysis, etc.). Large-scale functional genomics studies have provided us with comprehensive data on networks of protein–protein interactions (e.g. Refs 1–3), and bioinformatic approaches have been used to analyse and clean this extensive and complex data in order to build comprehensive, curated representations of the interaction network within a living cell (e.g. Refs 4 and 5).

The structure of a protein–protein complex provides important information about the function of the two proteins, and sets a starting point for targeted manipulation of the interaction. However, for most protein interactions, no structure has been solved. While thanks to significant efforts of large-scale Structural Genomics initiatives, most of the estimated 1,000 existing folds have been solved,[6] the coverage of protein–protein interactions by solved protein complex structures is significantly lower (~2000/10'000 [2]). Thus, successful docking protocols can be used to 'bridge the gap' and create structural models of an ever increasing number of protein–protein interactions, starting from the known monomer structures. In addition to improved functional characterization of specific interactions, a general approach for high-resolution docking would be very helpful in the elucidation of the structural basis for binding affinity and specificity, and allow the study of basic features that characterise protein–protein association and recognition.

8.1.2 Docking – The Creation of Protein Complex Structures Starting from the Monomers

Docking is defined as the modelling of a protein complex, starting from the free monomer structures. In principle, this procedure should be relatively easy, as only six degrees of freedom need to be evaluated: three translations and three rotations of a protein ligand relative to a protein receptor. And indeed, very efficient methods have been developed over the years to tackle this challenge in a quick and exhaustive way (see Chapters 6 of this book and Ref. 7), for example, by the use of grid-based Fast Fourier Transform Techniques (FFT)[8] or geometric hashing approaches.[9] However, even though the conformation space that needs to be searched to find the correct orientation between the two proteins is restricted, accurate selection of this orientation turns out to be complicated by the fact that proteins undergo conformational changes upon binding. Thus, although these methods do a pretty good job in finding conformations in which the two proteins create a large interface of two matching complementary surfaces, they will not necessarily be able to select the correct orientation,[10] since they are not able to account for the subtle changes that are due to conformational flexibility of each of the partners.[11–13]

8.1.3 Explicit Modelling of the Atomic Details of the Protein–Protein Interface Allows Distinguishing the Correct from Alternative Conformations

The underlying assumption of high-resolution docking approaches is that if the conformational changes are modelled explicitly, we should in principle be able to model the correct complex conformation to atomic detail, even if conformational changes have occurred during association, provided we use appropriate sampling techniques. It is assumed that the correct conformation provides the lowest free binding energy, and can thus be selected by energy criteria alone without the need of additional biological information. Obviously, the draw-back is the larger number of degrees of freedom that need to be sampled efficiently in order to locate the correct conformation, and therefore the increased running time.

Within the broader context of structure prediction *per se*, two fundamental questions of modelling arise: (a) can the approximate structure of a protein, or a protein complex, be determined in feasible time, by using a coarse-grained first step, and (b) can subsequent, extensive full-atom, high-resolution modelling create atom-level models that allow the selection of the correct structure among alternative decoys? In comparison to *ab initio* modelling for example, docking involves a relatively small number of degrees of freedom that need to be searched in the first step. Therefore, it allows judicious assessment of the second, high-resolution step. Indeed, high-resolution docking was one of the first examples that demonstrated that accurate modelling is indeed a feasible task, and that techniques that allow accurate packing can truly help in the distinction of the correct solution among alternative models, based on a general energy function alone and without prior information on the specific system.[14]

8.1.4 *The Scope of this Chapter*

In this chapter we will focus on the field of high-resolution protein–protein docking (we also refer the reader to other reviews, e.g. Refs 15, 16). By 'high-resolution', we refer to the criteria defined in the CAPRI community wide blind docking experiment[17, 18] (see below), but also in a broader sense to studies where a model provides atomic details of the protein interface that allows biological insights, similar to an experimentally solved protein structure. Modelling of flexibility in the monomers is crucial for a general high-resolution docking protocol, and different approaches in this difficult field will be briefly highlighted. Then we will introduce in more detail one of the successful high-resolution protocols, RosettaDock,[13, 19, 20] which is the docking protocol in the Rosetta programme suite, developed by several groups from the RosettaCommons (see https://www.rosettacommons.org/), mainly the Gray lab at John Hopkins University, the Baker lab at the University of Washington and our own lab at the Hebrew University in Jerusalem. Following this we will detail additional high-resolution approaches, and describe different schemes to combine low-resolution fast global searches with high-resolution local searches.

We will briefly summarise achievements of the described protocols within the CAPRI experiment. Then we will detail several 'real-world' examples of docking studies that have helped experimentalists define a protein–protein interaction at atomic detail, and therefore have made possible the detailed manipulation of that interaction. At the end of the chapter we formulate some of the current challenges in high-resolution docking that await new approaches by motivated and enthusiastic scientists.

8.2 High-resolution Docking, as Defined by CAPRI

CAPRI (Critical Assessment of PRotein Interactions) is a community wide blind experiment where protein docking groups are asked to create models for a protein complex just prior to its publication, given the free monomer conformations of the partners (or, if not available, using either homolog protein structures or the bound conformation).[17] The predictions can then be assessed based on the solved structure. Since its establishment in 2001, around 40 different targets have been assessed by CAPRI, and regular meetings have promoted discussions about performance and utility of different docking approaches for the creation of accurate docking models. CAPRI is described elsewhere in this book (see Chapter 6), here we would merely like to describe the criteria that have been defined by the CAPRI evaluation team to assess the submitted models:[18] they include three measures: (a) f_{nat} (Fraction of native contacts): the fraction of residue pair contacts across the interface in the native complex structure that is reproduced in the model; (b) L_RMS (Ligand C_α atom RMSD): Root mean square deviation of C_α atoms in the ligand protein (i.e. the smaller protein), upon superimposition of the receptor molecule (i.e. the larger protein) of the model onto the native structure; and (c) I_RMS (interface residue C_α atom RMSD): RMSD of the interface residues. Based on these measures, models are classified as 'high accuracy', 'medium accuracy', 'acceptable' and 'incorrect' models. The criterion for 'high accuracy' models is defined as a structure that reproduces more than 50% of residue-residue native contacts (i.e. $f_{nat} >$ 50%), and either lies within 1 Å ligand C_α RMSD, or within 1 Å interface residue C_α RMSD of the native complex structure (i.e. L_RMS

< 1 Å or I_RMS < 1 Å). 'Medium accuracy' models are defined as those with '0.5 > f_{nat} > 0.3 and (L_RMS < 5 Å or I_RMS < 3 Å)' or 'f_{nat} < 0.5 and L_RMS < 1 Å and I_RMS < 1 Å'. Indeed, such high-accuracy, and in particular also medium-accuracy models, are now routinely created.

The initial two rounds (Targets 1–7) established CAPRI as a coordinator of docking assessment and progress,[17] and defined the challenges that would accompany the docking field until now, namely the adequate modelling of conformational changes that occur upon binding, and the development of scoring functions that are able to select the correct models. The next set of rounds (rounds 3–5; Targets 8–19; summarised in Ref. 21) then experienced a series of targets without significant structural changes beyond the side chain, or where the bound backbone conformation of one of the partners was supplied. For these targets, many good models were created by different docking protocols, by either rigid body approaches, or approaches that limit the modelling of flexibility to the side chains. Many of the targets in these rounds were taken from enzyme–inhibitor or antibody–antigen interactions, which are the types of complexes that were used to benchmark many of the docking protocols.[22, 23] The high success rate for these targets is thus due to the fact that no major conformational changes beyond side chains occurred, and that the types of targets matched the types of complexes in the docking benchmarks. The latest summary of CAPRI (rounds 6–12;[24] Targets 20–28) shows that the overall model quality indeed depends very much upon the target, indicating that no generally successful high-resolution modelling protocol exists yet. Overall, only one high-accuracy model was submitted for these rounds, despite a significant increase in participating groups, and the availability of protocols that had gradually been improved during the previous challenges. The reason for this shift is most probably a consequence of the change in target types: Crystallographers have moved on to new types of protein complexes: while in the first rounds, most of the targets were antibody–antigen and enzyme–inhibitor complexes, recent targets include more complexes that are involved in a variety of cellular processes such as: signal transduction, membrane transport, and various aspects of transcription and translation. These targets show overall larger conformational changes upon binding and often contain small interfaces. Transient complexes show weak binding, and in turn, docking algorithms cannot always distinguish the correct orientation from others based on calculated binding energy only. The current challenge therefore lies in the adaptation of docking protocols to transient interactions with small interfaces, e.g. by including additional information about the interaction that will allow distinguishing the orientation from other models with interfaces of similar size and packing quality. It should be noted that these interactions can also be difficult to characterise by experiment, as was shown for Target 27 in which the experimentalists could not determine which of the interfaces in the crystal was biologically relevant.[24] Overall, it can be seen that certain protocols, such as ZDOCK[25–27] and HADDOCK,[28, 29] show a constant good performance (creating medium and acceptable resolution models for many targets). RosettaDock on the other hand allowed the generation of the most accurate high-resolution models – albeit only in specific cases. Since RosettaDock is very sensitive to clashes, it will only be successful if those clashes are efficiently removed (Baker and Gray groups[30–33]).

8.3 Accounting for Conformational Changes of Monomers is Crucial to High-resolution Modelling

8.3.1 *Modelling Side Chain Flexibility*

Conformational changes upon protein–protein binding range from side chains rotameric changes, through small backbone re-arrangements, up to large domain movements. We will refer here to movements at the side chain level and address methods to account for backbone flexibility in the docking process in the next section.

Several studies have established that the side chains at the interface of the unbound protein monomer tend to be pre-oriented to accommodate binding, perhaps due to the fact that there is not enough time during the formation of an encounter complex to allow for side chain re-arrangements.[34] Molecular Dynamics (MD) simulations showed that interface residues tend to be less mobile than other surface residues[35] and indeed incorporation of the unbound side chain conformation as an additional rotamer option into the RosettaDock protocol was able to increase the prediction quality.[13]

In some cases however, not only do the interface side chains undergo conformational changes, but the modelling of these changes is critical for the correct prediction of the bound complex. One example for such a change is the cohesin–dockerin complex presented as Targets 11 and 12 in the CAPRI blind docking experiment.[36] In this complex, a Leucine residue in the free monomer sticks out into the solvent. In the bound complex this residue undergoes a rotameric change to allow the binding of dockerin. Failure to model this change would not allow the prediction of the correct complex due to clashes (see Fig. 8.1a). Another example is CAPRI Target 21 – a complex of yeast origin recognition protein Orc1 and silent information regulator Sir1.[37] In this case three Sir1 interfacial side chains undergo a conformational change, and a small helical region of Orc1 undergoes a backbone conformational change of about 1.6 Å. Indeed, only three predictor groups were able to produce medium accuracy models for this target (DOT, HADDOCK and RosettaDock[24]).

8.3.2 *Taking it to the Next Step: Modelling Backbone Flexibility*

The next major challenge for high-resolution docking schemes is the correct modelling of complexes involving significant backbone conformational changes at the interface. This subject has been thoroughly covered in other reviews.[38,39] Incorporation of such flexibility into docking protocols can be divided into several categories, namely Ensemble Docking, Refinement and Minimization, and the Modelling of Hinge Motions.

8.3.2.1 *Ensemble Docking*

These methods create an ensemble of monomers with different backbones representing the flexibility of that monomer, prior to the docking process, thus reducing the search space to a feasible size.

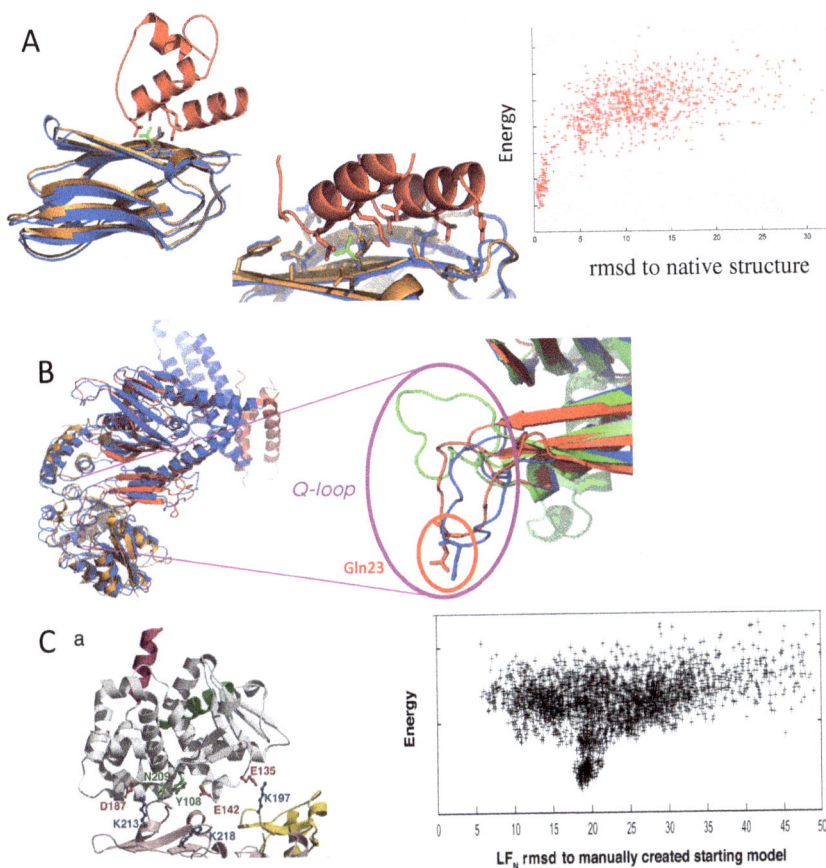

Fig. 8.1. Atom-resolution docking models of protein complexes. (A) Cohesin-dockerin interaction in CAPRI Targets 11 and 12. *Left panel:* the structural details of the bound structure (red + orange), and a high-resolution model produced with RosettaDock by the Baker group (blue). Note the central Leucine 87 side chain that undergoes a conformational change upon binding (free conformation in green). The blowup shows the accurate modelling of the interface side chain conformations. *Right panel:* energy funnel in the RosettaDock energy landscape (depicted as energy (in Rosetta Energy Units) vs distance to the crystal structure (in rmsd); adapted from Schueler-Furman *et al.*[30]). (B) HemK – RF1 interaction in CAPRI Target 20: the model predicts a large conformational change in the RF1 'Q-loop' upon binding to HemK. The loop is shown in a blowup on the right. Colouring as in (A); adapted from Wang *et al.*[32] (C) Model of the Anthrax toxin produced by RosettaDock: *Left panel:* the Protective Antigen (in yellow and pink) bound to the Lethal Factor (in grey). The experimentally verified residue pairs across the interface are shown in ball-and-stick presentation. *Right panel:* the energy landscape shows a clear funnel around the proposed orientation (according to Lacy *et al.*[87]).

This ensemble can be created by different approaches, such as using NMR ensembles,[40] different X-ray solved structures of the monomer, or alternatively, ensembles created by MD,[35] Normal Mode Analysis (NMA), or loop modelling. The different conformations can be docked sequentially (cross-docking as in Refs 35, 40, 41), which is computationally expensive, or collectively using algorithms such as the mean field approach. Examples for docking protocols which utilise ensemble docking are: RosettaDock,[42] HADDOCK[43] and ATTRACT (at the side chain[44] and loop levels[45]).

8.3.2.2 *Refinement and Minimization*

Different methods allow for flexible backbone modelling during a refinement and minimization step. In this step the backbone degrees of freedom might be minimized either according to a certain force field,[20,46] or along the lowest frequency normal modes (for protein–protein, protein–DNA and protein–ligand docking[47,48]). Many programmes, RosettaDock included, utilise Monte Carlo with Minimization (MCM[49]) to sample random rigid-body and backbone perturbations in pre-defined flexible regions. Local backbone refinement indeed allowed RosettaDock to successfully model CAPRI Target 18, while the original RosettaDock side chain flexibility alone could not select the correct conformation[20] (see below). Other docking methods ignore flexible loops during the docking and remodel them in a consecutive step. Such an approach allowed to create (using RosettaDock) the only acceptable model for CAPRI Target 20[32] (see Fig. 8.1b and below).

8.3.2.3 *Modelling Hinge Motion*

Many proteins consist of two or more globular domains connected by flexible hinges. Upon binding, these hinges may undergo conformational changes, which would change the rigid body orientation of the globular domains. Hinges can be detected computationally by methods such as HingeProt,[50] NMA, MD or by expert users. Modelling this kind of flexibility is usually performed by rigid body docking of the globular domains, and subsequent remodelling of the hinge such as in FlexDock.[51]

8.4 The High-resolution RosettaDock Protocol – Explicit Modelling of Full Side Chain Flexibility (and Beyond) Allows Accurate Modelling of Protein Complexes

The Rosetta modelling suite was originally developed for the prediction of protein structures, starting from the sequence only. Over the years it has been extended and adapted towards a wide range of different applications, taking advantage of an energy function that has been parameterized on a broad range of applications, and on search strategies that have been optimised on a broad range of different types of conformations.[52,53] RosettaDock is the docking protocol of Rosetta.[13, 20, 30,42,54] It applies a Monte Carlo-based sampling strategy that includes minimization prior to acceptance evaluation (MCM[49]) to find the optimal rigid body orientation of two proteins.

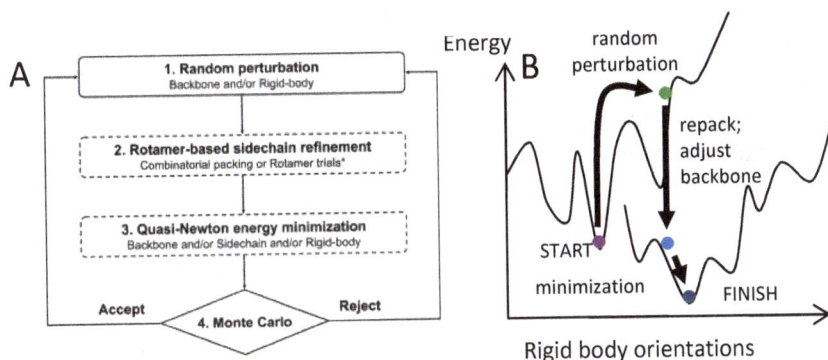

Fig. 8.2. The RosettaDock MCM protocol. (A) Flowchart of local optimization protocol. Monomer flexibility is introduced by perturbation (e.g. aggressive sampling of a specific loop; *Step 1*), or/and during minimization (e.g. removing of local clashes; *Step 3*). Figure according to Wang *et al.*[87] (B) The energy landscape in high-resolution MCM refinement, described as Energy (y-axis) vs Rigid body conformation (x-axis). After perturbation of the rigid body orientation, optimization removes local clashes, thereby changing the energy landscape to allow the detection of the minimum energy rigid body orientation. (Adapted from Gray *et al.*[19])

Figure 8.2 shows a schematic view of the minimization strategy: by optimising the side chain orientations, local clashes can be removed, and the correct orientation can be selected based on well-packed interfaces that result in low energy scores. RosettaDock uses the typical two-stage strategy, which includes first a fast, low-resolution search that optimises features not dependent on the detailed modelling of the protein side chains, such as the preference of residues to be at interfaces compared to other surface areas, and the preference for different residue pairs to contact each other across the interface.[54] For antibody–antigen docking, antibody residues that have been observed to contact antigens in known structures are given a bonus, while other residues are penalised, in order to bias towards the antigen binding region of the antibody.[19] In this first part of the protocol, side chains are approximated by a 'centroid' atom with a radius that simulates the size of the complete residue side chain, and side chain atoms are not explicitly modelled, thus this optimization step is fast. In the second step the resulting model is further optimised, side chains are added, and a full search that includes the optimization of both side chain conformations and rigid body orientation using a stringent energy function for evaluation is applied to locate the minimum energy conformation in the sampled region. Fifty steps of MCM are performed towards this goal: each step consists of a small perturbation that pushes the conformation out of its current minimum, the reorientation of clashing interface side chains and the optimization of the rigid body orientation through gradient-based minimization (see Fig. 8.2). The new conformation is only then evaluated and accepted based on the Metropolis criterion. Side chain sampling is done efficiently by using a rotamer library that contains a restricted number of possible conformations,[55] including the side chain conformation adapted in the free monomer structure.[13] For every eighth step, a full side chain repacking at the interface is performed, instead of the quick rotamer trial procedure that merely repositions individual side chains that clash with their environment.[56] In addition, an option called 'rtmin' allows to perform off-rotamer sampling by minimising each rotamer in the library prior to its selection.[13] Each simulation finds a local minimum conformation and therefore many independent simulations (10^4–10^5) are performed to guarantee adequate sampling. Clustering tools are then applied to locate heavily sampled energy minima basins[57] and cluster centres are selected.

The confidence in the selected model can be evaluated by further sampling of the local energy landscape around the model. With local perturbations, one samples densely around a candidate orientation (~500 samples). In many cases, the local landscape shows a deep funnel that centers around the native conformation (see right panels of Figs 8.1a and 8.1c for examples), and indeed, a model within 1–2 Å RMSD of the crystal structure can usually be found at the funnel tip. In general, the local energy landscape can be used to distinguish the correct orientation from alternative possibilities: the FunHunt classifier was developed to distinguish the correct orientation among a set of candidate orientations by evaluating the funnel around each of them.[58–60] Here again, we can see a two-step procedure that guides the protein towards its correct binding orientation: according to FunHunt, the main characteristics that distinguish near-native orientations from other low-energy candidate structures are (a) a first approximate encounter complex that is guided by optimization of a low-resolution feature, namely residue interface propensity (D_{env}), and (b) a subsequent specific binding that is supported by strong decrease in energy in the full-atom optimization step. This energy gap might prevent the two proteins from disassociating again, thereby creating a stable protein complex.

Since flexibility is explicitly modelled, an accurate, atom-based energy function can be used. It consists of a combination of physical terms, such as a van der Waals term, which contains a stringent repulsive term, and terms derived from statistical analyses of known protein structures. For example, hydrogen bond energy is derived from a statistical analysis of frequencies of hydrogen bond geometries,[61] as well as from quantum mechanical calculations of distance and angle preferences.[62] This is in contrast to molecular mechanics force fields such as in CHARMM, which model the hydrogen bond as a linear electrostatic interaction.[63] Thus, RosettaDock makes use of sophisticated sampling protocols and energy functions that have been calibrated previously on a range of different modelling tasks addressed by different Rosetta protocols.[52]

For a significant number of the CAPRI targets, we (i.e. the Baker and Gray groups) were able to create high accuracy models by using RosettaDock (see Fig. 8.1a; Targets 7, 8, 12, 14), or medium accuracy models (Targets 6, 11, 13, 21 and 26); (see above for the definition of accuracy measure). We submitted the best models to CAPRI for Target 19 (the only high-accuracy model for this target) and Target 20 (the only model of acceptable accuracy for this target; see below and Fig. 8.1b). As already noted above, these high and medium accuracy models were mostly submitted for targets where flexibility at the side chain level accounts for most of the conformational changes that occur upon binding (or where bound backbone conformations were provided by the CAPRI team).

8.4.1 *Adding Backbone Conformational Flexibility to the RosettaDock Protocol*

A reformulation of the representation of a protein as a 'fold-tree' in Rosetta allows the seamless integration of internal torsional degrees of freedom and rigid body degrees of freedom,[64] and can be used to extend the original RosettaDock protocol to include internal flexible regions.[20] Backbone flexibility can be introduced at different steps of the docking MCM protocol (see Fig. 8.2a). For example, inclusion of backbone flexibility during the minimization step in the MCM protocol improved modelling of some of the complexes in a benchmark (measured as the number of cases where a distinct energy funnel tip near the native orientation was observed, see below[20]). In the same study, it was demonstrated that increased conformational sampling of specified loops in the perturbation step of the MCM protocol improved the docking for other cases.

As an example, while the original RosettaDock protocol was too sensitive to local atomic clashes to produce a correct model for CAPRI Target 18 (Xylanase bound to the TAXI inhibitor), the new protocol succeeds: remodelling of the a loop during docking opens the structure to allow the accommodation of the inhibitor.[20] Note that Target 18 can successfully be modelled by rigid-body docking approaches without accounting for conformational changes, such as ZDOCK,[26] by using a softer energy function. However, larger conformational changes require their explicit modelling even with other, 'softer' methods. An example for the correct prediction of a large-scale conformational change in a loop is demonstrated in CAPRI Target 20:[65] the interaction of the HemK methyltransferase with Release Factor 1 (RF1) (see Fig. 8.1b). HemK catalyses the methylation of Q257 in RF1. The starting structure provided for HemK contained a methyl analogue which provided the information where the Q257 side chain of RF1 would be located in the complex. Using RosettaDock, we first modelled the interaction of RF1 with HemK without the flexible loops, and then added the loop back. Finally, conformations that allowed the remodelling of the loop to place Q257 onto the methyl analog in HemK were selected. The resulting model was the only CAPRI submission of acceptable quality for this challenging target. Figure 8.1c shows the significant conformational change that the loop undergoes upon binding to HemK, and the predicted RosettaDock conformation.

Even though these examples and others demonstrate the applicability of flexible docking, the significant increase in the degrees of freedom put off a general protocol, and can be only applied efficiently if backbone flexibility is restricted to specified regions.

8.4.2 *Ensemble Docking with RosettaDock*

Two different models have been suggested for the process of protein–protein recognition: Conformer Selection (CS) states that the bound conformation is sampled in the free monomer structure, albeit with very low frequency.[66] Upon binding, the equilibrium is shifted to the bound conformation. *Induced fit* postulates that the bound conformation is induced only upon binding to the partner.[67] Chaudhury *et al.*[42] try to implement *in-silico*, within RosettaDock, these different models. To capture CS during binding, an extra step was added to the RosettaDock low-resolution stage following the rigid-body perturbation, and preceding the Metropolis evaluation step. In this step, a pre-existing ensemble of conformers is superposed along the interface of the current conformer. The ensemble can be produced in a preliminary stage, using either solved X-ray/NMR structures, or using a computational modelling protocol, such as refinement with Rosetta,[53] to create an ensemble of structures. The centroid-mode binding energy is calculated for each conformer and used to generate a partition function. The current conformer is then replaced by a conformer selected from the ensemble based on its Boltzmann-weighted probability within the partition function. Once a conformer is selected, the Metropolis criterion is applied on the combined rigid-body/CS move. The induced-fit model is implemented in the high-resolution stage as an extended minimization scheme that includes the minimization of backbone torsion angles along local energy gradients at the interface (see previous paragraph[20]). This added flexibility at the interface improves the quality of predictions for cases with small to moderate conformation changes in the monomer backbone.

8.5 Additional High-resolution Docking Approaches

The ICM-DISCO software from Abagyan's group[11, 68] is another high-resolution method composed of a two-stage process. In the first step a rigid all-atom ligand molecule is docked onto a set of soft pre-calculated receptor potentials on a 0.5 Å grid. The sampling of rotational and translational degrees of freedom of the ligand starting from each grid position is performed by a pseudo-Brownian Monte Carlo minimization. The second step is needed to deal with induced conformational changes and includes a global optimization of the interface side chains of up to 400 of the best solutions from the previous step. This protocol profits from an efficient and extensive grid-search, and from smart softening of the energy function. The ICM-DISCO approach provided several high-accuracy submissions to CAPRI (for Targets 6, 12 and 14).

HADDOCK is an approach that makes use of experimental interaction data such as NMR titration experiments or mutagenesis.[43] This information is introduced as 'Ambiguous Interaction Restraints' to drive the docking process, and is formulated by an additional term in the energy function. The docking protocol consists of three stages: firstly a rigid body energy minimization is performed, followed by a semi-rigid simulated annealing in torsion angle space. During this stage the interface amino acids (both side chains and backbone) are allowed to move to optimise packing. Finally, a refinement in Cartesian space is carried out with explicit solvent. The final structures are clustered using the pairwise backbone RMSD at the interface and analysed according to their average interaction energies and their average buried surface area. HADDOCK provided several high accuracy submissions to CAPRI (for Targets 13 and 14), and notably the only medium-accuracy model for Target 10, a very large homo-trimer.

8.5.1 *High-accuracy Modelling with Rigid Body Docking*

Even though it is evident that in general it is crucial, or at least helpful, to include monomer flexibility to achieve high-accuracy models of protein–protein interactions, it should be noted that elegant applications of classical rigid-body docking protocols have in many cases created high-accuracy models for CAPRI as well. Notably, most of them are FFT-based approaches, including CLUSPRO,[69,70] ZDOCK[27,71] and PIPER[72,73] – all originating from Boston University, as well as DOT,[74,75] Hex (using spherical polar Fourier Transform correlation[76,77]), and Molfit.[78]

8.5.2 *A New Generation of Docking Protocols: Combining Successful Approaches of Low-resolution and High-resolution Searches*

While highly accurate models can in principle be created by RosettaDock, they are not always necessarily sampled, and depend on a very extensive, non-systematic global search.[54] On the other hand, a range of different docking approaches have been shown to quickly locate the region of correct orientations, such as by comprehensive searches using FFT[8,73,77,79] or geometric hashing.[9] In these cases, the correct solution is not always among the highest ranked models, and even if it is, further refinement is needed to create an atom-accuracy model. Several attempts have therefore been developed to create hybrid protocols that combine the advantages of a quick low-resolution protocol with the accuracy of a consequent high-resolution protocol. For example, Kozakov *et al.* have shown that by performing RosettaDock local perturbation searches on cluster centres obtained by PIPER, an FFT-based approach, correct orientations could be selected by evaluating the energy funnel landscape of the runs together with their divergence from the low-resolution cluster center.[80] This indicates that a combination of a broad low-resolution energy basin with a good full-energy score can distinguish near-native orientations. Along similar lines, ZDOCK, another FFT-based approach has been combined with Rosetta.[81] Finally, Andrusier *et al.* have devised FireDock,[82] which combines the fast PatchDock algorithm based on geometric hashing,[83] with a full-atom protocol that is largely based on the RosettaDock full-atom docking optimization protocol (see above). In principle, a general trend can be observed, where new protocols are developed that consist of combinations of different original protocols, which can lead at the end to greatly improved protocols, both regarding efficiency, accuracy and generality. Such progress has been possible thanks to the organisation of community-wide assessments and meetings within the CAPRI framework.[24, 84]

8.6 The Contribution of High-resolution Docking to the Understanding of Interactions of Biological Interest

The CAPRI experiment has had a major impact on the docking field, by making the assessment of performance easy: predictions are compared to the solved structure. In 'real-world' applications, 'success' is often defined in other ways, since the structure of the complex is not always solved. In this section we will detail a list of exciting applications of high-resolution docking, where atom-resolution model of the interaction has advanced our understanding of specific protein-protein interactions significantly.

8.6.1 *Entry Mechanism of Anthrax Toxin*

Anthrax toxin consists of three proteins: the Lethal Factor (LF), the Edema Factor (EF) and the Protective Antigen (PA). The latter heptamerizes and forms a pore in the cell membrane through which the two enzymatic factors enter the cell. The Collier laboratory identified several mutations that abolish the binding of LF to PA,[85, 86] and based on these mutations, two alternative principle orientations between LF and PA could be suggested; however, each of them explained only part of the mutational data. In an attempt to settle this discrepancy, we created a model of this interaction using RosettaDock.[87] The right panel of Fig. 8.1c shows the characteristic energy funnel picture obtained from the simulation run, which indicated that there was a third conformation distinct from either of the proposed orientations. Based on the details of the interface in this atom resolution model (see Fig. 8.1c, left panel), we designed a series of mutations that could validate our model: a disulfide link could be successfully introduced between residues Y108 in LF and N209 in PA, indicating their spatial vicinity in the complex. In addition, charge reversal in the residue pairs D187 in LF and K213 in PA, and E142 in LF and K218 in PA was applied, and in both cases, change of the charge in one partner only abolished the interaction, while a complementary charge in the other partner reconstituted the interaction. How then could the conflicting initial mutagenesis results be explained? Going back to their experimental protocols, the experimentalists recalled that they were working with an artificial PA dimer, and one set of mutations simply abolished the interaction between the PA monomers. This is a good example of how computational prediction can often advance experimental work and simplify the organisation of experimental evidence into a consistent model.

8.6.2 Antitumor Monoclonal Antibody 806 (mAb806) and the Epidermal Growth Factor Receptor (EGFR)

The antitumor antibody mAb806 binds the extracellular region of the EGFR in cancerous conditions (e.g. when the EGFR is over-expressed[88]), and the mAb806 epitope was mapped to a disulfide bonded loop (amino acids 287–302) in EGFR.[89] No molecular structure has been determined for the mAb806-EGFR complex. However, ample biochemical information was available to provide guidance. Sivasubramanian et al.[90] used RosettaDock to predict the structure of this complex. A homology model of mAb806 was docked against all the known structures of the EGFR epitope, and the resulting models were screened both for sterical hindrance with the entire EGFR complex (in different possible conformation) and for correlation between the results of computational mutagenesis performed on the models and experimental mutagenesis data. The model which best correlated with the experimental data was selected and was used to suggest new mutations that might affect the binding. These mutations were experimentally validated and indeed overall, the model proposed correctly predicts 33 of 40 (80%) mutations, including 14 of 16 (87%) new mutations suggested after the creation of the model. Based on the high resolution model, the authors could postulate that the steric hindrance created by the antibody near the EGFR dimer interface interferes with receptor dimerization, from which stems the antitumor effect of mAb806.

8.6.3 *High-resolution Docking in the Service of Biochemistry*

The atomic resolution of models produced by high-resolution docking protocols can shed light on the fine details of complex biochemical processes such as electron transfer and ion trafficking.

The relationship of two close redox (reduction–oxidation) complexes was examined by Medina *et al.*[91] using high-resolution docking simulations with pyDockRST[92] and ICM-dock.[68] Ferredoxin-NADP+ Reductase (FNR) interacts both with ferredoxin (Fd) and flavodoxin (Fld), in order to transfer two electrons, which will be used to reduce NADP+ to NADPH. While X-ray structures of the FNR:Fd complex have been reported, an experimental structure for the FNR:Fld interaction is highly elusive. In this study the authors conducted docking simulations of both complexes and proposed a model for the FNR:Fld interaction. This model, although displaying a different binding mode than the FNR:Fd complex, is in accordance with previous biochemical data and places the redox centres of both monomers within electron transfer range.

Another study, by Arnesano *et al.* shed light on the copper ion transfer mechanism by means of high-resolution docking. In yeast, Atx1 (a copper chaperone) delivers Cu(I) to the copper domain of Ccc2, an ATPase located in the trans-Golgi, which then transfers it to a cuproenzyme.[93] Using the HADDOCK programme[43] in combination with available NMR titration data,[94] an ensemble of high-resolution models was proposed for this interaction. The models provide a structural basis to discuss the mechanism of copper exchange. The copper binding motifs 'CxxC' of both monomers are put adjacent to each other in the models, with the copper ion almost at bond length (average of 3.7 Å between the copper ion and the sulfur atom of Cys13 at Ccc2). The interaction was found to be mainly of electrostatic nature, with a network of hydrogen bonds stabilising the complex. The implications of these models are relevant for a number of proteins homologous to Atx1 and Ccc2 and conserved from bacteria to humans.

8.6.4 *Applications of High-resolution Docking: Structure-based Prediction of Binding Specificity*

Protein–protein interaction networks underlie a new era of systems biology. The results of large-scale experiments determining interacting proteins allow for the construction of these networks, and from the networks arise new insights on the macro cellular level.[95] These experiments however are not very accurate and the coverage of the interactome is believed to be only partial.[96] Computational schemes to determine whether two proteins bind each other or not are therefore in dire need.[97]

The current approaches in this direction are based on the assumption that homologous proteins share the same binding mode, thus the sequences of the target proteins are threaded onto a template of the interaction and the energy of this model is evaluated.[98–101] However, when the sequence similarity is low, additional loops are usually inserted/deleted at the interface and overall structural differences occur that need to be modelled. High-resolution docking of these complexes while accounting for backbone conformational changes and loop modelling might be the answer to the holes in the systems biology interaction network. This field is however only at its beginning.

8.7 Conclusions and Outlook

8.7.1 *Impact of Docking on the Modelling Field*

Recent advances in large-scale genomic approaches have created an exciting era of abundant information regarding the sequences and structures of proteins, which in many cases can already broadly cover whole genomes. This poses interesting challenges in the field of systems biology, which attempts to create a comprehensive, macroscopic picture from this data. Importantly, this also pushes forward fields that are related to the microscopic basis of this complexity, namely the prediction of the structural basis of individual proteins and their interactions. It does not come as a coincidence that high-resolution structural modelling has advanced in correlation with the new era, as major breakthroughs have been possible by harnessing this information in the form of so-called 'knowledge-based' potentials that are based on statistical analyses of existing protein sequences and structures.

In turn, progress in docking procedures is expected to have impact beyond the docking field, and help improve local refinement strategies, e.g. *ab initio* modelling, where predictions within 2 Å RMSD from the native structure are also starting to appear more frequently.[102]

This review has focused on high-resolution docking, where atoms are explicitly modelled. In contrast to 'soft', rigid-body docking, these methodologies tend to be significantly slower due to the increased degrees of freedom that are sampled, but since conformational changes upon binding are explicitly modelled, the final structures can consistently be more accurate, and thus easier to select using stringent full-atom energy functions. As a consequence, several high-resolution models that reach experimental accuracy have been reported.

Despite impressive advances in docking, challenges definitely remain ahead. Most importantly, only anecdotal cases of high-resolution modelling of large conformational changes have been reported (e.g. the prediction of a large flip of a loop in the interface in CAPRI Target 20 using RosettaDock; see above and Fig. 8.1b[32]). Thus, in order to obtain an ultimate, generally applicable, high-resolution protocol, there is a need for new approaches to be included into a comprehensive modelling suite. As mentioned in the text, successful high-resolution modelling has many benefits: (a) understanding the basic principles that underlie protein–protein interactions: protocols that can model a protein complex at high level of detail, support the suggestion that the energy function describes accurately the actual binding of the partners; (b) structural support of large-scale protein–protein interaction networks: accumulation of solved monomer structures – mainly within the frame of the Structural Genomics project – is an excellent starting point for the modelling of a substantial fraction of the interactome, thereby supplementing the structural system level part; (c) fine-tuned manipulation of individual protein–protein interactions, by interface redesign and the design of inhibitors is dependent on an accurate structure of the protein complex, which can be provided by high-resolution docking protocols. In the future we will see an increasing number of functional assays of protein–protein interactions that are founded on structure-based manipulation of the complex.

Acknowledgements

This work was partially supported by The Israel Science Foundation founded by the Israel Academy of Science and Humanities, grant number 306/6, and the GIF (German–Israeli Foundation for Scientific Research and Development – 1709/2007).

References

1. Giot L., Bader J.S., Brouwer C., Chaudhuri A., Kuang B., Li Y., Hao Y.L., Ooi C.E., Godwin B., Vitols E., *et al.* (2003). *Science* 302: 1,727–1,736.
2. Russell R.B., Alber F., Aloy P., Davis F.P., Korkin D., Pichaud M., Topf M., Sali A. (2004). *Curr Opin Struct Biol* 14: 313–324.
3. Uetz P., Giot L., Cagney G., Mansfield T.A., Judson R.S., Knight J.R., Lockshon D., Narayan V., Srinivasan M., Pochart P., *et al.* (2000). *Nature* 403: 623–627.
4. Cusick M.E., Klitgord N., Vidal M., Hill D.E. (2005). *Hum Mol Genet* 14: R171–181.
5. Vidal M. (2005). *FEBS Lett* 579: 1,834–1,838.
6. Vitkup D., Melamud E., Moult J., Sander C. (2001). *Nat Struct Biol* 8: 559–566.

7. Halperin I., Ma B., Wolfson H., Nussinov R. (2002). *Proteins* 47: 409–443.
8. Katchalski-Katzir E., Shariv I., Eisenstein M., Friesem A.A., Aflalo C., Vakser I.A. (1992). *Proc Natl Acad Sci USA* 89: 2,195–2,199.
9. Fischer D., Lin S.L., Wolfson H.L., Nussinov R. (1995). *J Mol Biol* 248: 459–477.
10. Camacho C.J., Vajda S. (2002). *Curr Opin Struct Biol* 12: 36–40.
11. Fernandez-Recio J., Totrov M., Abagyan R. (2002). *Protein Sci* 11: 280–291.
12. Lorber D.M., Udo M.K., Shoichet B.K. (2002). *Protein Sci* 11: 1,393–1,408.
13. Wang C., Schueler-Furman O., Baker D. (2005). *Protein Sci* 14: 1,328–1,339.
14. Schueler-Furman O., Wang C., Bradley P., Misura K., Baker D. (2005). *Science* 310: 634–638.
15. Gray J.J. (2006). *Curr Opin Struct Biol* 16: 183–193.
16. Ritchie D.W. (2008). *Curr Protein Pept Sci* 9: 1–15.
17. Janin J., Henrick K., Moult J., Eyck L.T., Sternberg M.J., Vajda S., Vakser I., Wodak S.J. (2003). *Proteins* 52: 2–9.
18. Mendez R., Leplae R., Lensink M.F., Wodak S.J. (2005). *Proteins* 60: 150–169.
19. Gray J.J., Moughon S.E., Kortemme T., Schueler-Furman O., Misura K.M., Morozov A.V., Baker D. (2003). *Proteins* 52: 118–122.
20. Wang C., Bradley P., Baker D. (2007). *J Mol Biol* 373: 503–519.
21. Janin J. (2005). *Protein Sci* 14: 278–283.
22. Mintseris J., Wiehe K., Pierce B., Anderson R., Chen R., Janin J., Weng Z. (2005). *Proteins* 60: 214–216.
23. Chen R., Mintseris J., Janin J., Weng Z. (2003). *Proteins* 52: 88–91.
24. Lensink M.F., Mendez R., Wodak S.J. (2007). *Proteins* 69: 704–718.
25. Chen R., Tong W., Mintseris J., Li L., Weng Z. (2003). *Proteins* 52: 68–73.
26. Wiehe K., Pierce B., Mintseris J., Tong W.W., Anderson R., Chen R., Weng Z. (2005). *Proteins* 60: 207–213.
27. Wiehe K., Pierce B., Tong W.W., Hwang H., Mintseris J., Weng Z. (2007). *Proteins* 69: 719–725.
28. de Vries S.J., van Dijk A.D., Krzeminski M., van Dijk M., Thureau A., Hsu V., Wassenaar T., Bonvin A.M. (2007). *Proteins* 69: 726–733.
29. van Dijk A.D., de Vries S.J., Dominguez C., Chen H., Zhou H.X., Bonvin A.M. (2005). *Proteins* 60: 232–238.
30. Schueler-Furman O., Wang C., Baker D. (2005). *Proteins* 60: 187–194.
31. Daily M.D., Masica D., Sivasubramanian A., Somarouthu S., Gray J.J. (2005). *Proteins* 60: 181–186.
32. Wang C., Schueler-Furman O., Andre I., London N., Fleishman S.J., Bradley P., Qian B., Baker D. (2007). *Proteins* 69: 758–763.
33. Chaudhury S., Sircar A., Sivasubramanian A., Berrondo M., Gray J.J. (20070. *Proteins* 69: 793–800.
34. Kimura S.R., Brower R.C., Vajda S., Camacho C.J. (2001). *Biophys J* 80: 635–642.
35. Smith G.R., Sternberg M.J., Bates P.A. (2005). *J Mol Biol* 347: 1,077–1,101.

36. Carvalho A.L., Dias F.M., Prates J.A., Nagy T., Gilbert H.J., Davies G.J., Ferreira L.M., Romao M.J., Fontes C.M. (2003). *Proc Natl Acad Sci USA* 100: 13,809–13,814.

37. Hou Z., Bernstein D.A., Fox C.A., Keck J.L. (2005). *Proc Natl Acad Sci USA* 102: 8,489–8,494.

38. Andrusier N., Mashiach E., Nussinov R., Wolfson H.J. (2008). *Proteins* 73: 271–289.

39. Bonvin A.M. (2006). *Curr Opin Struct Biol* 16: 194–200.

40. Grunberg R., Leckner J., Nilges M. (2004). *Structure* 12: 2,125–2,136.

41. Krol M., Chaleil R.A., Tournier A.L., Bates P.A. (2007). *Proteins* 69: 750–757.

42. Chaudhury S., Gray J.J. (2008). *J Mol Biol* 381: 1,068–1,087.

43. Dominguez C., Boelens R., Bonvin A.M. (2003). *J Am Chem Soc* 125: 1,731–1,737.

44. Zacharias M. (2003). *Protein Sci* 12: 1,271–1,282.

45. Bastard K., Prevost C., Zacharias M. (2006). *Proteins* 62: 956–969.

46. Fitzjohn P.W., Bates P.A. (2003). *Proteins* 52: 28–32.

47. May A., Zacharias M. (20080. *Proteins* 70: 794–809.

48. Lindahl E., Delarue M. (2005). *Nucleic Acids Res* 33: 4,496–4,506.

49. Li Z., Scheraga H.A. (1987). *Proc Natl Acad Sci USA* 84: 6,611–6,615.

50. Emekli U., Schneidman-Duhovny D., Wolfson H.J., Nussinov R., Haliloglu T. (2008). *Proteins* 70: 1,219–1,227.

51. Schneidman-Duhovny D., Inbar Y., Nussinov R., Wolfson H.J. (2005). *Proteins* 60: 224–231.

52. Rohl C.A., Strauss C.E., Misura K.M., Baker D. (2004). *Methods Enzymol* 383: 66–93.

53. Das R., Baker D. (2008). *Annu Rev Biochem* 77: 363–382.

54. Gray J.J., Moughon S., Wang C., Schueler-Furman O., Kuhlman B., Rohl C.A., Baker D. (2003). *J Mol Biol* 331: 281–299.

55. Dunbrack R.L. Jr, Cohen F.E. (1997). *Protein Sci* 6: 1,661–1,681.

56. Kuhlman B., Baker D. (2000). *Proc Natl Acad Sci USA* 97: 10,383–10,388.

57. Shortle D., Simons K.T., Baker D. (1998). *Proc Natl Acad Sci USA* 95: 11,158–11,162.

58. London N., Schueler-Furman O. (2007). *Proteins* 69: 809–815.

59. London N., Schueler-Furman O. (2008). *Biochem Soc Trans* 36: 1,418–1,421.

60. London N., Schueler-Furman O. (2008). *Structure* 16: 269–279.

61. Kortemme T., Morozov A.V., Baker D. (2003). *J Mol Biol* 326: 1,239–1,259.

62. Morozov A.V., Kortemme T., Tsemekhman K., Baker D. (2004). *Proc Natl Acad Sci USA* 101: 6,946–6,951.

63. Brooks B., Bruccoleri R., Olafson B., States D., Swaminathan S., Karplus M. (1983). *J Comp Chem* 4: 187–217.

64. Bradley P., Baker D. (2006). *Proteins* 65: 922–929.

65. Janin J. (2007). *Proteins* 69: 699–703.

66. Monod J., Wyman J., Changeux J.P. (1965). *J Mol Biol* 12: 88–118.
67. Koshland D. (1958). *Proc Natl Acad Sci USA* 44: 98–104.
68. Fernandez-Recio J., Totrov M., Abagyan R. (2003). *Proteins* 52: 113–117.
69. Comeau S.R., Gatchell D.W., Vajda S., Camacho C.J. (2004). *Nucleic Acids Res* 32: W96–99.
70. Comeau S.R., Kozakov D., Brenke R., Shen Y., Beglov D., Vajda S. (2007). *Proteins* 69: 781–785.
71. Li L., Chen R., Weng Z. (2003). *Proteins* 53: 693–707.
72. Shen Y., Brenke R., Kozakov D., Comeau S.R., Beglov D., Vajda S. (2007). *Proteins* 69: 734–742.
73. Kozakov D., Brenke R., Comeau S.R., Vajda S. (2006). *Proteins* 65: 392–406.
74. Law D.S., Ten Eyck L.F., Katzenelson O., Tsigelny I., Roberts V.A., Pique M.E., Mitchell J.C. (2003). *Proteins* 52: 33–40.
75. Mandell J.G., Roberts V.A., Pique M.E., Kotlovyi V., Mitchell J.C., Nelson E., Tsigelny I., Ten Eyck L.F. (2001). *Protein Eng* 14: 105–113.
76. Ritchie D.W. (2003). *Proteins* 52: 98–106.
77. Ritchie D.W., Kemp G.J. (2000). *Proteins* 39: 178–194.
78. Ben-Zeev E., Kowalsman N., Ben-Shimon A., Segal D., Atarot T., Noivirt O., Shay T., Eisenstein M. (2005). *Proteins* 60: 195–201.
79. Chen R., Weng Z. (2002). *Proteins* 47: 281–294.
80. Kozakov D., Schueler-Furman O., Vajda S. (2008). *Proteins* 72: 993–1,004.
81. Pierce B., Weng Z. (2008). *Proteins* 72: 270–279.
82. Andrusier N., Nussinov R., Wolfson H.J. (2007). *Proteins* 69: 139–159.
83. Duhovny D., Nussinov R., Wolfson H. (2002). *Efficient Unbound Docking of Rigid Molecules*: 185–200.
84. Janin J., Wodak S. (2007). *Structure* 15: 755–759.
85. Lacy D.B., Mourez M., Fouassier A., Collier R.J. (2002). *J Biol Chem* 277: 3,006–3,010.
86. Cunningham K., Lacy D.B., Mogridge J., Collier R.J. (2002). *Proc Natl Acad Sci USA* 99: 7,049–7,053.
87. Lacy D.B., Lin H.C., Melnyk R.A., Schueler-Furman O., Reither L., Cunningham K., Baker D., Collier R.J. (2005). *Proc Natl Acad Sci USA* 102: 16,409–16,414.
88. Mishima K., Johns T.G., Luwor R.B., Scott A.M., Stockert E., Jungbluth A.A., Ji X.D., Suvarna P., Voland J.R., Old L.J., *et al.* (2001). *Cancer Res* 61: 5,349–5,354.
89. Johns T.G., Adams T.E., Cochran J.R., Hall N.E., Hoyne P.A., Olsen M.J., Kim Y.S., Rothacker J., Nice E.C., Walker F., *et al.* (2004). *J Biol Chem* 279: 30,375–30,384.
90. Sivasubramanian A., Sircar A., Chaudhury S., Gray J.J. (2008). *Proteins* 74: 497–514.
91. Medina M., Abagyan R., Gomez-Moreno C., Fernandez-Recio J. (2008). *Proteins* 72: 848–862.
92. Cheng T.M., Blundell T.L., Fernandez-Recio J. (2007). *Proteins* 68: 503–515.

93. Yuan D.S., Dancis A., Klausner R.D. (1997). *J Biol Chem* 272: 25,787–25,793.

94. Arnesano F., Banci L., Bertini I., Bonvin A.M. (2004). *Structure* 12: 669–676.

95. Russell R.B., Aloy P. (2008). *Nat Chem Biol* 4: 666–673.

96. Yu H., Braun P., Yildirim M.A., Lemmens I., Venkatesan K., Sahalie J., Hirozane-Kishikawa T., Gebreab F., Li N., Simonis N., *et al.* (2008). *Science* 322: 104–110.

97. Aloy P., Russell R.B. (2006). *Nat Rev Mol Cell Biol* 7: 188–197.

98. Kiel C., Serrano L. (2007). *Bioinformatics* 23: 2,226–2,230.

99. Lu L., Lu H., Skolnick J. (2002). *Proteins* 49: 350–364.

100. Lu L., Arakaki A.K., Lu H., Skolnick J. (2003). *Genome Res* 13: 1,146–1,154.

101. Aloy P., Russell R.B. (2003). *Bioinformatics* 19: 161–162.

102. Bradley P., Misura K.M., Baker D. (2005). *Science* 309: 1,868–1,871.

CHAPTER 9

Scoring and Refinement of predicted Protein–Protein Complexes

Martin Zacharias

Physics Department, Technical University Munich,
85747 Munich, Germany
E-mail: martin.zacharias@ph.tum.de

Docking approaches typically result in a large number of putative protein–protein complexes. The selection of the most realistic predicted complex by an appropriate scoring function is an important step of the modelling process. In addition, an initially selected complex structure often requires structural refinement to arrive at an accurately structural model of the complex. A variety of scoring and refinement methods have been developed in recent years. The chapter is intended to introduce and discuss the most relevant methods for evaluating and refining predicted protein–protein complex structures.

9.1 Introduction

Knowledge of the structure of protein–protein complexes is of major importance in understanding the biological function of protein–protein interactions. Experimental structure determination of protein complexes, for example by X-ray crystallography, requires purification of large amounts of proteins and the ability to crystallise the protein–protein complex which may not be feasible for all known interacting proteins. In particular complexes of weakly or transiently interacting protein partners may not be stable enough to allow experimental structure determination at high (atomic) resolution. However, frequently such transient protein–

protein interactions are of functional importance for the cell during signal transduction or regulation of metabolism.[1,2] Many cellular functions are mediated by multi-protein complexes in a dynamic equilibrium with the isolated components or sub-complexes.[3,4] Each protein of a cell may potentially interact with many other proteins so that the number of potential complexes greatly exceeds the number of single proteins. In order to fully understand the function of these protein–protein interactions, knowledge of the three-dimensional (3D) complex structure is desirable. The prediction of protein–protein complex structures is therefore of increasing importance to obtain at least realistic structural models of protein–protein complexes.[5–10] If the structure of the isolated protein partners is available, it is possible to use a variety of computational docking methods to generate putative complex structures (reviewed in Refs 11–19). A short overview of the most common methods will be given in Section 9.2 and docking methods are reviewed in detail in Chapter 6.

Alternatively, in case of sufficient sequence similarity of proteins forming a putative complex with respect to a protein complex with experimentally known structure, it is also possible to use comparative modelling approaches to build a structural model of the complex[20–22] (Section 9.3). It is assumed that not only do the partner proteins adopt similar structures but that also that the interface is similar to the interface in a template complex of known structure.

Often, protein–protein complex structures obtained from protein–protein docking but also in case of comparative modelling are of limited accuracy and require further structural refinement to achieve the generation of a realistic structural model. The structural refinement of such complex is the subject of Section 9.4. In fact, the majority of current protein–protein docking approaches distinguish between a first exhaustive systematic docking search followed by a second refinement step of pre-selected putative complexes.[11–19] Docking protocols may even consist of several consecutive refinement and rescoring steps.[18,19]

In order to limit the computational demand during the first systematic search, typically a simple and rapid scoring is employed (e.g. based on surface complementarity of the docking interface) to initially rank and pre-select putative complexes. An important next step after the refinement of a subset of complexes (often still several hundreds or even thousands of complexes) is to select the most realistic model out of the set of solutions. The various options for scoring or re-scoring (after refinement) of a set of modelled complexes are discussed in Section 9.5 of this chapter.

Both scoring and refinement are frequently considered as separate issues of modelling protein–protein complex structures. However, the majority of scoring functions are parameterised using experimental structures as input. Hence, highly accurate scoring of a protein–protein docking geometry is only possible if the complex structure has also been predicted with high precision. The more errors a scoring function may tolerate, the less specific it will become. Consequently, there is a direct relation between the robustness or softness of a scoring function and the number of false–positive solutions. It is likely that the design of more rigorous and more specific scoring functions requires at the same time an improvement of the prediction accuracy of binding modes in terms of deviation from the experimental binding geometry.

9.2 Generation of Protein–Protein Complexes by Docking Methods

The purpose of computational protein–protein docking methods is to predict the structure of a protein–protein complex based on the structure of the isolated protein partners. Several efficient approaches have been developed in recent years to efficiently generate a large number of putative binding geometries (see Chapter 6).

Among the most common methods are Fast Fourier transform (FFT) correlation techniques.[23-28] The FFT correlation technique allows the rapid calculation of the optimal overlap of functions that describe the boundaries of proteins. The two protein partners are represented by cubic grids, the grid points are assigned discrete values for inside, outside and on the surface of the protein. For various relative orientations of the two binding partners a geometric complementarity score can be calculated by computing the correlation of the two grids. Instead of summing up all the pair products of the grid entries one can make use of the Fourier correlation theorem. The correlation integral can easily be computed in Fourier space. The discrete Fourier transform for the receptor grid needs to be calculated only once. Due to the special shifting properties of Fourier transforms the different translations and orientations of the ligand grid with respect to the receptor grid can be computed by a simple multiplication in Fourier space.[23]

The extension of the FFT correlation method employing polar variables in stead of Cartesian coordinates has also been described and successfully applied in the field of protein–protein docking.[29,30]

Geometric hashing is a computer visualization technique to match complementary substructures of one or several data sets.[31] Typically, each data set is broken down to triangles, which are stored in a hash table. By means of a hash key similar triangles can be found very quickly. During docking, these triangles comprise points on a molecular surface, having a certain geometrical (concave, convex) or physico-chemical (polar, hydrophobic) character. By matching triangles belonging to different molecules and being of complementary character, putative complex geometries can be generated.[32–34]

A third class of methods uses either Monte Carlo, Brownian Dynamics or multi-start energy minimization to generate large sets of putative protein–protein docking geometries.[35–42] Since these methods are computationally more expensive compared to FFT based correlation methods or geometric hashing, a search is often limited to pre-defined regions of the binding partners.[36] Alternatively, it is possible instead of atomistic models to employ coarse-grained (reduced) protein models to perform systematic docking searches. With such reduced protein models it is possible to optimise docking geometries starting from tens of thousands of protein start configurations.[38–40] In order to limit the number of putative complex structures generated during an initial docking search cluster analysis is typically employed to reduce the number to a subset of representative complex geometries.

9.3 Protein–Protein Complexes Based on Homology to Known Complexes

Many proteins of unknown structure share sequence similarity to proteins of known experimental structure. It has also been recognised that sequence similarity of natural protein sequences often implies structural similarity. Based on the sequence similarity it is often possible to create a structural model of a protein with unknown structure by using the corresponding known structure as template.[43,44] In case of sufficient sequence similarity (> 40% identical residues) such structural models can be quite accurate frequently with a root mean square deviation (RMSD) from the real native structure of < 2 Å.[44]

In addition to comparative modelling of single proteins, it is also possible to build comparative models of whole protein complexes.[20–22] Comparative modelling of complexes benefits from the growing number of experimentally determined protein–protein complexes but also from the observation that a significant number of protein interactions involves recurrent interfaces or the interactions are mediated by similar domains or regions of proteins.[21,22,45] Russel and co-workers[20,21] as well as Sali and co-workers[45,46] succeeded in comparative modelling of a significant fraction of protein–protein complexes of the yeast protein interactome.[47,48] The interactome is defined as a set of experimentally verified protein–protein interactions in a cell or organism. Sequence-to-structure threading methods have also been successfully applied for comparative modelling of protein–protein interactions.[49] In the approach of Sali *et al.*[46] alternative interaction geometries were considered and evaluated using an empirical scoring function (see below).

Nevertheless, a prerequisite of comparative modelling in general is that the complex structure of the unknown target is indeed structurally similar to the template complex. In case of a very limited sequence similarity (homology) of the complexes this may not be the case and the generated complex will deviate from the native complex structure. In addition, the accuracy of the predicted protein interaction geometries depends on additional factors such as the accuracy of the modelled protein partners. Similar to protein–protein complexes obtained by docking searches, comparative models of complexes may require further structural refinement and further evaluation (scoring) if putative alternative interaction geometries have been predicted.

9.4 Structural Refinement of Modelled Protein–Protein Complexes

9.4.1 *Force Field Description of Proteins and Protein Complexes*

Computational methods to refine putative protein–protein complex structures obtained from docking searches or comparative modelling are based on a molecular mechanics force field description of the participating molecular structures. A force field employs as variables the positions of atoms (not electron coordinates as in quantum mechanical methods) and consists of several additive terms that control the bonded and non-bonded geometry of molecules,[50,51]

$$V_{tot} = \sum_{i=1}^{Nbonds} \tfrac{1}{2} k_{b\_i} (b_i - b_{oi})^2 + \sum_{i=1}^{Nangles} \tfrac{1}{2} k_{\theta\_i} (\theta_i - \theta_{oi})^2 +$$
$$\sum_{i=1}^{Ndihedral} \sum_{n=1}^{N\tau} k_n (1 + \cos(n\tau_i + \delta_{ni})) +$$
$$\sum_{i \neq j}^{Npairs} \left(\frac{A_{ij}}{r_{ij}^{12}} - \frac{B_{ij}}{r_{ij}^6} + \frac{q_i q_j}{r_{ij}} \right) \tag{9.1}$$

The bonded terms (first three summations in the above equation) contain a sum over all bonds, all bond angles and dihedral angles of the protein structures. Usually quadratic penalty terms with appropriate force constants (k_b and k_θ, respectively) are used to control the bond lengths (b) and bond angles (θ) of the molecules. A linear combination of periodic functions is employed to control torsion angles (τ). Additional non-bonded terms describe van der Waals and Coulomb interactions (as a double sum over all non-bonded pairs of atoms). The form of the energy function allows a rapid evaluation of the potential energy of a molecule and calculation of gradients necessary for energy optimization and molecular dynamics simulations based on a numerical solution of the classical equations of motion.

9.4.2 *Optimization Based on Energy Minimization*

A standard procedure to refine the interface region of predicted protein–protein complexes is to perform energy minimization. The start structure corresponds to a pre-selected complex obtained as a putative solution from homology modelling or docking searches. A variety of energy minimization procedures exists mostly based on the gradient of the molecular force field employed to describe atomic interactions. It usually leads to the closest local energy minimum with respect to the start structure. Energy minimization can be performed in Cartesian coordinates of the protein atoms. Several molecular mechanics programme packages are available for Cartesian energy minimization of biomolecules (e.g. Amber,[52] Charmm,[53] Gromacs[54]). The programme RDOCK is an example of refinement of docked complexes based on multi-start energy minimization.[55] Due to the stiff nature of some of the energy terms in the force field, Cartesian minimization typically results only in small adjustments of atoms and is used to remove any sterical atom overlap at the predicted interface. Typically, the movement of atoms during Cartesian minimization is limited to a few tenths of Angstrom from the initial positions.

Alternatively, it is also possible to perform the minimization in internal degrees of freedom (e.g. dihedral angles) of the protein molecules combined with variables that describe the relative position and orientation of the protein partners. Docking programmes such as RosettaDock[40] (described in detail in Chapter 8) or the ICM (Internal Coordinate Molecular mechanics) programme[36,56] employ energy minimization in dihedral angles to adjust side chains and the protein backbone at the protein–protein interface. This limits the minimization to fewer variables. An additional advantage is also that the optimization can easily be limited to a subset of relevant conformational variables (for example only side chain dihedrals at the interface). The possible conformational adjustments are often larger compared to optimization in Cartesian coordinates.[36] Energy minimization can be combined with other conformational search techniques such as Monte Carlo methods to introduce random perturbations prior to minimization (see Section 9.4.4). It is also possible to perform multiple energy minimizations starting for example from different side chain placements at a protein–protein interface.[57]

9.4.3 *Accounting for Global Conformational Changes*

Protein partner structures can undergo not only local adjustments (e.g. conformational adaptation of side chains and backbone conformation at the interface) during association but also more global conformational changes that involve for example large loop movements or domain opening-closing motions. Several methods have been designed to detect hinge regions in proteins that are potentially involved in mediating global changes.[58–60]

Normal mode (NM) analysis is a molecular mechanics based method to identify global soft collective degrees of freedom of protein structures. A normal-modes analysis involves the calculation of the curvature of the molecular mechanics energy function at an energy minimum. The curvature of the energy function corresponds to the second derivate with respect to the atom coordinates of the system. Based on the calculation of the eigenvectors and eigenvalues of the second derivative matrix it is possible to identify soft (and hard) collective degrees of freedom of the protein molecules. Soft degrees of freedom represent possible collective motions of many atoms in the system that result in only small corresponding energy changes. Such motions may represent for example opening-closing motions of enzyme active sites or relative displacements of domains in proteins (illustrated in Fig. 9.1).

Normal mode analysis has been used to characterise the global mobility of many classes of biomolecules[60–63] and has also been employed to identify hinge regions in proteins.[60] A drawback of the method is the computationally expensive calculation of normal modes of large receptor molecules due to the large number of coordinates and the requirement for extensive energy minimization. Furthermore, the energy minimization is often performed in the absence of solvent and can lead to significant deviations from the experimental protein geometry.

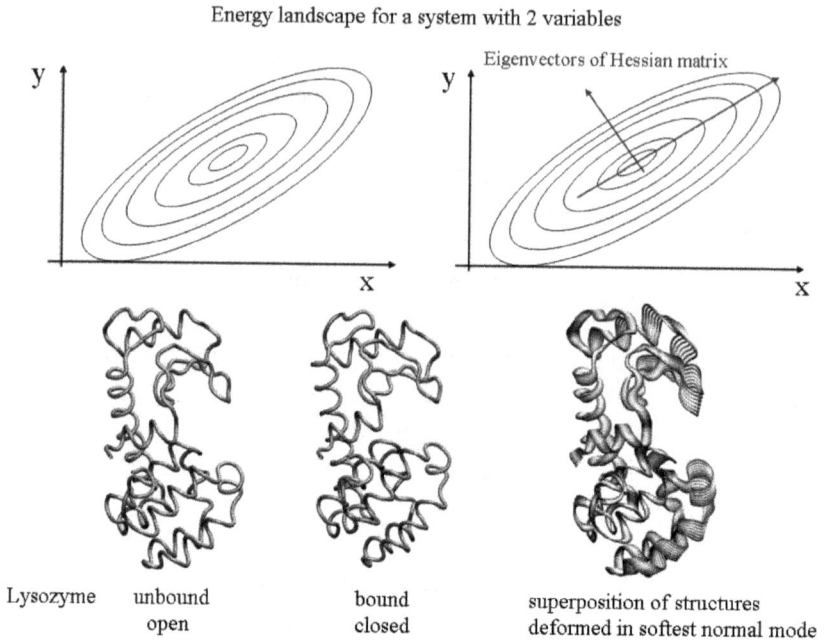

Fig. 9.1. Illustration of normal mode directions that point towards the eigenvector directions of the second derivative matrix (Hessian matrix) of the energy function at an energy minimum (upper two panels). The lower three panels show the backbone tube representation of lysozyme in an unbound and substrate-bound conformation. The substrate binding region is localised at the centre between the upper and lower domains of the protein. A superposition of lysozyme structures deformed along the direction of the softest normal mode illustrates the collective character of the motion (lower panel on the right) with different grey levels of deformed structures. It also illustrates that the softest collective mode describes an opening-closing motion of two domains that encompass the substrate binding site.

It has been shown that soft normal modes obtained from Elastic Network Models (ENMs) of proteins frequently overlap with experimentally observed global conformational changes.[64–69] In an ENM the experimental structure serves as a reference (energy minimum) structure and the mobility of atoms is determined by harmonic springs that control the distances between atoms.[64–69]

The force constant for restraining the distance close to the distance in the reference structure decreases in ENMs with the distance (either continuously or by introducing a cut-off distance). Consequently, the relative mobility of atoms depends on the local density and a larger number of short range contacts restricts locally the relative mobility of atoms.[66,67] Normal modes derived from an ENM of a protein can be calculated very rapidly because of the simple form of the energy function and typically only the protein backbone is considered. Also, compared to normal mode analysis using a molecular mechanics force field the calculation of ENMs does not require a costly energy minimization and the reference structure is identical to the experimental protein structure.

Instead of employing NM analysis to just identify putative directions of largest mobility in protein molecules it is also possible to employ the identified soft degrees of freedom directly as variables during docking searches or during refinement of docked complexes. The inclusion of such collective variables during docking has been used by Zacharias and Sklenar for docking of ligands to DNA.[70] In combination with normal modes from ENMs it has later also been used for refinement of protein–protein and protein–DNA complexes as well as during systematic docking searches.[71–76] Alternatively, the soft directions can also be used to perform displacements followed by full Cartesian energy minimization to generate sterically possible large-scale deformations.[77] Use of soft modes as additional variables allows the rapid relaxation of protein structures on a global scale involving much larger collective displacements of atoms during minimization then conventional energy minimization using Cartesian or other internal coordinates. The eigenvalue of each normal mode is a measure of the energy required to deform the structure along the corresponding mode. It can be used to estimate the receptor deformation energy avoiding costly calculation of the internal receptor energy at every docking minimization step. An additional advantage compared to docking methods that employ ensembles of rigid receptor structures is that the receptor conformation can change continuously during docking in the pre-calculated soft degrees of freedom and has therefore a greater capacity for induced fit adaptation. The application of relaxation or refinement in normal mode variables has been applied successfully in a number of studies.[71–74] It has also been used during systematic docking searches accounting approximately for global conformational changes already during the first protein–protein docking stage of generating many putative complexes.[74,75] Instead of directly relaxing protein structures along normal mode directions to dock or refine docking solutions, it is also possible to generate ensembles of protein structures deformed along selected collective degrees of freedom.[77,78]

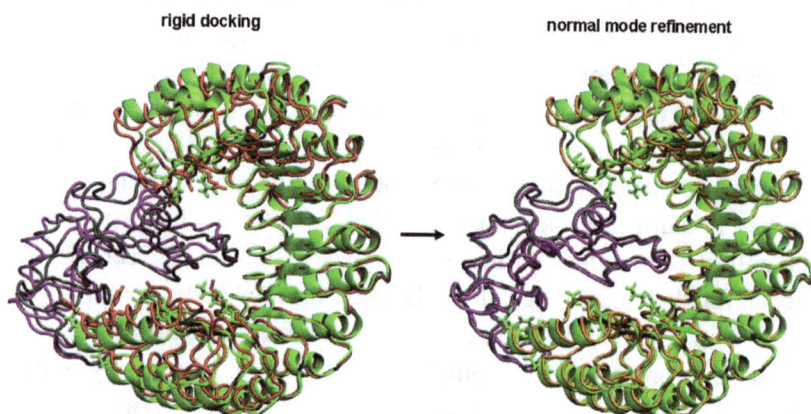

Fig. 9.2. Comparison of rigid docking of RNAse A to the unbound RNAse inhibitor (horse shoe type structure) and docking including flexible adjustment along five soft normal modes of the inhibitor protein. The RNAse inhibitor in the bound form is shown for comparison as green cartoon whereas the unbound structure is illustrated as red tube (left panel). The yellow tube model (in closer agreement with the bound structure) indicates the inhibitor structure after normal mode relaxation and docking, and bound receptor structures in green. Mobile side chains at the interface are also included. The smaller RNAse protein at the experimental binding position is shown in grey (tube representation) and the placement after docking as purple tubes. It illustrates that both the docked protein position and the inhibitor conformation can markedly improve upon normal mode refinement when starting from unbound partner structures. Docking was performed with the ATTRACT software[75] that allows simultaneous optimization of docking placement and conformational adjustment in normal mode variables.

For example, Mustard and Ritchie[78] generated protein structures deformed along directions compatible with a set of distance constraints reflecting large scale sterically allowed deformations. Subsequently, the structures were used in rigid body docking searches to identify putative complex structures.[78] Other authors used principal component of motions derived from molecular dynamics simulations to evaluate global conformational changes in proteins and to generate ensembles of (rigid) structures which can then be used for docking searches.[79,80] Ensemble docking approaches are increasingly popular because it does not require modification of the rigid docking approaches.[80] A drawback is, however, that it requires many more rigid docking searches (one for each generated protein structure) which may also increase the number of false–positive solutions.

9.4.4 *Molecular Dynamics Simulation of Protein–Protein Complexes*

Molecular dynamics (MD) simulations are frequently employed to achieve larger conformational adjustments compared to energy minimization during docking refinement. MD simulations are based on numerically solving Newton's equation of motion in small time steps (1– 2 fs = 1–2 10^{-15} s) based on forces derived from a force field description of the molecular system.[81] In contrast to energy minimization and due to the kinetic energy of every atom of the molecule, it is in many cases possible to overcome energy barriers and to move the structure significantly further away from the initial placement. Depending on the simulation length and temperature displacements up to several Angstroms from the initial atom positions are possible. Due to the dynamics of the complex it is hoped that the structure can overcome energy barriers and may move closer to a realistic binding geometry.

Whether the displacements achieved during MD simulations indeed move the structures towards a more realistic complex structure depends on the accuracy of the force field and on how appropriately the aqueous environment has been accounted for. In principle, refinement simulations on a given protein–protein complex require the inclusion of surrounding aqueous solvent and ions. However, the inclusion of a sufficiently large number of explicit waters increases the computational demand. In addition, the equilibration of explicit solvent molecules around a solute molecule requires significant simulation times (currently limited to tens of nanoseconds).[81] Nevertheless, during the final stages of some protein–protein docking protocols explicit water molecules being added to the simulation system.[82]

MD simulations including surrounding waters and ions have been used to investigate the flexibility of protein structures prior to docking.[79,83,84] It is possible for example to identify the alternative or most likely side chain conformations.[83,84] Global conformational flexibility can be analysed by principal component analysis of the motions extracted from MD simulations.[79]

The computational demand of MD simulations including many thousands of explicit solvent molecules can limit the maximum simulation time and number of complexes used for refinement. Instead of including explicit solvent molecules during MD refinement of complexes it is also possible to implicitly account for solvent effects. This allows longer simulation times and refinement of more predicted complexes in a given amount of time. A variety of implicit solvation models has been developed (reviewed in Refs 85–87) and only a brief description of the most relevant concepts for protein–protein docking and scoring will be given.

Hydration shell models are typically based on the exposed solvent accessible surface of the proteins and can be used to rapidly estimate solvation effects approximately. The contribution to the solvation of the molecule depends on the type of atom, and in most cases linearly on the amount of exposed surface area of the atom.[88,89] Solvation parameters or surface tension parameters of each atom are often derived from the transfer free energy of chemical groups between a non-polar solvent (e.g. octanol) and water.[88,89] However, the environment of buried atoms at protein–protein interfaces is not necessarily well represented by an organic alcohol. There have been attempts to readjust surface tension parameters to better represent the change in solvation upon transfer of atoms from an aqueous solution to the environment at protein–protein interfaces[90,91] (see sub-section on force field scoring of docked complexes). Solvation models based on accessible surface area have been used not only during refinement but also for scoring docked complexes (see below) and for identifying putative binding regions on the surface of proteins.[90-94]

A short coming of accessible surface area based solvation models is the fact that polar and charged chemical groups, or atoms once completely buried in the interior of a biomolecule, do not influence the hydration properties of the molecule. It is, however, well known that the position of a charged group inside a protein relative to the surface can significantly affect the hydration properties and stability of a protein or protein–protein complex.[95,96]

Another macroscopic solvation concept describes the protein interior as a medium with a low dielectric permitivity embedded in a high dielectric continuum representing the aqueous solution.[96-98] The effect of the solvent is then calculated as a reaction field from a solution of Poisson's equation for the charges assigned to each atom of the molecule. The mean effect of a salt atmosphere can be included by solving the Poisson–Boltzmann equation. Within such model the total energy of a molecule in solution is given by its force field energy and a polar solvation contribution calculated from a solution of the Poisson- or Poisson–Boltzmann equation.[96-98] The polar solvation part is usually supplemented with a surface area dependent nonpolar solvation contribution (uniform for the whole surface of the molecule). Typically, the PB equation is numerically solved by a finite-difference method (FDPB) or by finite-element or boundary-element techniques to represent the molecular surface. For the FDPB method and with an appropriate choice of parameters for atomic radii a very reasonable correlation between calculated and experimental hydration free energies can be achieved. However, the method cannot easily be combined with MD refinement due to the difficulty to extract accurate solvation forces from grid solutions of the Poisson–Boltzmann equation.[99]

Instead of the FDPB approach it is possible to use more approximate methods like the Generalized Born (GB) method.[100-105] In the GB approach an effective solvation radius is assigned to each atom. This effective radius can be thought of as an average distance of the selected atom from the solvent or from the solvent accessible surface of the molecule. With the effective Born radii calculated for each atom the electrostatic solvation and its derivative (solvation forces) can be calculated very rapidly.[100–102] It is therefore possible to apply this method in molecular mechanics calculations such as energy minimization, conformational searches and molecular dynamics simulations. The GB method and related implicit solvent approaches are frequently used during refinement of docked protein–protein complexes.

9.4.5 *Refinement of Docked Complexes by Molecular Dynamics Simulation*

The HADDOCK docking programme (see Chapter 7)[37] is entirely based on MD simulations to perform docking searches and to also use it during the final docking stage (refinement). During the initial search it employs a simple distance-dependent dielectric constant during docking and includes a surface area dependent solvation term for the final evaluation of predicted complexes.[37,106] Alternatively, during the final refinement step the approach also allows inclusion of an explicit solvent shell around the docked protein structures. This approach has shown promising results on a set of ten test complexes.[82]

Krol *et al.*[107] employed short MD simulations in combination with a distance-dependent dielectric constant or an implicit solvent model to refine docked complexes obtained from rigid docking using the FFT correlation method. The protocol was fast enough to refine 1,000 putative docking start geometries, however in an application to several targets from the CAPRI docking challenge it was able to improve the docking geometry only for one out of ten target complexes.[107] Similar docking MD protocols have been used to refine complexes obtained from other systematic docking searches.[74]

The application of MD simulations for refinement of complexes is limited by the accuracy of the force field but also by the limited ability to overcome energy barrier during the relatively short simulation time scales. It is possible to increase the simulation temperature or to use advanced sampling methods such as simulated annealing or replica exchange methods to overcome these limitations.[108–110] However, higher simulation temperatures may also result in undesired conformational changes of the protein partners. An alternative technique to overcome energy barriers is to soften the energy function for residues at the protein interface. This can be achieved by introducing an energy cut-off for the repulsive Lennard–Jones term[36,37,40] but also by a continuous deformation of the potential (potential scaling approach). Riemann and Zacharias[111] described a method for interface refinement where the interaction potential of interface side chains is scaled down to zero at an initial stage of an MD refinement simulation and then gradually rescaled to the full potential. The approach showed promising results on several test cases.[111]

9.4.6 *Monte Carlo and Brownian Dynamics Refinement of Docked Complexes*

Monte Carlo (MC) search methods introduce random conformational changes in the predicted protein–protein complex and the changes are accepted or rejected according to a Metropolis criterion. The advantage of Monte Carlo (MC) methods is that no derivatives of the potentials are required and the search can be limited to the most relevant variables affecting the docking geometry. It is possible to use the dihedral angles of side chains located at the interface of protein–protein complexes as variables and keep the rest of the protein structures fixed. Monte Carlo methods are employed by several docking approaches at the refinement step. For example, the ICM-method employs a biased MC approach in side chain dihedrals and subsets of backbone dihedrals in combination with a truncated van der Waals potential for refinement[36] (see also Chapter 6). Other approaches employ combinations of random Monte Carlo coordinate changes followed by energy minimization, implemented for example in the programmes FireDock[112] and RosettaDock[40,41] (described in detail in Chapter 8). A combination of Monte Carlo side chain optimization and optimal potential scaling to improve the ability to overcome barriers has also been described.[113] The approach employs a combination of rigid body moves and side chain adjustments and the gradual transition from a smoothed potential to the full interaction potential at the protein–protein interface. The approach was successfully tested on a large set of putative complex structures and starting from unbound partner structures.[113] Another way of smoothing the interface region during docking is to apply multi-copy mean field approaches for side chains or protein loop structures.[114–116] Instead of a smoothed potential these methods employ a set of alternative side chain or loop conformations, and a mean field of all copies is used during docking employing energy minimization[116] or MC searches.[114,115] Beside to the placements of the partner structures, the relative weights of each copy are adjusted during the docking process leading to the selection of the best fitting side chain or loop copies at the interface.[114–116]

9.5 Scoring of Modelled Protein–Protein Complexes

Realistic scoring of putative protein–protein complexes is of critical importance at the systematic search and the refinement stages of docking. The design of an appropriate and optimal scoring function for the realistic evaluation of protein–protein complexes is still an issue that has not been solved satisfactorily for all docking applications.

As indicated above, the scoring function needs to realistically account for several enthalpic and entropic contributions that may influence the binding affinity of a receptor–ligand complex. Since computational speed is an important issue for ligand-receptor docking, it is necessary to identify a reasonable compromise between accuracy and speed to calculate a score for a protein–protein complex. Current scoring functions to evaluate predicted complexes range from simple schemes that just account for sterical complementarity used for example during the early systematic docking stage to complete force field functions that may be used during refinement. The scoring function can include terms that account for sterical and electrostatic interactions but also account for desolvation and hydrophobic contributions to ligand–receptor interaction.[117–134]

Instead of scoring functions based on a molecular mechanics force field, it is also possible to design knowledge-based potentials to evaluate complexes that are extracted from known protein–protein complexes.[135–143] The basic idea of such knowledge-based scoring is to relate the observed frequency of atom–atom (or group-group) contacts to the expected contact frequency at protein–protein or other receptor–ligand interfaces to extract favourable and unfavourable atom–atom interactions. Hereby, expected contact frequency means contacts that are obtained if atoms would be distributed randomly at the interface.

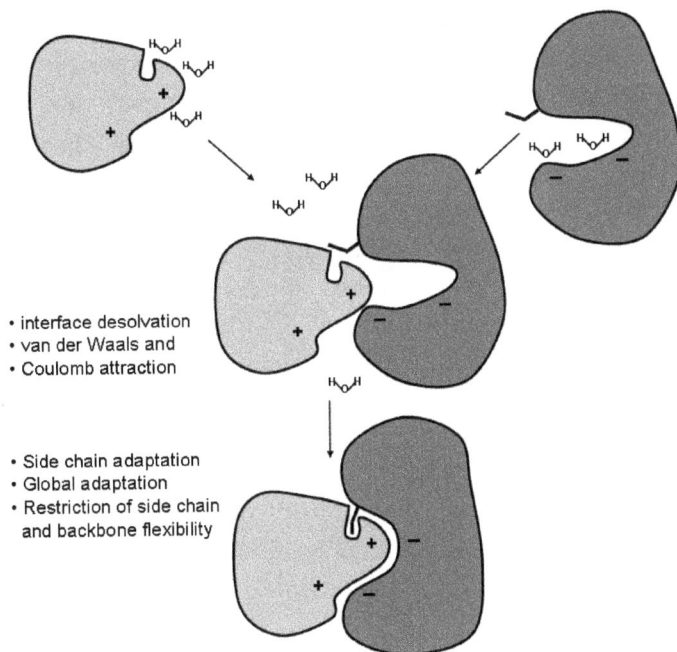

Fig. 9.3. Schematic illustration of the binding process of two proteins and of associated changes in the binding partners. Association involves changes in the interactions of the binding partners but also the release of water molecules (desolvation) from the interface region. It may also involve adaptive conformational changes and changes in the mobility of backbone and side chains (one side chain is illustrated as a stick model).

9.5.1 *Driving Forces for Molecular Association and the Scoring Problem*

In order to design an optimal scoring of docked complexes it is important to consider in detail the process of association and the changes associated with the binding event. The driving force for the association of two proteins to form a complex corresponds to the free energy change associated with the binding reaction (see Chapter 3 on the thermodynamics of protein–protein association). Protein–protein association can involve changes in electrostatic interactions, van der Waals interactions and hydrogen bonding interactions. The type of interactions can include contributions within and between the proteins (intra- and intermolecular contributions) but also between proteins and the surrounding solvent (typically aqueous solution). In addition, complex formation can also result in changes of interactions between solvent molecules.

The binding process may not only involve changes in internal energy or enthalpy as indicated above but can also involve changes in the entropy of the protein partners and the surrounding water.[144] For example, the conformational entropy of both interacting protein molecules can be affected by changes in flexibility of the binding partners. Complex formation can also lead to a change in the ordering of solvent molecules around the binding partners that influences the association process (hydrophobic effect).[144] Contributions that can affect the binding of two proteins are schematically illustrated in Fig. 9.3.

9.5.2 Scoring Based on Physical Force Fields

Scoring based on a physical force field refers to the use of a molecular mechanics force field described above (Section 9.4.1) for evaluating or scoring a predicted complex structure. The binding partners are treated at atomic resolution and the force field is typically supplemented by a solvation model. The solvation model corresponds typically to one of the implicit solvation models described in Section 9.4.3. In principle, if the same solvation description has been used during the docking refinement stage the final calculated interaction energy (+ solvation contribution) can also be used to score a complex.

The force field energy scoring can be decomposed into several contributions of the following typical form,

$$\Delta E_{score} = \Delta E_{vdW} + \Delta E_{Coulomb} + \Delta E_{Solvation} + \Delta E_{Adaptation} + T \cdot \Delta S_{Conf}$$

$$(9.2)$$

The first three terms in the score correspond to the change in van der Waals, Coulomb and solvation free energy, respectively, upon complex formation. These contributions are calculated by subtracting the energy contributions of the isolated partners (in the same conformation as in the complex) from the corresponding energy contributions of the complex. This contribution can be termed interaction energy of the protein partners in the complex. The $\Delta E_{Adaptation}$ energy terms refers to the change in internal energy of each protein partner upon adapting from the unbound form to the conformation in the bound complex structure. This term formally includes several other contributions such as possible changes in the bonded and non-bonded energy terms of each partner protein that need to change upon forming the bound structures.

The $\Delta E_{Adaptation}$ terms are often neglected during scoring. Hence, only the intermolecular (interaction) contributes to the score and one may wonder if this could dramatically affect the quality of the scoring. However, accounting for the conformational adaptation energy is difficult in case of scoring predicted complexes represented by a single structure because a single bound protein conformation may differ from a representative unbound conformation, not only at the interface but also in many regions that are irrelevant for the binding process (e.g. side chains or loops that are far away from the interface region). It is likely that each of the complexes that one wishes to score differs from the representative unbound form as well as from other predicted complexes in regions not relevant for binding. Thus, any straightforward estimation of the adaptation energy by calculating the difference between a partner structure in the bound form minus the energy of a representative unbound structure may include contributions due to accidental differences in regions irrelevant for binding. These irrelevant contributions can be significant and differ in each of the evaluated complexes and may even dominate the score. The problem of course occurs less in cases where the conformational changes in partner structures can be limited to the region of interest.

The conformational entropy term ($T\Delta S_{Conf}$) represents all contributions due to restriction (or enhancement) of the mobility of the partners upon complex formation. Approximate methods can be used to estimate changes in conformational entropy of binding based for example on normal mode analysis of the complex vs isolated partners.[63,145] Here, one assumes that the energy function close to a given structure can be approximated by a quadratic function and one can calculate entropies by treating the system as a system of harmonic oscillators.[145] Alternatively, it is also possible to evaluate the number of accessible stable rotameric states of side chains in the complex vs isolated partners to estimate the restriction of the side chain flexibility upon protein–protein association.[146,147] However, in most force field scoring approaches it is assumed that the change in conformational entropy is similar for all evaluated complexes. Hence, the term might be very significant for calculating absolute binding free energies. In case of scoring one is primarily interested in identifying the most realistic complex geometry relative to alternative geometries and the conformational entropy term is therefore often neglected.

As outlined in previous sections, the molecular mechanics force field based approach bears a number of approximations and several attempts have been made in recent years to further improve force field based scoring. The score consists of several contributions and a frequently applied strategy to improve scoring is to give each force field contribution an adjustable weight. The weights can be optimised on a set of known complex structures compared to a set of incorrectly docked complexes (termed decoy complexes). Examples of scoring with optimal weights of energy components are implemented in the HADDOCK approach,[36] the RosettaDock programme,[40] the ICM docking software,[36] the ZRANK programme[124] and the EMPIRE programme[148] (as well as other approaches).

The force field scoring of putative complexes can be supplemented by a pseudo-energy term that accounts for available experimental data on a given system.[36,117] This can, for example, be a pseudo energy term for the distance between residues that are known to interact in the complex or a term that accounts for the presence of a residue at the interface based for example on experimental mutagenesis data.

9.5.2.1 *Scoring Based on Ensembles of Structures*

Protein molecules and protein–protein complexes undergo conformational fluctuations such that the representation by one static structure may not be realistic. In principle, a more accurate evaluation of a given complex can be achieved if one calculates an average score (force field energy) over an ensemble of conformations compatible with a protein–protein complex structure. Such ensemble can for example be generated by a molecular dynamics simulation of the complex and the isolated partners. Typically, during the evaluation an implicit solvent model such as the finite-difference Poisson–Boltzmann approach (see above and Chapter 11) or the GB approach are employed to account for solvation effects. The methodology is frequently referred to as MM/PBSA (Molecular Mechanics Poisson Boltzmann Surface Area) approach[149] and can result in improved scoring compared to force field based scoring based on single representative structures for each docking geometry.[149,150] However, due to the additional demand to generate and evaluate an ensemble of conformations (often several hundreds of conformations) around each docked complex it is limited to the scoring of only a few putative protein–protein complexes.

9.5.3 *Knowledge-based Scoring of Docked Complexes*

In addition to scoring functions based on a molecular mechanics force field, it is also possible to design knowledge-based or statistical potentials to evaluate complexes. As the term 'knowledge-based' indicates such potentials are usually extracted from known protein–protein complexes or other receptor–ligand complexes. In knowledge-based scoring one relates the observed frequency of atom–atom (or group–group) contacts (or distances) to the expected contact frequency at receptor–ligand interfaces. Over- or under-representation of a given pair of atoms or residues at an interface, then relates to favourable or unfavourable interactions, respectively.[135–143] Expected contact frequency indicates contacts that are obtained if atoms would be distributed randomly at the interface.

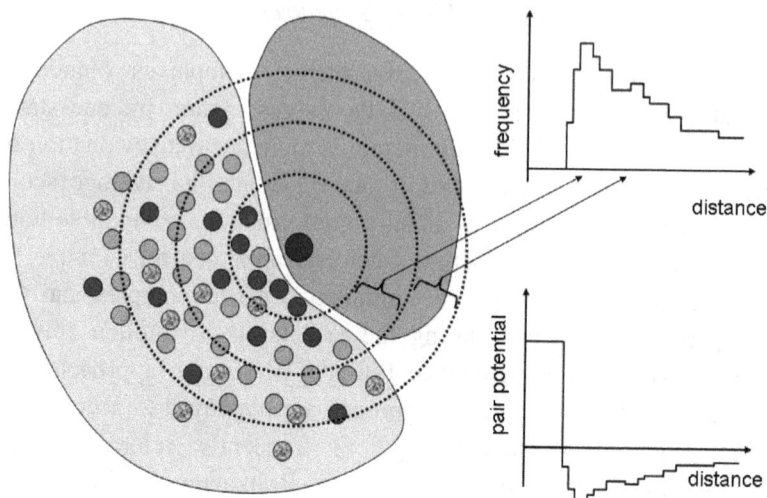

Fig. 9.4. Scheme for the design of a distance-dependent statistical potential from amino acid distributions at interfaces. For a selected atom or residue (large grey sphere on the smaller protein partner) the frequency of other residues (or atoms) in the second protein is evaluated at various distances. The ratio of observed frequencies (from many protein–protein interfaces) relative to expected frequencies (assuming a uniform distribution of atoms or residues) determines an effective statistical pair potential for a pair of residues or atoms (see text).

9.5.3.1 *Principles of Statistical Potentials to Score Predicted Complexes*

The idea to extract interaction of free energies between amino acid residues or atoms from the frequency of contacts in known protein structures has been pioneered by Tanaka and Scheraga[151] and further developed by Miyazawa and Jernigan,[152,153] and Sippl and co-workers.[154,155] Since then many statistical potentials have been developed in most cases depending either on distances or on only contacts between atoms or whole residues in proteins.[135–143] A few statistical potentials also contain orientation dependent terms.[154] The main application area of statistical potentials is the assessment of predicted or experimentally determined protein structures. Statistical potentials are used in several programme packages to predict protein structures using fold recognition and threading methods,[155–158] *ab initio* folding of proteins[159–161] and prediction of the stability of proteins.[162,163] Similar approaches can also be applied to derive a statistical potential for evaluating protein–protein complexes.[135–143] In statistical mechanics for a set of n particles the n-body correlation function is defined as,

$$g^{(n)}(\vec{r}_1,\vec{r}_2....\vec{r}_n) = \frac{p(\vec{r}_1,\vec{r}_2........\vec{r}_n)}{p(\vec{r}_1)\cdot p(\vec{r}_2)...\cdot p(\vec{r}_n)} \qquad (9.3)$$

with $p(r_1,r_2...r_n)$ corresponding to the probability of finding the n particles in configuration r_1, $r_2...r_n$ and $p(r_i)$ is the probability to find particle i at r_i. The correlation function is a measure of the interdependence of particles (would be 1 in case of no correlation or no interdependence of particles to adopt a certain configuration). From the n-body correlation function one can calculate a potential of mean force for a given configuration of particles,

$$w^{(n)}(\vec{r}_1,\vec{r}_2....\vec{r}_n) = -R\cdot T\cdot ln(g^{(n)}(\vec{r}_1,\vec{r}_2....\vec{r}_n) \qquad (9.4)$$

where R is the gas constant, T is the temperature. The basis for deriving a statistical potential for protein structure prediction or evaluation of docked complexes is a set of native structures. Ideally, one would use the n-body correlation function (or probability distribution) of atoms or residues in protein structures or at protein–protein interfaces. However, in practice the available data is insufficient for accurately evaluating multi-body correlations and one restricts in most cases the derivation of statistical potentials to pair correlations. That is most statistical potentials are based on contacts or distances between residues or atoms in the set of experimentally determined protein structures or protein–protein complexes. Commonly the inverse-Boltzmann equation is used[135] to derive a distance-dependent potential V of the form,

$$W(a,b,d) = -RT\, ln\left(\frac{N_{observed}(a,b,d)}{N_{expected}(a,b,d)}\right) \qquad (9.5)$$

where $N_{observed}(a, b, d)$ corresponds to the number of observed atom pairs a, b at a distance d (within a distance interval) in the database and $N_{expected}(a, b, d)$ is the expected number of atom pairs at the same distance interval if atoms would be randomly distributed (e.g. no interactions between atoms).

Differences in statistical potentials based on the inverse-Boltzmann equation can be due to the use of different data bases but the main difference is due to differences in the calculation of the number of expected atom pairs in case of no interaction between atoms or residues (termed reference state). Several different assumptions have been made to choose an appropriate reference state. In case of an infinite system the statistical potential corresponds to the potential-of-mean force derived from the pair correlation function of the two atoms a, b. In this case the probability to find non-interacting particles at an interval at a distance d is proportional to the square of the distance. Hence, the statistical potential is given by

$$W(a,b,d) = -RT \ln\left(\frac{V \cdot N_{observed}(a,b,d)}{4\pi d^2 \cdot N_a \cdot N_b}\right) \tag{9.6}$$

Here V is the volume and N_a and N_b correspond to the number of atoms of type a and b in the protein, respectively. This potential would be correct in case of using a uniform density of atoms as reference state. However, proteins are finite and the number of atoms in a shell at a distance does not necessarily increase with the square of the distance. This has been recognised by Zhou and co-workers[135,164] who introduced an exponent $\alpha = 1.57$ (later adjusted to 1.61) instead of 2 when calculating the expected number of pairs at distance d. The smaller exponent accounts more realistically for the expected distance distribution of atoms in proteins.

It should also be emphasised that statistical potentials derived from dense systems by the inverse-Boltzmann method do not necessarily reflect true interactions between particles. For example, oscillations in the distance dependence may simply be due to the size of the atoms and the shell or atom-layer-like packing structure around atoms.

9.5.3.2 *Application of Statistical Potentials to Score Predicted Complexes*

A number of knowledge-based scoring functions for specifically evaluating docked protein–protein complexes have been designed in recent years and only an overview of a few developments can be given. As indicated above only very few knowledge-based potentials go beyond accounting for pairwise interactions or using variables other than contact or distance to evaluate an interaction.[156] The potentials differ in the resolution of the interface description (e.g. atomic resolution or pseudo atoms representing parts or whole residues) but also in the number of atom or pseudo atom types or the reference state from which the expected contact or distance probability for atom-pairs are derived (see above).

For example, Jiang *et al.*[165] developed an atomic-resolution statistical potential based on physico-chemical properties and propensities of interface atoms which contained just four different atom types. Despite the small number of atom types the potential reproduced the known binding energies of several complexes in good agreement with experiment. However, the ability to distinguish between native docking geometry and alternative solution was not tested. Skolnick and co-workers developed atomic-resolution and residue-resolution statistical contact potentials based on a large set of experimental protein–protein complex structures.[166] The potential has been used successfully within the TASSER protein structure prediction programme[166] and can also be applied to evaluate docked protein–protein complexes.[167]

One of the best performing statistical potential is the DFIRE (Distance-scale Finite Ideal Gas Reference state) potential by Zhou and Zhou.[135] The potential involves 19 different atom types. The potential has been applied both for evaluating modelled protein structures and protein–protein complexes.[136,168] Although original designed as an all-atom statistical potential, a variant residue-level potential was also developed (DFIRE-SCM).[169] Interestingly, the coarse-grained potential showed very similar performance in discriminating native protein structures from non-native decoy structures compared to the all-atom potential and other statistical potentials.

Another distance-dependent statistical potential has been developed by Sali and co-workers.[46,142] The derivation of the potential is based on an Ansatz that factorises the probability distribution to find any configuration of atoms or pseudo atoms at an interface.[42] Most other statistical potentials assume a Boltzmann law for relating probabilities to effective interaction energies. Instead the potential by Sali and co-workers uses a factorization of the full probability function in terms of pairwise probabilities and other terms depending on higher order probabilities. This potential has been used for comparative modelling of protein structures[170] but also for evaluating protein–protein complexes that have been built based on sequence homology to known complexes.[46]

Chuang et al.[171] developed an alternative knowledge-based potential that uses as reference state docking decoy structures (incorrectly docked complexes). For this potential the pairwise atom occurrences in native complexes are related to the corresponding frequencies in the non-native decoy complexes to derive a score for the actual atom pair. The method has been combined with an FFT docking programme and improved the scoring performance compared to a previous scoring function.[171]

Tobi and Bahr[172] used a different approach to derive a knowledge-based potential. Most existing statistical potentials are based on a set of native structures or complexes. Except for the concepts like the above discussed method by Chuang et al.[171] this leaves out the information on complexes that are incorrect. In order to optimally account for the additional information in non-native complexes that include which interactions should not occur in native complexes, the authors used linear programming to optimise a scoring function. The optimization started with a pairwise additive contact-based scoring function, a number of native protein–protein complexes and a large number of incorrect complex geometries. The parameters of the contact scoring function were then systematically optimised to achieve the largest possible gap between rankings of the native complexes vs non-native complexes. The residue-based scoring function was successfully applied to reproduce the native complex as best ranking solution on a set of test systems.[172]

It is well known that for example solvation forces between molecules are non-additive and can, in general, not be accurately described by simple pairwise potentials. For example, a charged residue is usually effectively repelled from a non-polar side chain. However, the effective repulsion is likely to be stronger (at the same distance) if the interaction occurs in a fully buried environment like the centre of a protein–protein interface compared to a partially solvated rim region of an interface.

A few attempts have been made so far to extend the concept of knowledge-based potentials to interaction patterns beyond pairs. In the field of protein tertiary structure prediction 3-body and 4-body potentials have been developed. A drawback of multi-body statistical potentials is the increased number of parameters compared to pairwise potentials which depends on how many atom or residue types the model may distinguish. Application of 3-body or 4-body potentials in the area of protein structure prediction resulted so far only in modest improvement of tertiary structure prediction compared to pairwise potentials alone.[173–175] However, Majeweski[176] could show that a multi-body potential consisting of up to 17 residues representing a whole interaction pattern was effective in tertiary structure prediction efforts. It remains to be seen if multi-body potentials can also be effective in scoring docked protein–protein complexes.

9.6 Conclusions and Outlook

Realistic modelling of protein–protein interactions and the accurate prediction of putative complex structures is of increasing importance for molecular life sciences. Proteins in the cell have many potential binding partners and undergo a large number of transient interactions with other proteins. For many of these interactions information on the 3D structure is desirable but experimentally often difficult or impossible to elucidate. Improvement of the refinement of docked complexes as well as better scoring of putative complexes to identify the most realistic complex structure is of key importance for modelling protein–protein complexes.

A variety of scoring functions have been designed in recent years and many are effective for the identification of near-native complex structures if docking was performed using bound structures or in case of only minor conformational changes upon complex formation. This observation indicates the strong coupling of scoring and appropriate treatment of conformational changes during docking. A key issue of future prediction efforts should be to solve both issues simultaneously. Ideally, scoring and refinement of docked complexes should not only identify the native or near-native protein binding geometry but also provide a realistic prediction of the binding affinity. This is also a prerequisite for applying docking methods to predict for a given protein the correct binding partner among several other proteins (cross docking). Attempts to use protein–protein docking methods for cross-docking have not been fully satisfying likely due to the limitations of current scoring functions.[177]

It has been recognised that many protein–protein interactions are mediated through recurrent use of similar interaction interfaces or through domain–domain interactions that occur in many proteins.[178–185] A likely route for future directions will be the increased use of comparative modelling methods to model not only single proteins but also complexes of proteins. This direction will benefit from methodological improvements to detect even remote sequence similarity between a target sequence and a template complex. However, improvements in refinement techniques are necessary to eliminate possible errors in the complex structure due to the limited accuracy of comparative modelling methods in case of low target-template similarity.

The focus of the majority of docking efforts in recent years was on modelling of homo- or heterodimeric protein–protein complexes. However, many biological processes involve not only the interaction of two proteins but multiple simultaneous interactions. In many of these cases the isolated pairwise interaction may not be strong enough for a high-resolution experimental characterization. The prediction of multi-protein complexes and assemblies has the potential to provide structural models for such structures and could be helpful to elucidate the associated biological function. Only few methods have been designed to tackle the multi-protein docking problem.[186–189] Such docking efforts can be combined with experimental methods that provide experimental constraints on the structure of the assemblies. For example, electron microscopy and atomic force microscopy do not require crystals of biological structures and can provide low resolution electron density distributions of large macromolecular assemblies. Frequently, docking methods have been used to place atomic resolution structures into the low resolution electron density obtained from electron microscopy and related techniques.[190–192] Depending on the resolution of the experimental data scoring and refinement of possible alternative protein arrangements will be of increasing importance to obtain realistic models of large macromolecular assemblies.

In recent years the rational design of several new or modified protein–protein interactions has become possible.[193–195] The achievement offers the future opportunity to use designed protein–protein interactions to build functional nanomolecular structures. Improved docking refinement and scoring methods can help to better evaluate suggested protein modifications with respect to a desired binding mode and may help to limit the number of redesigned proteins that need to be evaluated experimentally.

Acknowledgements

I would like to thank Drs A. May and R. Bahadur for helpful discussions, and the Deutsche Forschungsgemeinschaft and the European Community for financial support.

References

1. Eisenberg D., Marcotte E.M., Xenarios I., Yeates T.O. (2002). *Nature* 405: 823–826.
2. Bahadur R.P., Zacharias M. (2008). *Cell Mol Life Sci* 65: 1,059–1,072.
3. Rousseau F., Schymkowitz J. (2005). *Curr Opin Struct Biol* 15: 23–30.
4. Charbonnier S., Gallego O., Gavin A.C. (2008). *Biotechnol Annu Rev* 14: 1–28.
5. Sali A., Kuriyan J. (1999). *Trends Cell Biol* 9: 20–24.
6. Valencia A., Pazos F. (2002). *Curr Op Struct Biol* 12: 368–373,
7. Franzot G., Carugo O. (2003). *J Struct Funct Genomics* 4: 245–255.
8. Kim P.M., Lu L.J., Xia Y., Gerstein M.B. (2002). *Science* 314: 1,938–1,941.
9. Sawinski L., Eisenberg D. (2003). *Curr Op Struct Biol* 13: 377–382.
10. Aloy P., Russel R.B. (2006). *Nature* 188: 188–197.
11. Smith G.R., Sternberg M.J.E. (2002). *Curr Op Struct Biol* 12: 28–35.
12. Halperin I., Ma B., Wolfson H., Nussinov R. (2002). *Proteins* 47: 409–443.

13. Deremble C., Lavery R. (2005). *Curr Op Struct Biol* 15: 171–175.
14. Vajda S., Vakser I.A., Sternberg M.J.E., Janin J. (2002). *Proteins* 47: 444–446.
15. Bonvin A.M.J.J. (2006). *Curr Op Struct Biol* 16: 194–200.
16. Gray J.J. (2006). *Curr Op Struct Biol* 17: 183–193.
17. Vakser I.A., Kundrotas P. (2008). *Curr Pharm Biotechnol* 9: 57–66.
18. Ritchie D.W. (2008). *Curr Protein Pept Sci* 9: 1–15.
19. Andrusier N., Mashiach E., Nussinov R., Wolfson H.J. (2008). *Proteins* 73: 271–289.
20. Russell R.B., Alber F., Aloy P., Davis F.P., Korkin D., Pichaud M., Topf M., Sali A. (2004). *Curr Opin Struct Biol* 14: 313–319.
21. Aloy P., Russell R.B. (2004). *Nature Biotechnology* 22: 1,317–1,321.
22. Sanchez R., Pieper U., Melo F., Eswar N., Mart-Renom M.A., Madhusudhan M.S., Mirkovi N., Sali A. (2000). *Nat Struct Biol* 7: 986–990.
23. Katchalski-Katzir E., Shariv I., Eisenstein M., Friesem A.A., Aalo C., Vakser I.A. (1992). *Proc Natl Acad Sci USA* 89: 2,195–2,200.
24. Vakser I.A., Matar O.G., Lam C.F. (1999). *Proc Natl Acad Sci USA* 96: 8,477–8,482.
25. Vakser I.A. (1995). *Protein Eng* 8: 371–377.
26. Chen R., Li L., Weng Z. (2003). *Proteins: Struct Func Genet* 52: 80–87.
27. Mandell J.G., Roberts V.A., Pique M.E., Kotlovyi V., Mitchell J.C., Nelson E., Tsigelny I., Ten Eyck L.F. (2001). *Protein Eng* 14: 105–113.
28. Gabb H.A., Jackson R.M., Sternberg M.J.E. (1997). *J Mol Biol* 272: 106–120.
29. Ritchie D.W., Kemp G.J.L. (2000). *Proteins: Struct Func Bioinfo* 39: 178–194.
30. Ritchie D.W. (2005). *J Appl Cryst* 38: 808–818.
31. Norel R., Lin S.H., Wolfson H., Nussinov R. (1994). *Biopolymers* 34: 933–940.
32. Sandak B., Nussinov R., Wolfson H.J. (1998). *J Comput Biol* 5: 631–654.
33. Sandak B., Wolfson J., Nussinov R. (1998). *Proteins: Struct Funct Bioinfo* 32: 159–167.
34. Shatsky M., Nussinov R., Wolfson H.J. (2004). *J Comput Biol* 11: 83–106.
35. Cherfils J., Duquerroy S., Janin J. (1991). *Proteins* 11: 271–280.
36. Fernandez-Recio J., Totrov M., Abagyan R. (2002). *Protein Sci* 11: 280–291.
37. Dominguez C., Boelens R., Bonvin A.M.J.J. (2003). *J Am Chem Soc* 125: 1,731–1,737.
38. Zacharias M. (2003). *Prot Sci* 12: 1,271–1,282.
39. Zacharias M. (2005). *Proteins: Struct Func Bioinf* 60: 252–256.
40. Gray J.J., Moughan S., Wang C., Schueler-Furman O., Kuhlman B., Rohl C.A., Baker D. (2003). *J Mol Biol* 331: 281–299.
41. Wang C., Schueler-Furman O., Baker D. (2005). *Prot Sci* 14: 1,328–1,339.
42. Motiejunas D., Gabdoulline R., Wang T., Feldman-Salit A., Johann T., Winn P.J., Wade R.C. (2008). *Proteins* 71: 1,955–1,969.
43. Sali A., Blundell T.L. (1993). *J Mol Biol* 234: 779–815.

44. Pieper U., Eswar N., Webb B.M., Eramian D., Kelly L., Barkan D.T., Carter H., Mankoo P., Karchin R., Marti-Renom M.A., Davis F.P., Sali A. (2009). *Nucleic Acids Res* 37: 347–354.
45. Devos D., Russel R.B. (2007). *Curr Opin Struct Biol* 17: 370–377.
46. Davis F.P., Braberg H., Shen M.Y., Pieper U., Sali A., Madhusdhan M.S. (2006). *Nucl Acids Res* 34: 2,943–2,952.
47. Uetz P., Giot L., Cagney G., Manseld T.A., Judson R.S., Knight J.R., Lockshon D., Narayan V., Srinivasan M., Pochart P., Qureshi-Emili A., Li Y., Godwin B., Conover D., Kalbeisch T., Vijayadamodar G., Yang M.J., Johnston M., Fields S., Rothberg J.M. (2000). *Nature* 403: 623–671.
48. Ito T., Chiba T., Ozawa R., Yoshida M., Hattori M., Sakaki Y. (2001). *Proc Natl Acad Sci* 98: 4,569–4,574.
49. Lu L., Arakaki A.K., Lu H., Skolnick J. (2003). *Genome Res* 13: 1,146–1,154.
50. Leach A.R. (2001). *Molecular Modelling, Principles and Applications*. Pearson Education Limited, Dorchester.
51. Schlick T. (2006). *Molecular Modeling and Simulation*. Springer Press, New York, USA.
52. Case D., Pearlman D.A., Caldwell J.W., Cheatham III T.E., Ross W.S., Simmerling C.L., Darden T.A., Merz K.M., Stanton R.V., Cheng A.L., Vincent J.J., Crowley M., Tsui V., Radmer R.J., Duan Y., Pitera J., Massova I., Seibel G.L., Singh U.C., Weiner P.K., Kollman P.A. (2003). *Amber 8*. University of California, San Francisco.
53. Brooks B.R., Bruccoleri R.E., Olafson B.D., States D.J., Swaminathan S., Karplus M. (1983). *J Comp Chem* 4: 187–217.
54. Van Der Spoel D., Lindahl E., Hess B., Groenhof G., Mark A.E., Berendsen H.J. (2005). *J Comput Chem* 26: 1,701–1,718.
55. Li L., Chen R., Weng Z. (2003). *Proteins* 53: 693–707.
56. Fernandez-Recio J., Totrov M., Abagyan R. (2003). *Proteins* 52: 113–117.
57. Liang S., Wang G., Zhou Y. (2009). *Proteins* 76: 309–316.
58. Shatsky M., Nussinov R., Wolfson H.J. (2004). *J Comput Biol* 11: 83–106.
59. Flores S.C., Keating K.S., Painter J., Morcos F., Nguyen K., Merritt E.A., Kuhn L.A., Gerstein M.B. (2008). *Proteins* 73: 299–319.
60. Keating K.S., Flores S.C., Gerstein M.B., Kuhn L.A. (2009). *Protein Sci* 18: 359–371.
61. Hayward S., de Groot B.L. (2008). *Methods Mol Biol* 443: 89–106.
62. Zacharias M., Sklenar H. (2000). *Biophys J* 78: 2,528–2,542.
63. Van Wynsberghe A.W., Cui Q. (2005). *Biophys J* 89: 2,939–2,949.
64. Tirion M. (1996). *Phys Rev Lett* 77: 1,905–1,908.
65. Bahar I., Atilgan A.R., Erman B. (1997). *Folding Design* 2: 173–181.
66. Hinsen K. (1998). *Proteins* 33: 417–429.
67. Tama F., Sanejouand Y.H. (2001). *Protein Eng* 14: 1–6.
68. Bahar I., Rader A.J. (2005). *Curr Opin Struct Biol* 15: 589–592.

69. Van Wynsberghe A.W., Cui Q. (2006). *Structure* 14: 1,647–1,653.
70. Zacharias M., Sklenar H. (1999). *J Comput Chem* 20: 287–300.
71. Zacharias M. (2004). *Proteins* 54: 759–767.
72. May A., Zacharias M. (2005). *Biochim Biophys Acta* 1754: 225–231.
73. Lindahl E., Delarue M. (2005). *Nucleic Acids Res* 33: 4,496–4,506.
74. May A., Zacharias M. (2007). *Proteins* 69: 774–80.
75. May A., Zacharias M. (2008). *Proteins* 70: 794–809.
76. May A., Zacharias M. (2008). *J Med Chem* 51: 3,499–3,506.
77. Cavasotto C.N., Kovacs J.A., Abagyan R.A. (2005). *J Am Chem Soc* 127: 9,632–9,640.
78. Mustard D., Ritchie D.W. (2005). *Proteins* 60: 269–274.
79. Smith G.R., Sternberg M.J.E., Bates P.A. (2005). *J Mol Biol* 347: 1,077–1,101.
80. Carlson H.A., McCammon J.A. (2000). *Mol Pharmacol* 57: 213–218.
81. Karplus M., McCammon J.A. (2002). *Nat Struct Biol* 9: 646–652.
82. van Dijk A.D.J., Bonvin A.M.J.J. (2006). *Bioinformatics* 22: 2,340–2,347.
83. Rajamani D., Thiel S., Vajda S., Camacho C.J. (2004). *Proc Natl Acad Sci* 101: 11,287–11,292.
84. Camacho C.J. (2005). *Proteins* 60: 245–251.
85. Bashford D., Case D.A. (2000). *Annu Rev Phys Chem* 51: 129–152.
86. Baker N.A. (2005). *Curr Opin Struct Biol* 15: 137–143.
87. Chen J., Brooks C.L. 3rd, Khandogin J. (2008). *Curr Opin Struct Biol* 18: 140–148.
88. Holm L., Sander C. (1992). *J Mol Biol* 225: 93–105.
89. Juffer A.H., Eisenhaber F., Hubbard S.J., Walter D., Argos P. (1995). *Protein Sci* 4: 2,499–2,509.
90. Fernandez-Recio J., Totrov M., Skorodumov C., Abagyan R. (2005). *Proteins* 58: 134–143.
91. Fernandez-Recio J., Abagyan R., Totrov M. (2005). *Proteins: Struct Func Bioinf* 60: 308–313.
92. Heifetz A., Eisenstein M. (2003). *Protein Eng* 16: 179–185.
93. am Busch M.S., Lopes A., Amara N., Bathelt C., Simonson T. (2008). *BMC Bioinformatics* 13: 148–155.
95. Honig B., Sharp K., Yang A.S. (1993). *J Phys Chem* 97: 1,101–1,109.
96. Sheinerman F.B., Norel R., Honig B. (2000). *Curr Opin Struct Biol* 10: 153–159.
97. Koehl P. (2006). *Curr Opin Struct Biol* 16: 142–151.
98. Grochowski P., Trylska J. (2008). *Biopolymers* 89: 93–113.
99. Gilson M.K., Davis M.E., Luty B.A., McCammon J.A. (1993). *J Phys Chem* 97: 3,591–3,600.
100. Still W.C., Tempczyk A., Hawley R.C., Hendrikson T. (1990). *J Am Chem Soc* 112: 6,127–6,129.
101. Hawkins G.D., Cramer C.J., Truhlar D.G. (1995). *Chem Phys Lett* 246: 122–129.
102. Schaefer M., Karplus M. (1996). *J Phys Chem* 100: 1,578–1,599.
103. Bashford D., Case D.A. (2000). *Annu Rev Phys Chem* 51: 129–152.

104. Onufriev A., Case D.A., Bashford D. (2002). *J Comput Chem* 23: 1,297–1,304.
105. Baker N.A. (2005). *Curr Opin Struct Biol* 15: 137–143.
106. de Vries S.J., van Dijk A.D., Krzeminski M., van Dijk M., Thureau A., Hsu V., Wassenaar T., Bonvin A.M. (2007). *Proteins* 69: 726–733.
107. Krol M., Chaleil R.A.G., Tournier A.L., Bates P.A. (2007). *Proteins* 69: 750–757.
108. Swendsen R.H., Wang J.S. (1986). *Phys Rev Lett* 57: 2,607–2,609.
109. Gnanakaran S., Nymeyer H., Portman J., Sanbonmatsu K.Y., Garcia A.E. (2003). *Curr Opin Struct Biol* 15: 168–174.
110. Kannan S., Zacharias M. (2007). *Proteins* 66: 697–706.
111. Riemann N.R., Zacharias M. (2005). *Prot Eng* 18: 465–476.
112. Andrusier N., Nussinov R., Wolfson H.J. (2007). *Proteins* 69: 139–159.
113. Lorenzen S., Zhang Y. (2007). *Protein Sci* 16: 2,716–2,725.
114. Jackson R.M., Gabb H.A., Sternberg M.J. (1998). *J Mol Biol* 276: 265–285.
115. Bastard K., Thureau A., Lavery R., Prevost C. (2003). *J Comput Chem* 24: 1,910–1,920.
116. Bastard K., Prevost C., Zacharias M. (2006). *Proteins* 62: 956–969.
117. Gabb H.A., Jackson R.M., Sternberg M.J.E. (1997). *J Mol Biol* 272: 106–120.
118. Camacho C.J., Gatchell D.W., Kimura S.R., Vajda S. (2000). *Proteins: Struct Func Genet* 40: 525–537.
119. Mandell J.G., Roberts V.A., Pique M.E., Kotlovyi V., Mitchell J.C., Nelson E., Tsigelny I., Eyck L.F.T. (2001). *Protein Eng* 14: 105–113.
120. Heifetz A., Katchalski-Katzir E., Eisenstein M. (2002). *Protein Sci* 11: 571–587.
121. Chen R., Weng Z. (2002). *Proteins* 47: 281–294.
122. Cerutti D.S., Ten Eyck L.F., McCammon J.A. (2005). *J Chem Theory Comp* 1: 143–152.
123. Camacho C.J., Ma H., Champ P.C. (2006). *Proteins* 63: 868–877.
124. Pierce B., Weng Z.A. (2008). *Proteins* 72: 270–279.
125. Camacho C.J., Ma H., Champ P.C. (2006). *Proteins* 63: 868–879.
126. Audie J., Scarlata S. (2007). *Biophys Chem* 129: 198–211.
127. Zhao X., Liu X., Wang Y., Chen Z., Kang L., Zhang H., Luo X., Zhu W., Chen K., Li H., Wang X., Jiang H. (2008). *J Chem Inf Model* 48: 1,438–1,447.
128. Liang S., Meroueh S.O., Wang G., Qiu C., Zhou Y. (2008). *Proteins* 75: 397–403.
129. Moont G., Gabb H.A., Sternberg M.J.E. (1999). *Proteins: Struct Func Genet* 35: 364–373.
130. Martin O., Schomburg D. (2008). *Proteins* 70: 1,367–1,378.
131. Gottschalk K.E., Neuvirth H., Schreiber G. (2004). *Prot Eng Des Sel* 17: 183–189.
132. Duan Y., Reddy B.V.B., Kaznessis Y.N. (2005). *Prot Sci* 14: 316–328.
133. Murphy J., Gatchell D.W., Prasad J.C., Vajda S. (2003). *Proteins: Struct Func Genet* 53: 840–854.
134. Liu S., Li Q., Lai L. (2006). *Proteins: Struct Func Bioinf* 64: 68–78.
135. Zhou H., Zhou Y. (2002). *Protein Sci* 11: 2,714–2,726.
136. Zhang C., Liu S., Zhou Y. (2005). *Proteins: Struct Func Bioinf* 60: 314–318.

137. Kozakov D., Brenke R., Comeau S.R., Vajda S. (2006). *Proteins* 65: 392–406.
138. Huang S.Y., Zhou X. (2008). *Proteins* 72: 557–579.
139. Sippl M. (1990). *J Mol Biol* 213: 859–883.
140. Jernighan R.L., Bahar I. (1996). *Curr Opin Struct Biol* 6: 195–209.
141. Samudrala R., Moult J. (1998). *J Mol Biol* 275: 895–916.
142. Shen M.Y., Sali A. (2006). *Protein Sci* 15: 2,507–2,524.
143. Rojnuckarin A., Subramanian S. (1999). *Proteins* 36: 54–67.
144. Gilson M.K., Given J.A., Bush B.L., McCammon J.A. (1997). *Biophys J* 72: 1,047–1,069.
145. Ma J. (2004). *Curr Protein Peptid Sci* 5: 119–123.
146. Mintseris J., Pierce B., Wiehe K., Anderson R., Chen R., Weng Z. (2007). *Proteins* 69: 511–520.
147. Pierce B., Weng Z. (2007). *Proteins* 67: 1,078–1,086.
148. Liang S., Liu S., Zhang C., Zhou Y. (2007). *Proteins* 69: 244–253.
149. Massova I., Kollman P.A. (1999). *J Am Chem Soc* 121: 8,133–8,142.
150. Golke H., Case D.A. (2004). *J Comput Chem* 25: 238–250.
151. Tanaka S., Scheraga H.A. (1976). *Macromolecules* 9: 945–950.
152. Miyazawa S., Jernigan R.L. (1999). *Proteins* 36: 357–369.
153. Moult J. (1997). *Curr Opin Struct Biol* 7: 194–199.
154. Sippl M.J., Weitckus S. (1992). *Proteins* 13: 258–271.
155. Sippl M.J. (1995). *Curr Opin Struct Biol* 5: 229–235.
156. Lu M., Dousis A.D., Ma J. (2008). *J Mol Biol* 376: 288–301.
157. Lu H., Skolnick J. (2001). *Proteins* 44: 223–232.
158. Melo F., Sanchez R., Sali A. (2002). *Protein Sci* 11: 430–448.
159. Makino Y., Itoh N. (2008). *BMC Struct Biol* 8: 46–57.
160. Fitzgerald J.E., JHA A.K., Colubri A., Sosnick T.R., Freed K.F. (2007). *Prot Sci* 16: 2,123–2,139.
161. Skolnick J., Kolinski A., Ortiz A.R. (1997). *J Mol Biol* 265: 217–241.
162. Tobi D., Elber R. (2000). *Proteins* 41: 40–46.
163. Zhang C., Kim S.H. (2000). *Proc Natl Acad Sci* 97: 2,550–2,555.
164. Zhou H., Zhou Y. (2003). *Proteins* 54: 15–22.
165. Jiang L., Gao Y., Mao F., Liu Z., Lai L. (2002). *Proteins: Struct Func Genet* 46: 190–196.
166. Lu H., Lu L., Skolnick J. (2003). *Biophys J* 84: 1,895–1,901.
167. Lu L., Skolnick J. (2002). *Proteins* 49: 350–364.
168. Zhang C., Liu S., Zhu Q., Zhou Y. (2005). *J Med Chem* 48: 2,325–2,335.
169. Zhang C., Liu S., Zhou H., Zhou Y. (2004). *Protein Sci* 13: 400–411.
170. Eswar N., Eramian D., Webb B., Shen M.Y., Sali A. (2008). *Methods Mol Biol* 426: 145–59
171. Chuang G.Y., Kozakov D., Brenke R., Comeau S.R., Vajda S. (2008). *Biophys J* 95: 4,217–4,227.
172. Tobi D., Bahar I. (2006). *Proteins* 62: 970–981.

173. Munson P.J., Singh R.J. (1997). *Protein Sci* 6: 1,467–1,481.

174. Carter C.W., LeFebvre B.C., Cammer S.A., Tropsha A., Edgell M.H. (2001). *J Mol Biol* 311: 625–638.

175. Gan H.H., Tropsha A., Schlick T. (2001). *Proteins* 43: 161–174.

176. Mayewski S. (2005). *Proteins* 59:, 152–169.

177. Sacquin-Mora S., Carbone A., Lavery R. (2008). *J Mol Biol* 382: 1,276–1,289.

178. Stein A., Russell R.B., Aloy P. (2005). *Nucleic Acids Res* 33: D413–D417.

179. Davis F.P., Sali A. (2005). *Bioinformatics* 21: 1,901–1,907.

180. Gong S., Park C., Choi H., Ko J., Jang I., Lee J., Molser D.M., Oh D., Kim D.S., Bhak J. (2005). *BMC Bioinformatics* 6: 207–218.

181. Levy E.D., Pereira-Leal J.B., Chothia C., Teichmann S.A. (2006). *PLOS Comp Biol* 2: 155–182.

182. Mintz S., Shulman-Peleg A., Wolfson H.J., Nussinov R. (2006). *Proteins: Struct Func Bioinf* 61: 6–20.

183. Winter C., Henschel A., Kim W.K., Schroeder M. (2006). *Nucleic Acids Res* 34: D310–D314.

184. Aloy P., Russell R.B. (2003). *Bioinformatics* 19: 161–162.

185. Berchanski A., Eisenstein M. (2003). *Proteins: Struct Func Genet* 53: 817–829.

186. Comeau S.R., Camacho C.J. (2004). *J Struct Biol* 150: 233–244.

187. Inbar Y., Benyamini H., Nussinov R., Wolfson H.J. (2005). *J Mol Biol* 349: 435–447.

189. Inbar Y., Benyamini H., Nussinov R., Wolfson H.J. (2005). *Phys Biol* 2: S156–S165.

190. Wriggers W., Milligan R.A., McCammon J.A. (1999). *J Struct Biol* 125: 185–195.

191. Roseman A.M. (2000). *Acta Cryst* D56: 1,332–1,340.

192. Rossmann M.G. (2000). *Acta Cryst* D56: 1,341–1,349.

193. Kortemme T., Joachimiak L.A., Bullock A.N., Schuler A.D., Stoddard B.L., Baker D. (2004). *Nat Struct Mol Biol* 11: 371–379.

194. Kortemme T., Baker D. (2004). *Curr Opin Chem Biol* 8: 91–97.

195. Kossiakoff A., Koide S. (2008). *Curr Opin Struct Biol* 18: 499–506.

CHAPTER 10

Motif-Mediated Protein Interactions and their Role in Disease

Holger Dinkel and Heinrich Sticht

Institute of Biochemistry, Erlangen University,
Fahrstraße 17, 91054 Erlangen, Germany
E-mail: h.sticht@biochem.uni-erlangen.de

Many important protein–protein interactions are mediated by relatively small recognition domains which bind to peptides exhibiting specific sequence motifs. In this type of interaction, only the interaction domain adopts a globular three-dimensional structure while the interaction motif is mostly linear. These interactions are of crucial importance to signalling mechanisms, cell compartment targeting and post-translational modification. During recent years it became clear that mutations in linear sequence motifs also represent the cause of several hereditary diseases like Noonan Syndrome, Liddle's Syndrome and Usher's Syndrome. In addition, genome sequencing and functional analyses revealed that viral effector proteins frequently rely on linear sequence motifs to ensure viral persistence and replication. Well-characterised examples are the HIV-1 Nef protein or the Tip protein from Herpesvirus saimiri. This book chapter will review the principles of motif-mediated interactions, their role in disease and computational approaches to identify functional sequence motifs.

10.1 Introduction

Proteins play crucial roles in virtually all biological processes. They catalyse chemical reactions, oligomerise to form filaments, control progression through the cell cycle, regulate other proteins' activity and integrate into the cell membranes. Originally, it was assumed that protein function was exclusively governed by the three-dimensional structure and that proteins consisted of a single large spherical shape with unique fold and activity. Today, this view has changed significantly in several aspects.

It is now known that a large number of proteins consist of multiple distinct domains. Domains are defined as compact, spatially distinct units, which fold separately into their three-dimensional structure and which exhibit a distinct function. Therefore, the traditional rule of 'one protein – one function' has been extended into 'one domain – one function'.

It also became apparent that some segments of the peptide chain do not adopt a globular three-dimensional structure at all, and were consequently termed 'disordered protein regions': approximately 20%–50% of all eukaryotic proteins have some disorder in their sequence and up to 17% of the proteins in a eukaryotic cell are completely disordered.[1] It is important to note that these disordered regions do not represent simple spacers connecting globular domains, but instead contain linear sequence motifs important for protein functions. These sequence motifs mediate protein–protein interaction, cell compartment targeting and represent the sites of post-translational modification.[2]

Today, approximately 200–300 motif patterns are known and are catalogued by several resources including the Eukaryotic Linear Motif (ELM) database,[3] PROSITE[4] and Minimotif-Miner.[5] Binding via linear motifs was estimated to account for 15–40% of the interactions in the human proteome,[6] suggesting that there is a significant number of novel motifs which still need to be discovered. McEntyre and Gibson estimate that there are most likely more linear motif instances for signalling-pathway regulation and cell-compartment targeting than there are globular domains in the proteome.[7]

Most linear sequence motifs interact with modular protein interaction domains. One prominent protein that exploits motif-mediated interactions for its regulation is the lymphocyte-specific tyrosine-kinase Lck (Fig. 10.1). While the kinase domain exhibits enzymatic activity, the two other domains (SH2, SH3) are adaptor-domains, responsible for regulatory processes: they are able to recognise distinct short sequence motifs, which contain phosphorylated tyrosine residues in the case of the

SH2 domain and proline-rich sequence stretches in the case of the SH3 domain. These interactions are important for the activation of the kinase activity. Additional motif-mediated interactions of Lck include the interaction between substrates and the catalytic domain as well as binding of CD4 to the amino-terminal region of Lck[8] (Fig. 10.1).

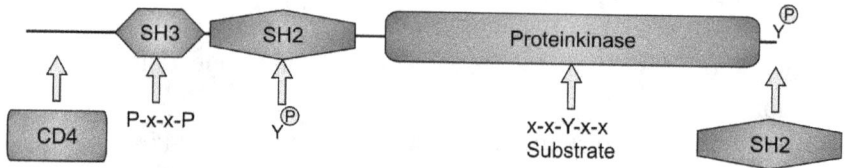

Fig. 10.1. Schematic drawing of human tyrosine-kinase Lck illustrating the principle of motif-mediated interactions. The top line shows the domain architecture of Lck and the C-terminal tyrosine, which can become phosphorylated. The second line shows the domains and linear sequence motifs binding to distinct regions of Lck: the amino-terminal region is responsible for CD4/CD8 binding. The SH3 interaction domain binds to a P-x-x-P consensus motif, while the SH2 domain interacts with motifs containing a phosphorylated tyrosine residue. Tyrosines are also the substrate for the kinase domain, which is able to phosphorylate these amino acids.

Linear sequence motifs, however, are not only involved in physiological cellular processes, but also play a role in disease. Recently it became clear that numerous hereditary diseases can be attributed to mutations within linear motifs, and there is increasing evidence that viral effector proteins have evolved to exploit motif-mediated interactions to interfere with the signalling machinery of their host.[9]

This book chapter will review motif-mediated interactions by first giving an introduction to the properties of interaction domains (Section 10.2) and of linear motifs (Section 10.3). Subsequent sections will deal with the role of linear motifs in disease (Section 10.4) and computational tools to identify sequence motifs (Section 10.5).

10.2 Protein Interaction Domains

Regulatory proteins frequently exhibit a modular arrangement of different domains that mediate molecular interactions or have enzymatic activity. Interaction domains can target proteins to a specific subcellular location, provide a module for recognition of protein post-translational modifications or chemical second messengers, initiate the formation of multiprotein signalling complexes, and control the conformation, activity and substrate specificity of enzymes.[10]

SH2 and SH3 domains were the first protein modules described that mediate interactions with short linear motifs. They were originally discovered by comparing members of the Src tyrosine kinase family and exist as conserved sequences of 50–100 amino acids.[11,12]

Subsequently, these domains were also found in other proteins and termed 'Src homology 2' (SH2) and 'Src homology 3' (SH3) domains. They can be seen as prototypical examples of interaction domains with the following characteristic properties:

- Interaction domains are relatively small (about 100 amino acids in length), which allows easy genomic shuffling and integration into new protein sequences.
- Interaction domains are of modular nature, with amino- and carboxy-termini in close spatial proximity. This allows easy integration into existing proteins without major reorganisation.
- Interaction domains can be found in many different kinds of proteins (Fig. 10.2), which underscore their importance for diverse molecular tasks.
- Interaction domains mediate interactions with other proteins by recognising short sequence motifs (3–10 amino acids). These interactions are of transient nature with high dissociation rates.

Figure 10.2 provides an overview of the diversity of different protein families in which SH2 and SH3 domains can be found.

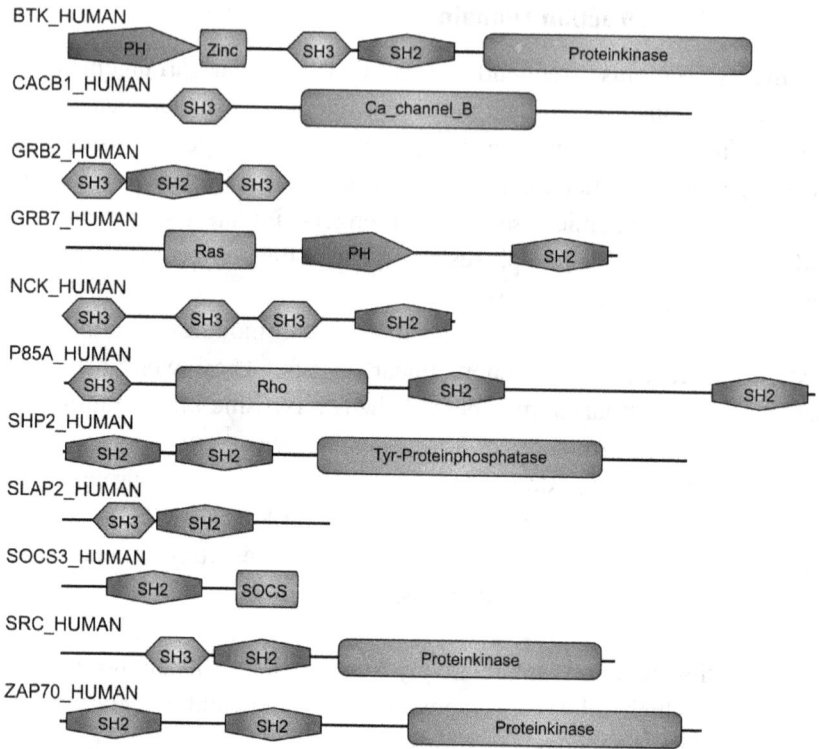

Fig. 10.2. Schematic drawing of different proteins that contain SH2 and/or SH3 domains, highlighting the versatility of these elements. Of special interest is the diverse nature of the different proteins: depicted are adaptor proteins (GRB2, NCK), enzymes such as tyrosine kinases (BTK, SRC, ZAP70) and protein phosphatases (SHP2), ion channels (CACB1), scaffolding proteins (GRB7) and regulatory proteins (P85A, SLAP2, SOCS3).

Several interaction domains are present in multiple or even hundreds of copies in the human proteome (Table 10.1) and these domains are used in many different contexts to regulate distinct tasks of the cell.[13,14,15] Due to the enormous number of individual domains, a comprehensive review of their function is far beyond the scope of this article. Therefore, only some key molecules of the cell, which exhibit SH3 and/or SH2 domains are discussed below.

Table 10.1. Overview of the most prominent types of human protein interaction domains, which recognise short linear peptide sequences. For each domain type, the number of human proteins containing the respective domain is given, as well as the total number of domain instances. Differences between both numbers result from the fact that numerous proteins contain more than one interaction domain of the same type. Numbers were obtained from Pfam,[16] version 23. For each domain, the patterns of interacting motifs available from the ELM database[3] are given. According to the ELM nomenclature, variable pattern positions are denoted by a '.' instead of an 'x'. Bold letters indicate amino acids that need to be phosphorylated to become functional.

Domain	Proteins	Instances	Interacting patterns
SH3	597	839	[RKY]..P..P P..P.[KR] ...[PV]..P
PDZ	328	563	.[ST].[VIL]$.[VYF].[VIL]$.[DE].[IVL]
SH2	268	297	**Y**.N. **Y**[IV].[VILP] **Y**[QDEVAIL][DENPYHI][IPVGAHS] **Y**..Q **Y**[VLTFIC].. G**Y**[KQ].F
WW	124	214	PP.Y PPLP .PPR. ...[**ST**]P.
FHA	90	90	..**T**..[ILV]. ..**T**..[DE].
14-3-3	22	22	R[SFYW].**S**.P R.[SYFWTQAD].[**ST**].[PLM] [RHK][STALV].[**ST**].[PESRDIF]
PTB	22	22	NP.**Y** NP.Y.
EVH	21	21	FP.[PAILSK]P PP..F
GYF	7	7	[QHR].{0,1}P[PL]PP[GS]H[RH]

10.2.1 SH3 Domains

Many different classes of proteins exhibit one or more SH3 domains, e.g. tyrosine kinases (Src, Btk, Zap-70), adaptors (Hsb-1, Grb2, Slap2), guanine nucleotide exchange factors (Arhg7, Vav, Vav2), motor molecules (Myosin) and ion channels (Cab1). Peptides bound by SH3 domains are proline-rich (Table 10.1). The ligand binding surface of SH3 domains is relatively flat and hydrophobic and consists of three shallow grooves defined by conserved aromatic residues.[17] The ligand adopts an extended left-handed helical conformation termed type-II polyproline helix. SH3 ligands can bind in two different orientations: Class I ligands have the consensus pattern [RK]-x-x-P-x-x-P, whereas Class II ligands bind in the opposite direction with their canonical pattern P-x-x-P-x-[RK]. For both classes of ligands, binding affinity is quite low with dissociation constants in the range of 1 µM–100 µM.[18] This finding is in accordance with the fact that most signalling events are highly dynamic; if such interactions were too strong, they would no longer allow a fine-tuned control of signalling events.

10.2.2 SH2 Domains

SH2 domains are present in tyrosine kinases (Src, Btk, Zap-70), tyrosine phosphatases (Shp1 and Shp2), transcription factors (STATs), adaptors (Grb2, Nck, Crk), negative regulators (SOCS, SH21A), scaffolding proteins (Shc, Grb7, SLP76) and guanine nucleotide exchange factors (Vav, Vav2). All SH2 domains recognise short linear motifs including a phosphorylated tyrosine residue (Table 10.1; Fig. 10.3b). The interaction with the phosphotyrosine alone conveys approximately half of the binding energy and thereby increases the binding affinity about 1,000–fold compared to a non-phosphorylated peptide.[19] Experiments have shown that the selectivity of SH2 binding is mainly moderated by the three amino acid residues immediately carboxy-terminal to the crucial tyrosine residue.[20] Generally, SH2 domains show dissociation constants of 500 nM–1 µM.[21] This affinity is higher than in the case of SH3-mediated interactions, but phosphorylation and dephosphorylation of the phosphotyrosine still allow a transient and fine-tuned control of the respective interactions.

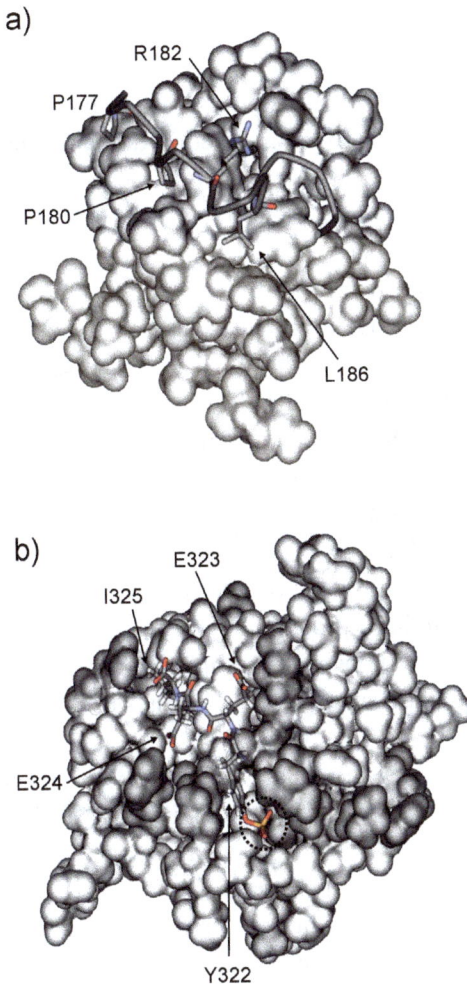

Fig. 10.3. Structural properties of motif-mediated interactions. (a) Structure of the Lyn SH3 domain (space-filled presentation) in complex with a linear sequence motif from the herpesviral Tip protein (PDB code: 1WA7).[22] Key residues of the P-x-x-P-x-R binding motif are shown in stick presentation and are labelled. L186 is highlighted because it forms additional contacts outside the classical recognition motif (see Section 10.4 for more details on this interaction). (b) Structure of the Lck SH2 domain (space-filled presentation) in complex with a linear sequence motif from middle-T-antigen of polyoma virus (PDB code 1LKK[23]). Residues of the interacting Y-E-E-I motif are labelled and the phospho-group of the tyrosine is marked by a circle. Basis residues of the SH2-domain are shaded in dark grey to highlight the charge complementarity between SH2-domain and ligand.

10.2.3 *Signalling Adaptors: Proteins Containing Multiple Interaction Domains*

The importance of signalling domains is corroborated by the existence of proteins which consist solely of interaction domains and motifs, i.e., do not show any enzymatic activity. Such proteins are called 'signalling adaptors', examples of which are Grb2, Nck, SLP76 and Crk (Fig. 10.2). Such adaptor proteins play an important and highly dynamic role in signalling, especially in transmitting signals from different types of receptors at the cell surface to intracellular networks.[24] The human Grb2 protein, for example, has a length of 217 amino acids and consists of two SH3 domains flanking a central SH2 domain. The latter domain binds preferentially to Y-x-N motifs on activated receptor tyrosine kinases (RTKs), while the SH3 domains interact with proteins such as Sos[25] and the Gab1/2 docking proteins.[26] Sos is a guanine nucleotide exchange factor for the Ras-GTPase that stimulates the MAP-Kinase pathway. The Gab proteins become subsequently phosphorylated and are targeted by SH2-domain containing proteins like phosphatidylinositol-3-kinase (PI3K) and the tyrosine phosphatase Shp2. By these means, Grb2 is able to couple RTKs to pathways such as the Ras-MAP-Kinase pathway and PI3K pathways.[27]

Signalling adaptors were also shown to be essential for T-cell signalling. For instance, the adaptor proteins SLP76, Nck and Grb2 are involved in signal transduction events in activated T cells: SLP76 promotes recruitment of phospholipase C1 (PLC1) to the membrane with subsequent Ca^{2+} increase, influences transcriptional activation via the MAP-kinase cascade, and together with Sos and Ras induces cytoskeletal re-arrangements in conjunction with Nck and Vav.[28]

In summary, the main task of signalling adaptors is to physically couple activated receptors to downstream targets. They represent flexible docking modules that allow a dynamic activation of signalling pathways. Although being relatively simple adaptors, they are able to mediate sophisticated cooperative functions, which are necessary for complex cellular behaviour.

10.3 Properties and Regulation of Motif-mediated Interactions

In motif-mediated interactions, globular domains in one protein recognise short stretches of approximately 3–10 residues in their binding partner. These regions often show a particular sequence pattern termed 'linear motif', which contains the key residues involved in function or binding. These key residues may be connected by variable residues (denoted as 'x'), which ensure the proper spacing of the interacting amino acids. This kind of interaction differs from domain–domain interactions, as no pre-formed binding interface exists. Examples for linear motifs include the classical P-x-x-P motif for binding to SH3 domains or the N-P-x-Y motif for the interaction with PTB domains (Table 10.1).

Another important aspect of motif-mediated protein interactions is the tendency of interaction motifs to be found in protein regions without regular three-dimensional structure, so called disordered regions.[29] Conversely, interaction motifs are usually not found inside globular regions of proteins.[30] The main properties of short linear motifs can be summarized as follows:

- Interaction motifs are relatively short (3–10 amino acid residues).
- Interaction motifs are preferentially found in disordered protein regions.
- In addition to mediating protein–protein interactions, linear motifs also represent the target sites of post-translational modifications and they mediate cell compartment targeting.
- Interaction motifs can evolve convergently in different organisms as they frequently arise and disappear by point mutations, thereby conferring extreme adaptability to the interactome.[29]

Short linear motifs can have distinct regulatory properties, e.g. to increase specificity or to regulate certain pathways efficiently; some of the key features are described below.

10.3.1 *Inducible Interactions*

Many linear motifs need certain key residues to be post-translationally modified in order to specifically bind to their cognate domain. The most prominent example is the binding of SH2 domains to phosphorylated tyrosine motifs: a non-phosphorylated tyrosine residue shows a binding affinity several orders of magnitude lower than a phosphorylated one. Phosphorylation at serine or threonine is also required for linear motifs to interact with forkhead-associated (FHA) or 14-3-3 domains. In case of WW and phosphotyrosine-binding (PTB) domains at least a subset of the respective domains requires phosphorylation of the interaction motif for tight complex formation (Table 10.1).

10.3.2 *Cooperative Effects*

In most cases a single motif is sufficient to mediate a certain interaction; sometimes, however, a greater affinity or specificity is required, which might not be provided by a single motif. For instance, T-cell receptors contain multiple characteristic motifs in their cytoplasmic tail, termed Immunoreceptor Tyrosine-based Activation Motifs (ITAMs). Upon ligand binding and subsequent clustering of T-cell receptors, tyrosine kinases of the Src-family are recruited to phosphorylate tyrosine residues within these motifs. The spacing of the double motif provides docking sites for the tyrosine kinase Zap-70 containing two SH2 domains next to each other, to bind specifically to these tandem motifs. One critical downstream target of Zap-70 is PLC1, which hydrolyses phosphatidylinositol-4,5-bisphosphate, generating second messengers responsible for mediating intracellular Ca^{2+} release and protein kinase C activation, respectively. These pathways are important for the activation of transcription factors, such as NF-AT and AP-1, which are required for T-cell proliferation.[31]

Another example of a cooperative effect of interaction motifs is the interplay between the adaptor Nck and Nephrin. In this case, Nephrin has several Y-D-x-V motifs, which can bind to the SH2-domain of Nck adaptors. Nck in turn contains several SH3 domains (Fig. 10.2), which can bind to proline-rich regions in N-Wasp and thereby promote actin reorganisation.[32] Although a single SH2-motif is sufficient for binding to Nck and a single Nck SH3 domain is sufficient for actin polymerization, it is the cooperative effect of multiple copies of the same interaction motif that leads to high local concentration of effector molecules, ultimately resulting in potent actin polymerization induced by Nephrin.[33]

10.3.3 *Mutually Exclusive Interactions*

One single amino acid residue can be subject to different post-translational modifications. However, in most cases, only one modification can be applied at a time, e.g. an acetylated lysine cannot be methylated. An example, where several different modifications compete for the same residues, are histones: when Lys9 of Histone 3 is trimethylated, it binds to Chromodomains of heterochromatin protein 1, implicated in gene silencing.[34] Conversely, when the residues Lys9 and Lys14 are acetylated, Histone 3 binds to the Bromodomains of histone acetyltransferases such as TAFII250, which in turn lead to an increased acetylation of further lysine residues within Histone 3,[35] marking transcriptional activation. These interactions are mutually exclusive, mediated by mutually exclusive modifications.[36]

10.3.4 *Intra- Versus Intermolecular Interactions*

A linear motif can be recognised by an interaction domain within the same protein, thereby changing the conformation of the whole protein. This is exemplified in kinases of the Src type, where an SH2 domain binds to a phosphorylated carboxy-terminal tyrosine residue in an intramolecular fashion. This 'closed' conformation is further sustained by an additional interaction between an SH3 domain and an intramolecular linker, which blocks the SH3 ligand-binding site. This prevents the enzymatic domain of the protein from engaging its substrates and thereby keeps the enzyme in an inactive state. Activation of Src kinases can be achieved by dephosphorylation of the tyrosine in the C-terminal region and/or the presence of SH2/SH3 competing ligands that bind with a similar or stronger binding affinity.

10.4 The Role of Linear Motifs in Disease

The enhanced knowledge about motif-mediated interactions that was gained during recent years has also allowed assigning the molecular origin of several diseases to mutations in linear motifs. In addition, the function of numerous viral effector proteins could be attributed to the presence of linear motifs which interfere with cellular signalling pathways. Known examples of diseases resulting from mutated cellular motifs or form viral infections will be discussed below in Sections 10.4.1 and 10.4.2.

10.4.1 *Diseases Caused by Mutated Motifs*

There are several diseases known which are caused by one or more mutations in linear motifs mediating important interactions. Examples include Noonan Syndrome, Liddle's Syndrome and Usher's Syndrome: Noonan Syndrome is caused by mutations in Raf-1 which impede the motif-mediated interaction with 14-3-3 proteins thereby deregulating Raf-1 kinase activity.[37] Usher's Syndrome is the most frequent cause of hereditary deaf-blindness in humans,[38] affecting one child in 25,000. This disease can be caused by mutations in either PDZ domains in Harmonin or the corresponding PDZ interaction motifs in SANS protein.[39,40] Another example of a disease involving PDZ domains is 'Familial Hypomagnesemia with Hypercalciuria and Nephrocalcinosis' (FHWHN), an autosomal recessive wasting disorder of renal Mg^{2+} and Ca^{2+} that leads to progressive kidney failure. Here, motifs mediating interaction to PDZ domains are mutated in Claudin16, abolishing important interactions to the scaffolding protein ZO-1 resulting in lysosomal mislocalization of the protein.[41,42]

Liddle's Syndrome has been described as a consequence of autosomal dominant activating mutations in the WW interaction motif in the β- and γ-subunits of the epithelial sodium channel ENaC.[43] These mutations abrogate the binding to the ubiquitin ligase NEDD4-2, thereby inhibiting channel degradation and prolonging the half-life of ENaC, ultimately resulting in increased Na^+ resorption, plasma volume extension and hypertension.[44,45]

Up to present, there are just a few examples of diseases known, in which a mutation was pinpointed to amino acids residing within a linear motif. However, only little attention has been paid to small motifs as critical mediators of protein interactions in the past. Therefore, it is likely that more interaction motifs involved in disease will be detected after a comprehensive motif annotation is available for the human proteome.

10.4.2 *Linear Motifs Involved in Viral Infection*

Viruses are known to be limited in genome size[46,47] and often use overlapping genes to achieve maximum compression of information.[48] The independent evolution of overlapping reading frames for instance in the Paramyxoviridae and Rhabdoviridae 'P genes', as well as in other RNA viruses, can be seen as a response to selective pressure in order to maximise genomic information content while maintaining a small genome.[49] In accordance with these properties is the observation, that viruses frequently use linear motifs to interact with their host, rather than incorporating complete modular domains or whole proteins into their genome.[9]

In principle almost every type of protein domain can be targeted by viral sequence motifs, but the majority of the interactions yet known are formed to SH2, SH3, PDZ, and 14-3-3 domains of host proteins.[9] These domains all constitute large protein families of 22–839 individual members (Table 10.1), which raise the question about the specificity of the respective interactions. Inspection of viral regulatory proteins reveals two interesting principles, which might be particularly important to ensure viral infection:

• The presence of multiple linear interaction motifs.
• The presence of high-affinity binding motifs.

10.4.2.1 *Viral Proteins Containing Multiple Interaction Motifs*

Herpesvirus saimiri codes for a tyrosine kinase interacting protein (Tip) that interacts with the SH3 domain and the kinase domain of the T-cell-specific tyrosine kinase Lck via two separate sequence motifs.[50,51] The resulting activation of Lck by Tip is considered as a key event in the transformation of human T-lymphocytes during herpesviral infection. In addition, Tip contains tyrosine residues, which become phosphorylated and can interact with the SH2 domains of STAT factors and Src-family kinases.[52] These multiple motifs thus allow both tight binding and linking of different signalling pathways.

Another example of a viral protein containing multiple linear interaction motifs is the middle-T-antigen of polyoma virus:[53] This membrane-bound protein is able to regulate several signalling pathways of its murine host. For instance, upon phosphorylation of tyrosine residue Y250, it binds to SHCA, which subsequently becomes phosphorylated and activates the MAP-Kinase pathway via Grb2 and Sos. Tyrosine residue Y315 mediates binding to SH2 domains of PI3K, which in turn influences transformation, cytoskeletal re-arrangements and survival signals. Other signalling events stimulated by middle-antigen include binding to PLC1 via Y322, and to 14-3-3 proteins via a phosphoserine motif at S257.

10.4.2.2 *Principles to Enhance Binding Affinity and Specificity*

The Y-E-E-I motif present at position Y322 in the middle-T-antigen of the polyoma virus is also interesting for structural reasons: this motif was shown to bind SH2 domains with a very high affinity of approximately 100 nM,[52] which is significantly higher than most physiological SH2-ligand interactions. This tight binding is mainly due to optimised electrostatic interactions of the glutamates with basic residues of the SH2 domain (Fig. 10.3b) and ensures that the viral protein can efficiently compete with cellular ligands for binding.

This principle of high-affinity binding of viral interaction motifs was also observed for two herpesviral proteins (Tip, Tio) that bind to the SH2 domain of the protein tyrosine kinase Lck.[52] The viral proteins showed a higher binding affinity than the intramolecular interaction formed by the regulatory C-terminus of Lck, suggesting that the viral proteins are capable of disrupting the intramolecular inhibition of Lck, thereby constitutively activating the kinase.[52]

As outlined above in Section 10.4.2.1, Tip contains an additional linear motif that binds to the SH3 domain of Src-family kinases. The structure determination of the Tip-LynSH3 complex[22] revealed that Tip forms additional contacts outside its classical proline-rich recognition motif and, in particular, a strictly conserved leucine (L186) of the C-terminally adjacent sequence stretch packs into a hydrophobic pocket on the Lyn surface (Fig. 10.3a). These additional interactions were shown to contribute to a significantly enhanced binding affinity for Lyn compared to other Src-family kinases.[54,22]

The regulatory Nef-protein from HIV, which is critical for viral pathogenesis, uses a slightly different principle to achieve enhanced binding specificity to SH3 domains. Again, the canonical interaction to SH3 domains is formed by a P-x-x-P motif present in the extended N-terminus of the Nef-protein. A key determinant is for the discrimination between different SH3-domains, however, an isoleucine residue present in Hck, but not in Fyn. This isoleucine forms hydrophobic contacts with two α-helices in Nef that cannot be formed by the respective arginine present in Fyn. The importance of this interaction for binding specificity was confirmed by site-directed mutagenesis, showing that a R96I mutant of Fyn had a comparable affinity for Nef compared to Hck.[55–57]

Another high-affinity SH3-interaction has been recently reported by Shelton and Harris for the Hepatitis C virus 'non-structural 5A protein' (NS5A): this viral protein contains a highly conserved P-x-x-P-x-R motif that is able to interact with the SH3 domain of Fyn with high affinity.[58] By this means, the Hepatitis C virus NS5A protein effectively competes with cellular ligands for binding to SH3 domains in the infected cell and influences important signal transduction interactions.

10.4.2.3 *Conclusions*

The examples above demonstrate that viral proteins frequently exploit variations of cellular interaction motifs to interfere with signalling pathways of the host. As shown above, these viral motifs frequently exhibit a higher affinity than cellular motifs thus ensuring an efficient reprogramming of signal transduction in the host. It is also important to note that the respective viral proteins do not share any overall sequence homology to a protein of the host cell, strongly suggesting that the viral interaction motif has resulted from convergent evolution. The concept of reprogramming cellular processes by the interaction via linear motifs appears highly attractive for microbial pathogens for the following reasons: the short length of 3–10 amino acids allows easy integration even in small regulatory proteins, thereby meeting the requirement for a small overall genome size in numerous pathogens. In addition, multiple motifs may evolve within the same effector, which ensures either enhanced binding affinity to one protein or to link different cellular pathways by interacting with multiple proteins. Since few amino acids within a linear motif are sufficient for the interaction, convergent evolution can occur much faster compared to a classical domain–domain interaction between two proteins, because the latter type of interaction requires replacement of numerous amino acids within the interface, while maintaining the overall three-dimensional fold. Several linear motifs rely on phosphorylation to become functional, which allows triggering the effects of viral action in a time-dependent fashion (e.g. by activation through a cellular kinase) instead of exerting constitutive effects.

10.5 Computational Approaches Addressing Motif-mediated Interactions

With respect to the computational approaches in the field of linear motifs, there are at least two major challenges, both of which are briefly outlined here using the P-x-x-P motif as an example and are described in more detail in the following paragraphs.

Identification of Linear Interaction Motifs: This goal requires dissecting those instances of a particular pattern that occur by chance from the functional instances that are actually used for interaction. This type of prediction aims to answer the question whether a P-x-x-P pattern is actually used for SH3 binding or whether it is non-functional, e.g. because it is buried in the interior of the protein.

Prediction of Binding Specificity: After a P-x-x-P motif is identified, the prediction of distinct cellular SH3 domain(s) which bind to the P-x-x-P motif is highly desirable. In this context, computational approaches for the characterization of the binding specificity of cellular adapter domains represent a valuable complement of time-intensive experiments.

10.5.1 *Identification of Linear Motifs*

The widespread occurrence of motif-mediated interactions in a large variety of signalling processes make a computational prediction of the respective interactions highly desirable. Today, there are several web-based tools available for the detection of linear motifs in query protein sequences. Most of these tools like ELM[3] or Minimotif-Miner[5] are directly linked to a motif database containing patterns describing different types of linear motifs. In addition, tools like Dilimot[59] and SlimDisc[60] were developed for the discovery of novel types of interaction motifs. A common key problem in all methods for motif identification is the short length of linear motifs resulting in a large number of instances, of which only a small fraction is functional. This large number of non-functional ('false positive') instances makes it difficult to decide which motifs to select for further experimental characterization. Different strategies have been suggested to reduce this problem.

Context filters mask those parts of the sequence space in which little or no linear motifs are expected to occur. Masked regions include globular domains, in which no motifs are expected since they would be buried and therefore not accessible for interaction. Other filters take into account that some motifs are only functional within particular cellular compartments.[3,61] Scoring schemes, which measure the conservation of linear motifs among homologs, proved to be particularly powerful in the reduction of false positive hits.[5,62,61,63]

10.5.2 *Methods for Determining Binding Specificity*

The pattern notation of sequence motifs allows an initial assignment of the type of interacting domain (e.g. SH3), but is not well suited to describe the binding specificity of individual protein interaction domains (e.g. Lck-SH3). The specificity of individual domains can preferably be represented by position specific scoring matrices (PSSMs), which describe the relative importance of the individual sequence positions of a motif, for binding to a particular protein domain.

The respective PSSMs can be generated based on experimentally binding data, which gives information about the ligand binding preference of individual protein interaction domains. In particular, the use of peptide libraries has allowed addressing the ligand binding specificity of numerous domains.[64,65,66,67] This information has been used to create PSSMs available for numerous individual members of the major types of interaction domains. Such protein-specific PSSMs are available in Scansite,[68] iSPOT,[69] PDZbase[70] and SMALI.[71]

Despite considerable methodical advances, experimental investigation of a significant number of domains has failed, due to low solubility or the formation of inclusion bodies during recombinant expression.[66] In addition, experimental approaches are time-consuming and do not necessarily provide an understanding of the structural properties that define the specificity of each domain. Numerous computational approaches that predict ligand binding specificity have therefore been developed to complement experimental studies.[72] Two recent promising approaches use general purpose energy force fields to predict the binding specificity of peptide-binding domains.[73,74] These methods have the great advantage that no domain-specific information is required aside from a model of the complex for the domain under investigation and it is possible to use these force fields to do *in silico* mutagenesis of the ligand. From this computational analysis, a PSSM is created containing the information on the preferred residues at each position in the ligand. These scoring matrices can help to identify physiological interaction partners by scoring a linear motif with different PSSMs (e.g. those derived for different members of the SH3-domain family).

10.5.3 *WWW-resources for the Investigation of Linear Motifs*

As described above, computational methods provide valuable tools to supplement experimental binding data and to gain insight into the structural determinants governing protein affinity. Some recent tools suitable for the respective analyses are summarised below in Table 10.2.

Table 10.2. Overview of some computational tools available for the analysis of motif-mediated interactions. The tools are divided according to their major purpose: motif databases contain the pattern describing known types of motifs and can be used to search for linear motifs in a query sequence. Domain databases give information about the most prominent types of interaction domains that bind to linear motifs. Motif detection tools can be used to discover novel types of motifs in a group of proteins sharing common functional properties. The last group of tools is intended to identify a subset of interaction domains that specifically bind to a given linear motif.

NAME	Refs	WWW-link
Motif databases:		
ELM	3	http://www.elm.eu.org/
MMM	5, 75	http://mnm.engr.uconn.edu
PROSITE	76, 77	http://www.expasy.org/prosite/
Domain databases:		
CDD	78, 79	http://www.ncbi.nlm.nih.gov/Structure/cdd/cdd.shtml
InterPro	80, 81	http://www.ebi.ac.uk/interpro/
Pfam	82–85	http://pfam.sanger.ac.uk/
ProDom	86	http://prodom.prabi.fr/prodom/current/html/home.php
SMART	87, 88	http://smart.embl-heidelberg.de/
Motif detection tools:		
Dilimot	59	http://dilimot.embl.de
GLAM2	89	http://bioinformatics.org.au/glam2
MEME	90	http://meme.sdsc.edu
QuasiMotiFinder	91	http://quasimotifinder.tau.ac.il/
SlimDisc	60, 92	http://bioinformatics.ucd.ie/shields/software/slimdisc/
SlimFinder	62	http://bioinformatics.ucd.ie/shields/software/slimfinder/
Teiresias	93	http://cbcsrv.watson.ibm.com/Tspd.html
Motif binding specificity:		
iSPOT	69	http://cbm.bio.uniroma2.it/ispot/
PDZbase	70	http://icb.med.cornell.edu/services/pdz/start
Scansite	68	http://scansite.mit.edu/
SMALI	94	http://lilab.uwo.ca/SMALI.htm
Fold-X	95, 96	http://foldx.crg.es/

Acknowledgements

The authors would like to thank Heike Meiselbach for help with the figures and the Deutsche Forschungsgemeinschaft (SFB 796) for financial support.

References

1. Dunker A.K., *et al.* (2000). *Genome Inform Ser Workshop Genome Inform* 11: 161–171.
2. Diella F., *et al.* (2008). *Frontiers in Bioscience* 13: 6,580–6,603.
3. Puntervoll P., *et al.* (2003). *Nucleic Acids Res* 31: 3,625–3,630.
4. Hulo N., *et al.* (2006). *Nucleic Acids Res* 34: D227–D230.
5. Balla S., *et al.* (2006). *Nat Methods* 3: 175–177.
6. Neduva V., Russell R.B. (2006). *Curr Opin Biotechnol* 17: 465–471.
7. McEntyre J.R., Gibson T.J. (2004). *Trends Biochem Sci* 29: 627–633.
8. Palacios E.H., Weiss A. (2004). *Oncogene* 23: 7,990–8,000.
9. Kadaveru K., Vyas J., Schiller M.R. (2008). *Front Biosci* 13: 6,455–6,471.
10. Pawson T., Nash P. (2003). *Science* 300: 445–452.
11. Sadowski I., Stone J.C., Pawson T. (1986). *Mol Cell Biol* 6: 4,396–4,408.
12. Koch C.A., *et al.* (1991). *Science* 252: 668–674.
13. Pawson T., Nash P. (2000). *Genes Dev* 14: 1,027–1,047.
14. Zarrinpar A., Bhattacharyya R.P., Lim W.A. (2003). *Sci STKE* 2003: RE8.
15. Pawson T., Warner N. (2007). *Oncogene* 26: 1,268–1,275.
16. Sonnhammer E.L., Eddy S.R., Durbin R. (1997). *Proteins* 28: 405–420.
17. Mayer B.J. (2001). *J Cell Sci* 114: 1,253–1,263.
18. Musacchio A. (2002). *Adv Protein Chem* 61: 211–268.
19. Domchek S.M., *et al.* (1992). *Biochemistry* 31: 9,865–9,870.
20. Songyang Z., *et al.* (1994). *Curr Biol* 4: 973–982.
21. Ladbury J.E., *et al.* (1995). *Proc Natl Acad Sci USA* 92: 3,199–3,203.
22. Bauer F., *et al.* (2005). *Protein Sci* 14: 2,487–2,498.
23. Tong L., *et al.* (1996). *J Mol Biol* 256: 601–610.
24. Pawson T. (2007). *Curr Opin Cell Biol* 19: 112–116.
25. Rozakis-Adcock M., *et al.* (1993). *Nature* 363: 83–85.
26. Gu H., Neel B.G. (2003). *Trends Cell Biol* 13: 122–130.
27. Chazaud C., *et al.* (2006). *Dev Cell* 10: 615–624.
28. Norian L.A., Koretzky G.A. (2000). *Semin Immunol* 12: 43–54.
29. Neduva V., *et al.* (2005). *PLoS Biol* 3: e405.
30. Fuxreiter M., Tompa P., Simon I. (2007). *Bioinformatics* 23: 950–956.
31. Jordan M.S., Singer A.L., Koretzky G.A. (2003). *Nat Immunol* 4: 110–116.
32. Rivera G.M., *et al.* (2004). *Curr Biol* 14: 11–22.

33. Blasutig I.M., *et al.* (2008). *Mol Cell Biol* 28: 2,035–2,046.
34. Bannister A.J., *et al.* (2001). *Nature* 410: 120–124.
35. Jenuwein T., AllisC.D. (2001). *Science* 293: 1,074–1,080.
36. Seet B.T., *et al.* (2006). *Nat Rev Mol Cell Biol* 7: 473–483.
37. Pandit B., *et al.* (2007). *Nat Genet* 39: 1,007–1,012.
38. Eudy J.D., Sumegi J. (1999). *Cell Mol Life Sci* 56: 258–267.
39. Weil D. *et al.* (2003). *Hum Mol Genet* 12: 463–471.
40. Kalay E., *et al.* (2005). *J Mol Med* 83: 1,025–1,032.
41. Müller D., *et al.* (2003). *Am J Hum Genet* 73: 1,293–1,301.
42. Müller D., *et al.* (2006). *Hum Mol Genet* 15: 1,049–1,058.
43. Warnock D.G. (1998). *Kidney Int* 53: 18–24.
44. Furuhashi M., *et al.* (2005). *J Clin Endocrinol Metab*, 90 (2005), 340–344.
45. Wang Y., *et al.* (2007). *Clin Endocrinol (Oxf)* 67: 801–804.
46. Normark S., *et al.* (1983). *Annu Rev Genet* 17: 499–525.
47. Krakauer D.C. (2000). *Evolution Int J Org Evolution* 54: 731–739.
48. Pavesi A. (2007). *Gene* 402: 28–34.
49. Jordan I.K., Sutter B.A., McClure M.A. (2000). *Mol Biol Evol* 17: 75–86.
50. Jung J.U., *et al.* (1995). *J Biol Chem* 270: 20,660–20,667.
51. Hartley D.A., *et al.* (2000). *Virology* 276: 339–348.
52. Bauer F., *et al.* (2004). *Biochemistry* 43: 14,932–14,939.
53. Dilworth S.M. (2002). *Nat Rev Cancer* 2: 951–956.
54. Schweimer K., *et al.* (2002). *Biochemistry* 41: 5,120–5,130.
55. Lee C.H., *et al.* (1995). *Embo J* 14: 5,006–5,015.
56. Lee C.H., *et al.* (1996). *Cell* 85: 931–942.
57. Arold S., *et al.* (1997). *Structure* 5: 1,361–1,372.
58. Shelton H., Harris M. (2008). *Virol J* 5: 24.
59. Neduva V., Russell R.B. (2006). *Nucleic Acids Res* 34: W350–355.
60. Davey N.E., Shields D.C., Edwards R.J. (2006). *Nucleic Acids Res* 34: 3,546–3,554.
61. Dinkel H., Sticht H. (2007). *Bioinformatics* 23: 3,297–3,303.
62. Edwards R.J., Davey N.E., Shields D.C. (2007). *PLoS ONE* 2: e967.
63. Chica C., *et al.* (2008). *BMC Bioinformatics* 9: 229.
64. Tong A.H.Y., *et al.* (2002). *Science* 295: 321–324.
65. Landgraf C., *et al.* (2004). *PLoS Biol* 2: E14.
66. Huang H., *et al.* (2008). *Mol Cell Proteomics* 7: 768–784.
67. Tonikian R., *et al.* (2008). *PLoS Biol* 6: e239.
68. Obenauer J.C., Cantley L.C., Yaffe M.B. (2003). *Nucleic Acids Res* 31: 3,635–3,641.
69. Brannetti B., Helmer-Citterich M. (2003). *Nucleic Acids Res* 31: 3,709–3,711.
70. Beuming T., *et al.* (2005). *Bioinformatics* 21: 827–828.
71. Li L. *et al.* (2008). *Nucleic Acids Res* 36: 3,263–3,273.
72. Kiel C., Beltrao P., Serrano L. (2008). *Annu Rev Biochem* 77: 415–441.
73. Hou T., *et al.* (2006). *PLoS Comput Biol* 2: e1.

74. Kiel C., Serrano L. (2007). *Bioinformatics* 23: 2,226–2,230.
75. Rajasekaran S., *et al.* (2009). *Nucleic Acids Res* 37: D185–D190.
76. Bairoch A. (1991). *Nucleic Acids Res* 19 Suppl: 2,241–2,245.
77. Boeckmann B., *et al.* (2003). *Nucleic Acids Res* 31: 365–370.
78. Marchler-Bauer A., *et al.* (2002). *Nucleic Acids Res* 30: 281–283.
79. Marchler-Bauer A., *et al.* (2007). *Nucleic Acids Res* 35: D237–D240.
80. Apweiler R., *et al.* (2000). *Bioinformatics* 16: 1,145–1,150.
81. Mulder N.J., *et al.* (2007). *Nucleic Acids Res* 35: D224–D228.
82. Bateman A., *et al.* (1999). *Nucleic Acids Res* 27: 260–262.
83. Bateman A., *et al.* (2000). *Nucleic Acids Res* 28: 263–266.
84. Bateman A., *et al.* (2004). *Nucleic Acids Res* 32: D138–141.
85. Sammut S.J., Finn R.D., Bateman A. (2008). *Brief Bioinform* 9: 210–219.
86. Corpet F., *et al.* (2000). *Nucleic Acids Res* 28: 267–269.
87. Schultz J., *et al.* (1998). *Proc Natl Acad Sci USA* 95: 5,857–5,864.
88. Letunic I., *et al.* (2006). *Nucleic Acids Res* 34: D257–260.
89. Frith M.C., *et al.* (2008). *PLoS Comput Biol* 4: e1000071.
90. Bailey T.L., Elkan C. (1995). *Proc Int Conf Intell Syst Mol Biol* 3: 21–29.
91. Gutman R., *et al.* (2005). *Nucleic Acids Res* 33: W255–W261.
92. Davey N.E., Edwards R.J., Shields D.C. (2007). *Nucleic Acids Res* 35: W455–W459.
93. Rigoutsos I., Floratos A. (1998). *Bioinformatics* 14: 55–67.
94. Li L., *et al.* (2008). *Nucleic Acids Res* 36: 3,263–3,273.
95. Guerois R., Nielsen J.E., Serrano L. (2002). *J Mol Biol* 320: 369–387.
96. Schymkowitz J., *et al.* (2005). *Nucleic Acids Res* 33: W382–388.

CHAPTER 11

Prediction and Calculation of Protein–Protein Binding Affinities and Mutation Effects

Sébastien Fiorucci, Serge Antonczak, Jérôme Golebiowski

Molecular Modelling Team, UFR Sciences,
Centre National de la Recherche Scientifique,
Université de Nice-Sophia Antipolis,
UMR6001 LCMBA, 28 Avenue Joseph Vallot,
06108 Nice Cedex2, France
Email: Sebastien.Fiorucci@unice.fr

This chapter reviews the most popular theoretical methods to compute changes in protein–protein binding affinities and relative free energies due to substitution of amino acid residues. It includes techniques for computing free energy changes associated with alchemical mutations, like free energy perturbation or thermodynamic integration, as well as more approximate methods such as the linear interaction energy method and approaches to combine molecular mechanics calculations and continuum descriptions of the surrounding solvents and ions. The applicability of the methods for calculating protein–protein interactions is also discussed.

11.1 Introduction

Molecular sciences have reached a point where manipulation either experimentally or theoretically of simple molecular systems is often no more a severe limitation to the exploration of their functionality. However, for more complex systems, several issues still need to be addressed in order to ensure that theoretical results are in agreement with experimental data. This is particularly true in the framework of molecular modelling, where the size and complexity of studied structures are steadily increasing. The remarkable efforts and progress during the recent years have allowed us to gain detailed and relevant energetic descriptions associated with structural changes or interactions of multiple structures. Still, a lot needs to be done but many simulation approaches have matured to be applicable to a variety of biomolecular systems. In this chapter we will summarise the most common protocols that allow us to estimate protein–protein affinities and mutation effects.

Protein–protein association plays a crucial role in many biological processes including signal transduction, cell growth regulation, metabolism and adhesion, immune response and others.[1] Understanding how these macromolecular complexes are formed and what determines their specificity is not only fundamental for appreciating the underlying biological processes but also helpful in developing new therapeutic strategies. With the advent of the genomic and post-genomic era and the steady increase of computer capacity and speed, theoretical studies of highly complex systems become accessible. In combination with alanine scanning, single and multiple mutant cycles or saturation mutagenesis experiments, theoretical approaches allow the screening of a wide panel of amino acid sequences and provide energetic and structural information on the studied systems. In such context, computational protein design strategies have even been developed to engineer synthetic protein–protein interfaces.[2-5]

The tendency of molecular systems to react or to associate is represented by a thermodynamic quantity, the change in free energy. Predictions of ligand–receptor binding affinity, as well as mutation effects, remain a challenge for computational approaches, all the more so since two major difficulties hamper an accurate prediction: (a) despite spectacular progresses in both X-ray crystallography experiments and Nuclear Magnetic Resonance spectroscopy, it still remains unreasonable to hope for the experimental determination of the majority of protein structures in the near future. It is therefore necessary to build homology models[6-8] (if possible) for unknown protein structures; (b) Since many entries in the Protein Data Bank (PDB) do not describe macromolecular complexes but isolated proteins, theoretical approaches like docking methodologies[9,10] are needed to propose a binding mode for a protein–protein interaction. In cases where the structure of a protein–protein complex is known and the mutation of a single amino acid does not induce large structural changes, computational predictions of binding free energy becomes feasible.

The accurate estimation of thermodynamic quantities can in principle be obtained by so-called 'first principle' approaches. However, the corresponding calculations are often time-consuming and not straightforward to set up. In parallel, the use of more approximate methods, based on empirical rules, becomes a possible alternative in case of high-throughput *in silico* screening.

In this chapter, we present the currently most accurate force field based methods such as Thermodynamic Integration (TI) and Free Energy Perturbation (FEP) to describe the physical interactions of the system. Then more approximate methods like Linear Interaction Energy (LIE) and Molecular Mechanics/Poisson Boltzmann Surface Area (MM/PBSA) approaches will be described. Even simpler models based on empirical rules are also reviewed to complete the panel of computational approaches.

11.2 About Protein–Protein Interactions

The understanding of protein–protein interface organisation and composition contributes to identify the forces guiding the association of such macromolecular entities. The dissection of protein–protein binding sites has recently been the aim of many investigations in terms of geometry[11–14] and chemical nature of the interface (see Chapters 1 and 2 of this volume).[15–22] Considering different types of molecular assemblies such as heterodimers (protease-inhibitor, enzyme-inhibitor, antibody-antigen, etc.), homodimers or others, the distribution of amino acids at the interface differs from the rest of the surface exposed residues.[14,18] It is reflected by their interface propensities (defined as the ratio of the abundance of a given amino acid at the interface over its overall abundance on the surface) which are also quite different between the core and the rim of the interface.[16] It is commonly accepted that protein–protein interfaces are mainly composed of a buried hydrophobic core surrounded by a more hydrophilic ring partly exposed to the solvent.[12,23] However, depending on the type of interface, even the inner part of the interface can contain some polar or charged residues. Interacting via hydrogen bonds or salt-bridges, these residues generally act as strong anchor points and maintain the structural integrity of the complex.[22,24] Although electrostatic complementarity[25] is strongly involved in protein–protein recognition, non-polar interactions play a major role in the binding affinity[21,24] since the interface is to a large extent densely packed.[15]

However, structural analysis alone cannot predict whether all of these interface contacts are important for binding. Alanine scanning experiments on human growth hormone and the extracellular domain of its first bound receptor showed for the first time that few specific residues contribute dominantly to the binding free energy.[26] Some authors have defined a 'hotspot' as a residue contributing a significant part of the binding free energy ($\Delta\Delta G > 2$ kcal·mol^{-1}) as measured by alanine substitution. Bogan and Thorn[15] also showed that hotspots are located within densely packed areas, i.e. at the inner part of the interface. Substitutions of a hotspot residue by an alternative residue may create holes or results in steric hindrance preventing a perfect fit and, thus explaining the critical loss of affinity. Clearly, the capacity to give a rationale to such mutations or to protein–protein affinities in general requires a realistic description of the associated changes of the free energy.

11.3 The Free Energy of Binding

The free energy is a thermodynamic function of state that encodes information about the equilibrium state of a system. When a system (in which the temperature, the number of particles and the volume are constant) is at equilibrium, the free energy (here the Helmholtz free energy, F) is at a minimum. Depending on the thermodynamic conditions, one can speak of either the Helmholtz free energy (F) if the number of particles (N), the volume (V) and the temperature (T) are kept constant or of the Gibbs free energy (G) if N, T and P (pressure) are constant, respectively. See Ref. 27 for detailed explanations. The accurate calculation of the free energy of a molecular system is, however, difficult to perform. State-of-the-art approaches are rooted in a statistical thermodynamic treatment of the system. The free energy function (here called A) is directly connected to the partition function Q through the simple relation:

$$A = -kT \ln Q \qquad (11.1)$$

Here, k corresponds to the Boltzmann constant and T to the temperature of the system. The partition function Q fills the gap between the macroscopic properties of a system and its microscopic representations. It can be simply described as a sum of Boltzmann factors corresponding to the partition of the particles constituting the system throughout accessible states. In a simple system of well defined localised and indistinguishable particles partitioned amongst quantified energy levels (e_i), the partition function reads:

$$Q = \frac{q^N}{N!} ; q = \sum_i p_i e^{\frac{-e_i}{kT}} \qquad (11.2)$$

In a more complex system, with interacting particles, the concept is identical but requires the calculation of the energy of the system through its Hamiltonian (H(r,p)), which is a function of both the positions (r) and the momenta (p) of the particles (phase space). H(r,p) is a continuous function and the summation becomes an integral.

$$Q = \frac{1}{N!} \frac{1}{h^3 N} \int \ldots \int e^{\frac{-H(r,p)}{kT}} \, dr dp \qquad (11.3)$$

In principle, this equation can give access to the partition function, provided that the integral can be computed. This is, however, out of reach for typical biomolecular systems, where the phase space (the space of possible positions and momenta) is too large to be properly sampled with any kind of simulation to directly compute the partition function.

Fortunately, in the framework of molecular systems, one is often more interested in the difference of the free energies between two states than in the absolute free energy of a state. We indeed seek to estimate the free energy change throughout a given transformation that can be a chemical reaction, a protein folding process or the association of two protein molecules. In this case the free energy difference relies on a ratio of partition functions, which is in principle easier to estimate.

Figure 11.1 illustrates the typical thermodynamic cycle used to compute protein–protein binding affinity and the free energy change resulting from a mutation. The binding free energy ΔG_1 simply results from the association process between the two protagonists.

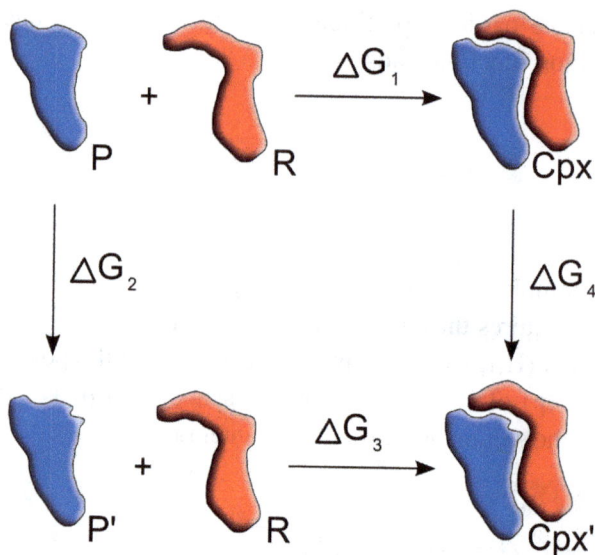

Fig. 11.1. Thermodynamic cycle used to estimate a protein–protein binding free energy and a free energy difference due to a mutation on one of the protein (P to P').

To compare the receptor (R) affinity between a wild-type protein and one of its mutant (called P') one can compute the free energy resulting from the mutation (ΔG_2), the free energy of binding of the mutant with the receptor (ΔG_3) and finally the free energy difference between the wild-type complex and the complex made up of the mutant and the receptor (ΔG_4).

The thermodynamic cycle indicates that $\Delta G_1 + \Delta G_4 - \Delta G_3 - \Delta G_2 = 0$. Consequently, the relative binding free energy is given by,

$$\Delta\Delta G_{\text{wt to mut}} = \Delta G_1 - \Delta G_3 = \Delta G_2 - \Delta G_4 \qquad (11.4)$$

11.4 First Principle Methods and End-point Approaches

11.4.1 *Alchemical Mutations. Free Energy Perturbation (FEP) and Thermodynamic Integration (TI)*

The theory of free energy perturbation and thermodynamic integration approaches was initially developed in the 1950s while the first application to a biomolecular system was performed during the 1980s. For a detailed explanation and the historical development of the methods, see Ref. 28.

Let us consider a typical chemical equilibrium between two states of a given system:

$$A \leftrightarrow B \qquad (11.5)$$

This equilibrium can be described by both the equilibrium constant and the corresponding free energy,

$$K = \frac{[B]}{[A]} \qquad (11.6)$$

and

S. Fiorucci, S. Antonczak, J. Golebiowski

$$\Delta A = -kT \ln\left(\frac{Q_B}{Q_A}\right), \qquad (11.7)$$

where Q_B and Q_A represent the partition functions of states B and A, respectively.

One can express the partition functions as shown above and the free energy expression reads

$$\Delta A = -kT \ln\frac{\int \ldots \int e^{\frac{-H_B}{kT}} \, drdp}{\int \ldots \int e^{\frac{-H_A}{kT}} \, drdp} \qquad (11.8)$$

Considering that the difference between the partition function of states A and B is small, one can write the Hamiltonian of B as a perturbation of the Hamiltonian representing state A:

$$H_B = H_A + \Delta H \qquad (11.9)$$

The change in free energy can be expressed as follows:

$$\Delta A = -kT \ln\frac{\int \ldots \int e^{\frac{-H_A}{kT}} . e^{\frac{-\Delta H}{kT}} \, drdp}{\int \ldots \int e^{\frac{-H_A}{kT}} \, drdp} \qquad (11.10)$$

This latter equation is just the ensemble average of the perturbation in the Hamiltonian, taken from a simulation obtained for the system in state A.

$$\Delta A = -kT \ln\left\langle e^{\frac{-\Delta H}{kT}} \right\rangle_A \qquad (11.11)$$

The Free Energy Perturbation (FEP) approach can be implemented either in Monte Carlo or Molecular Dynamics simulations. The approach is, in principle, exact (does not involve any approximations) and the accuracy of the calculated change in free energy depends on the sampling of configurations relevant for state B (under the control of the Hamiltonian representing state A).

From a more general point of view, even if the difference between states A and B is significant, one can decompose the path going from A to B into several smaller steps $(A..A_1...A_n...A_z...B)$ and then compute the sum of each associated small free energy change throughout the path from A to B.

The Thermodynamic Integration (TI) approach is also based on thermodynamic statistics of various simulations. In the TI scheme, one writes the free energy as a function of a coupling parameter, generally called λ. For the A to B transformation one has

$$\Delta A_{A \rightarrow B} = A(\lambda_A) - A(\lambda_B) = \int_{\lambda_A}^{\lambda_B} \frac{\partial A(\lambda)}{\partial \lambda} d\lambda \qquad (11.12)$$

It is possible to rewrite this free energy difference as a sum of averages of the Hamiltonian derivatives with respect to λ.

$$\Delta A_{A \rightarrow B} = \int_{\lambda_A}^{\lambda_B} \left\langle \frac{\partial H(\lambda)}{\partial \lambda} \right\rangle_\lambda d\lambda \qquad (11.13)$$

The free energy difference can then be obtained by a numerical integration of the ensemble average for the derivative of the Hamiltonian with respect to λ, obtained from various Monte Carlo runs or Molecular Dynamics simulations representing an alchemical transformation from A to B.

In both of these methods, the magnitude of the perturbation is critical for the accuracy of the calculated free energy difference. The free energy difference between two states might be calculated accurately, provided that state B can be considered as a small perturbation to state A. In this case, the partition functions of states A and B overlap and the simulation sampling can be properly achieved within reasonable computing times.

11.4.2 End-point Approaches: Molecular Mechanics Poisson–Boltzmann Surface Area (MMPBSA) and Linear Interaction Energy (LIE) Methods

11.4.2.1 MM-GB(PB)SA Approach

Since the free energy is a state function, it is in principle sufficient to only evaluate the initial and the final states for computing the binding free energy.[29] The Molecular Mechanics Poisson–Boltzmann Surface Area MM-GB(PB)SA method is based on the analysis of configurations obtained from equilibrated MD simulations with explicit solvent or other approaches treating the solvent as a continuum. The total free energy of the system can be expressed as the sum of several contributions:

$$G = E_{MM} + H_{rot/trans} + G_{sol} - TS \tag{11.14}$$

Where E_{MM} is the molecular mechanics energy

$$E_{MM} = E_{bond} + E_{angle} + E_{dihedral} + E_{vdW} + E_{Coulomb} \tag{11.15}$$

G_{sol} is the solvation free energy and $H_{rot/trans}$ corresponds to the contribution due to putative changes in the translational and rotational degrees of freedom of the binding partners.

The solvation free energy is computed as a sum of polar and non-polar contributions. The non-polar contribution corresponds to a cavity formation and van der Waals interactions between the solute and the solvent. This contribution is typically calculated from the solvent-accessible surface area of the molecule,

$$G_{SA} = \gamma SASA + \beta \tag{11.16}$$

SASA is the solvent accessible surface area estimated by rolling the a solvent-sized probe over the solute surface, γ and β are constants which were extracted from a least-squares fit to a plot of experimental alkane transfer free energies versus accessible surface area.

In the MM-PBSA approach, the polar contribution of G_{sol} is obtained by a calculation of the electrostatic potential $\phi(r)$ from a solution of the Poisson (or Poisson–Boltzmann) equation,

$$\nabla \varepsilon(r) \nabla \phi(r) + 4\pi \rho(r) = 0 \qquad (11.17)$$

here $\varepsilon(r)$ is a position dependent dielectric constant, and $\rho(r)$ is the charge distribution of the solute.

Alternatively, in the MM-GBSA approach the electrostatic component is calculated using the Generalized Born equation. In the GB equation, the protein atoms are represented by spheres with a dielectric constant different from that of the exterior of the protein (solvent). The electrostatic energy can be calculated by:

$$G_{sol(polar)} = \left(1 - \frac{1}{\varepsilon}\right) \sum \frac{q_i q_j}{\sqrt{r_{ij}^2 + \alpha_i \alpha_j \exp-\left(r_{ij}^2 / 2\alpha_i \alpha_j\right)}} \qquad (11.18)$$

with q_i and α_i are, respectively, the charges and the effective Born radii of atoms i and j.

The conformational entropic term TS is often estimated by a normal mode analysis of the complex and the isolated protein partners.

In order to calculate free energy changes associated with the complex formation, the free energies of isolated protein partners are subtracted from free energies calculated for the complex:

$$\Delta G_{bind} = \langle G_{complex} \rangle - \langle G_{protein1} \rangle - \langle G_{protein2} \rangle \qquad (11.19)$$

Where $\langle G_x \rangle$ corresponds to the average of the total free energy of the complex or isolated partner over snapshots taken from the MD trajectory.

An interesting feature of such a description of the free energy change is the possibility to decompose it on a per-residue basis.

Such a per-residue decomposition allows a semi-quantitative evaluation of alanine-scanning scoring by replacing a given residue side chain with an alanine side chain and performing the same free energy calculation. Often, this is performed using the trajectory obtained for the wild type proteins by replacing in each snapshot a given residue by alanine.[30,31]

11.4.2.2 Linear Interaction Analysis

The Linear Interaction Analysis (LIE) scheme is based on the idea that when a solute binds to a receptor, the change in free energy can be decomposed into polar and non-polar contributions.[32] The linear response theory is invoked to estimate the polar (electrostatic) component and the non-polar contribution is considered to scale proportional to the intermolecular van der Waals interaction energy, averaged over molecular dynamics simulations. The binding free energy can thus be written as:

$$\Delta G_{bind} = \alpha \Delta \left\langle V_{l-s}^{vdW} \right\rangle + \beta \Delta \left\langle V_{l-s}^{el} \right\rangle + \gamma \qquad (11.20)$$

Here, $\langle x \rangle$ denotes an average over a sampled trajectory (from MD or MC), for the van der Waals (vdW) or electrostatic (el) terms involving the ligand (l) and its surroundings. The Δ stands for the difference between the solute free in solution and bound to its partner. The parameters α, β and γ are generally fitted with respect to experimental results on a set of test cases with known binding free energy.

Several review articles on details and technical aspects of each approach to compute binding free energies in various systems have been published.[33,34] In this section the main aspects that have to be taken into account when setting up a free energy calculation will be discussed. These calculations can be applied to the prediction of both the total binding free energy or the effect of a residue mutation on binding free energy.

11.4.3 Applications on Protein–Protein Complex Structures

As for many computational approaches, free energy calculation techniques are subject to a trade-off between speed and accuracy. The methods are based on a statistical evaluation of various terms. The statistics are in general more accurately determined if a large number of configurations have been sampled. It is not surprising that in the framework of protein–protein association, the number of applications of the computationally of very expensive methods such as FEP and TI is still limited, since large numbers of configurations need to be evaluated.

Due to the large computational costs of alchemical transformations, applications typically involve single mutations often in model systems with known binding energies. TI was, for example, used to compute the relative binding free energy between a wild-type peptide and its Pro6 to Ala6 mutation on recognition by the T-Cell Receptor.[35] The TI calculation gave excellent agreement with experiment and allowed also the decomposition of the free energy change into various energetic contributions, which helps to explain the driving forces for binding. More generally, the decomposition into various components and more particularly on a per residue basis is a powerful tool to predict hotspots at protein binding sites and potential mutations that affect peptide or protein binding. Such prediction approaches include 'computational alanine scanning' or 'virtual alanine scanning' and are frequently used to predict hotspots in proteins that have a dominant effect on affinity for a given receptor. Variants of the MMPBSA calculation methods have been developed to explore the effect of alanine substitutions.[30,31] The MMPBSA approach involves more approximations but is computationally less expensive compared to FEP or TI methods and has been applied successfully to identify hotspot residues in proteins.[36]

Looking more closely at protein–protein interactions, the balance of hydrophobic and hydrophilic regions at the interfaces has to be properly considered. Systems dominated by hydrophobic contacts may have characteristics that are more difficult to predict.[37] This phenomenon is due to the limited directionality and often weak nature of hydrophobic interactions and due to the indirect role played by solvent molecules. The LIE approach is particularly sensitive to the hydrophobic character of a binding region and generally requires an additional term to produce accurate free energies for systems dominated by hydrophobic contributions.[33,38] The decomposition of the interaction free energy between two proteins indeed emphasized the importance of the non-polar terms.[35] The electrostatic contribution can either play the role of directional constraints[39] or can even make a repulsive contribution to binding.[40]

The LIE approach can give results in good agreement with experiment if the mutation does not involve residues sensitive to the electrostatic environment, for example due to changes in protonation states or the presence of counter ions. Indeed, although some LIE calculations involving aspartic or glutamic acids gave results in poor agreement with experimental data,[41] an appropriate treatment of possible changes in protonation states improved the results.[42] Additionally, the newly formed structure has to be examined with care since the mutated residue was shown to form steric clashes, either directly at the interface[42] or with the solvent just around the interface.[43]

In general, the binding free energy depends on buried water molecules present at the interface between the partner structures. During alchemical transformation in FEP or TI calculations, the removal of chemical groups or whole side chains can create large empty cavities at the interface which are likely to be filled with water molecules.[44] In such cases the FEP/TI protocol may need to include the creation or annihilation of water molecules. In case of employing implicit solvent methods one should be aware that the structural role played by explicit water molecules may not be appropriately accounted for.

The effect of the substitution of a putative hotspot at the interface can be decomposed into several parts: (a) the direct interaction energy of the new residue with its neighbours, (b) the influence of this residue on the overall structure of both the mutated protein and the complex, (c) the organisation of the water molecules around this residue, also both in the free protein and potentially in the complex.

An accurate calculation of these contributions requires sufficient sampling of conformations of the wild type and mutated complex structures. The adequate sampling of relevant states is indeed of primary importance since it represents the cornerstone to achieve convergence on the calculation of ensemble averages of a perturbation or free energy derivative. The large size of biomolecular systems implies that a thorough sampling is unfortunately not always possible. In case of screening a large number of possible substitutions it is often necessary to use much simpler empirically parameterized approaches.

11.5 Empirical Scoring Functions

The calculation of binding free energy changes using the methods discussed above is based on a statistical analysis of Monte Carlo or Molecular Dynamics simulations and usually requires the calculation of averages over many conformations. Specific force fields[45–48] have been developed to reduce the requirement of sampling many conformations. Simplifications of the energy function may, however, result in additional parameters to reproduce experimental data.

For instance the scoring of a predicted protein–protein complex usually involves the calculation of a scoring function for a single conformation of the complex (see the review of Halperin *et al.*[10] and references therein). Problems arising from docking simulations can be, in a general manner, split into two groups: (a) the generation of a pool of protein–protein complex conformations and (b) the scoring of the predicted structures. Searching and scoring methods can simply be based on geometric rules that tend to maximise the packing of interface atoms. Protein–protein docking algorithms can, however, use more sophisticated scoring functions taking into account solvation effects and flexibility of the residue side chains and the backbone,[9,49] and may provide a qualitative estimate of the binding free energy.

The development of a fast and reliable simplified protein force field is a difficult task due to the subtle balance between different energy terms. In the following the theoretical basis of empirical force fields used to estimate free energy changes in the context of protein folding[47,48,50] and protein design[2–5] will be discussed.

11.5.1 Empirical Force Fields

Several computational approaches[45–47,51] have been employed to approximately calculate binding energies and effects due to substitution of residues at the binding site. In each of these methods, several empirically parameterized terms are introduced in the energy function. A set of training structures is used to fit those terms followed by validation through blind test experiments. Among others, the FOLD-X[46] and Robetta[52] software packages are examples of such approaches and are described below.

11.5.1.1 FOLD-X

FOLD-X, developed by Guerois et al.,[46] provides a fast and quantitative estimation of mutational effects on protein stability and protein–protein association. The free energy can be decomposed into a combination of terms demonstrated to be important for protein stability:

$$\Delta G = W_{vdW}\Delta G_{vdW} + W_{solvH}\Delta G_{solvH} + W_{solvP}\Delta G_{solvP}$$
$$+\Delta G_{wb} + \Delta G_{hbond} + \Delta G_{el} + W_{mc}T\Delta S_{mc} + W_{sc}T\Delta S_{sc} \qquad (11.21)$$

ΔG_{vdW} is the van der Waals interactions term. ΔG_{solvH} and ΔG_{solvP} represent the hydrophobic and polar solvation energies, respectively. ΔG_{wb} and ΔG_{hbond} stand for stabilising effect of water bridges and hydrogen bonds, respectively, and ΔG_{el} is the electrostatic contribution of charged groups. The entropic term, described using fitted parameters,[53,54] accounts for the cost to restrict the backbone (ΔS_{mc}) and side chain (ΔS_{sc}) mobility in the folded state.

The set of parameters and weights (W_{xx} in the equation 11.21) are fitted using experimental free energy differences ($\Delta\Delta G_{wt/mut}$) of 339 single-point mutants. The predictive performance of the method was examined on a blind test database of 667 protein mutations as well as 82 protein–protein complex mutations. A good correlation was obtained between $\Delta\Delta G_{exp}$ and $\Delta\Delta G_{calc}$ with a standard deviation smaller than 0.85 kcal·mol^{-1}.

While hydrogen bonds, van der Waals and electrostatic interactions are explicitly evaluated, no computationally expensive simulations are needed by FOLD-X. The algorithm describes implicitly some specific properties of proteins (flexibility, the existence of an unfolded state) *via* the entropic terms and the fitted weights.

For issues like protein–protein interaction, where a critical point is to provide a fast and reasonably accurate estimation of the energetics of the system, the FOLD-X approach may be used to test the stability of a structural model.[55,56] For instance, Tur *et al.*[57] investigated the effect of mutations on protein–protein complexes (TRAIL-DR4 and TRAIL-DR5) involved in apoptosis using the FOLD-X method. A single amino acid mutation of TRAIL was predicted to have a favourable effect on the binding affinity.

11.5.1.2 *Robetta*

While the FOLD-X approach is primarily used to predict binding energies, the Robetta algorithm[52] covers different aspects of protein design. On one side, Rosetta is an *ab initio*[58] and comparative modelling[52] tool including NMR refinement[59] and side chain interface packing.[60] On the other side, the alanine-scanning module is of particular interest in the case of protein–protein interface analysis studies. Robetta approximates the binding free energy of protein–protein complexes accounting for shape complementarity at the interface as well as polar interactions and solvation effects.[45] As in the FOLD-X approach, the free energy function is a linear combination of various terms:

$$\Delta G = W_{attr} E_{LJattr} + W_{rep} E_{LJrep} + W_{HB(sc-bb)} E_{HB(sc-bb)}$$

$$+ W_{HB(sc-sc)} E_{HB(sc-sc)} + W_{Coul} E_{Coul} + W_{sol} G_{sol}$$

$$+ W_{\phi/\psi} E_{\phi/\psi}(aa) + \sum_{aa=1}^{20} n_{aa} E_{aa}^{ref} \qquad (11.22)$$

The free energy is decomposed into an attractive and a repulsive part of a Lennard–Jones potential (E_{LJattr} and E_{LJrep}). $E_{HB(sc-bb)}$ and $E_{HB(sc-sc)}$ stand for a side chain-backbone and a side chain-side chain hydrogen bonding terms, respectively. E_{Coul} is a Coulomb potential and G_{sol} an implicit solvation term. Two additional terms are only used for the alanine-scanning calculation: an amino acid type-dependent backbone torsion angle propensity ($E_{\phi/\psi}(aa)$) and a reference value (approximating the interactions in the unfolded state) for each amino acid (E_{aa}^{ref}). The weights (W_x) of the energy terms are parameterized using thermodynamic measurements of mutation effects[61] on both monomeric proteins and on protein interfaces. The effect of residue mutation is estimated for both the bound and the unbound states for the wild-type and the corresponding mutation, leading to the estimation of the binding free energy difference:

$$\Delta\Delta G_{bind} = \Delta G_{bind}^{MUT} - \Delta G_{bind}^{WT} \qquad (11.23)$$

For a test set of more than 2500 molecular systems, the unsigned error is roughly below 1.0 kcal·mol^{-1}.[45]

The influence of each term has been evaluated by removing its contribution to the free energy. The hydrogen bonding term (derived from an environment-dependent criterion) is a critical factor to realistically predict the existence of hotspots. As already discussed for MMPBSA, the lack of an explicit representation of the solvent is likely to result in an inaccurate description when water mediated hydrogen bonds are important for the protein-protein interactions. Applied to hotspot predictions, Robetta provided reasonable predictions of protein–protein binding free energies. Recent work reported the successful application of the model on the re-design of protein–protein interfaces with an experimentally verified significant improvement of the binding affinities.[62–64] For instance, Baker and co-workers[62] were able to increase the specificity of the colicin E7 DNase-Im7 immunity protein complex at least 300-fold.

11.5.2 *Knowledge-based Force Fields*

Knowledge-based (KB) potentials derive known proteins structure information to model interatomic interactions (see Refs 48, 50 and 65). The computational cost of such approaches allows them to estimate the free energy of a large amount of structures. According to the Boltzmann statistics, a contact potential can be related to the population of a given structural feature through the following equation:

$$V\left(P_{ij}\right) = -kT \ln P_{ij} \cong G_{ij} \tag{11.24}$$

Where P_{ij} is the probability of finding a pair of atom within a cut-off distance.

A free energy change can then be defined as a linear combination of individual contributions. The free energy can even be refined by introducing interaction contributions of higher order between a third residue (or atom) k and the coupling between i and j ($\Delta\Delta G_{ijk}$):

$$\Delta G_i = \Delta G_i^{intrinsic} + \sum_j \Delta G_{ij} + \sum_j \sum_k \Delta\Delta G_{ijk} + \dots \tag{11.25}$$

Here, $\Delta G_i^{intrinsic}$ represents the intrinsic change in free energy relative to the unfolded state.

Knowledge-based energy functions were used in protein folding or to evaluate the interaction energy in various receptor–ligand systems, including protein–protein,[51,66] protein–RNA[67] or protein–DNA[68] complexes.

11.6 About Computation Time

Computational demand is a critical issue for selecting an appropriate method to calculate or estimate binding free energy changes. Figure 11.2 gives a qualitative idea on the computational resources needed for the simulation approaches presented in the previous sections. The FEP and TI methods require an extensive and computationally demanding sampling of conformations to obtain accurate ensemble averages. Approaches based on few structures and using an empirically adjusted scoring function are largely devoid of sampling issues and are thus extremely useful for *in silico* high-throughput screening. In between these two extremes, the MMPBSA and LIE approaches usually require shorter simulation times than the FEP/TI methods and for many purposes represent a reasonable trade-off between speed and accuracy.

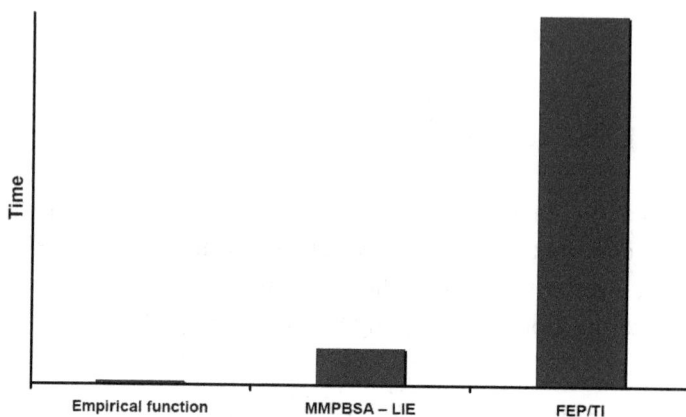

Fig. 11.2. Qualitative comparison of computation time needed to carry out a simulation using the different approaches discussed in this chapter.

11.7 Conclusions and Outlook

The ability to predict the binding mode of biomolecular complexes and to calculate the associated free energy change is a great challenge for computational chemists and biochemists. Several approaches exist that differ in computational demand and the necessary approximations to calculate free energy changes. The available approaches range from computationally very demanding free energy calculations including explicit solvent to empirically parameterized methods. The free energy simulation methods can in principle predict with a high accuracy the full structural and energetic features for biomolecular systems and are limited only by the accuracy of the force field and the sampling of relevant states. Due to the large computational demand, the application of free energy simulation approaches is often limited to the study of single mutations and its effect on protein–protein binding.

At the opposite end of the complexity scale, several approximate approaches are dedicated to produce qualitative results in very short time. These methods describe changes in the free energy of the system of interest using empirically derived terms and lack a detailed representation of the solvent and often also do not account for the flexibility of partner structures. The strength of these methods relies on a clever parameterisation and the capacity to deal with a large number of protein structures in reasonable computer time.

References

1. Jones S., Thornton J.M. (1996). *Proc Natl Acad Sci USA* 93: 13–20.
2. Jiang L., Althoff E.A., Clemente F.R., Doyle L., Rothlisberger D., Zanghellini A., Gallaher J.L., Betker J.L., Tanaka F., Barbas C.F. 3rd, Hilvert D., Houk K.N., Stoddard B.L., Baker D. (2008). *Science* 319: 1,387–1,391.
3. Kaplan J., DeGrado W.F. (2004). *Proc Natl Acad Sci USA* 101: 11,566–11,570.
4. Kuhlman B., Dantas G., Ireton G.C., Varani G., Stoddard B.L., Baker D. (2003). *Science* 302: 1,364–1,368.
5. Looger L.L., Dwyer M.A., Smith J.J., Hellinga H.W. (2003). *Nature* 423: 185–190.
6. Forster M.J. (2002). *Micron* 33: 365–384.
7. Marti-Renom M.A., Stuart A.C., Fiser A., Sanchez R., Melo F., Sali A. (2000). *Annu Rev Biophys Biomol Struct* 29: 291–325.
8. Moult J. (2008). *Structure* 16: 14–16.
9. Gray J.J. (2006). *Curr Opin Struct Biol* 16: 183–193.
10. Halperin I., Ma B., Wolfson H., Nussinov R. (2002). *Proteins* 47: 409–543.
11. Bahadur R.P., Chakrabarti P., Rodier F., Janin J. (2004). *J Mol Biol* 336: 943–955.
12. Bahadur R.P., Zacharias M. (2008). *Cell Mol Life Sci* 65: 1,059–1,072.
13. Janin J., Rodier F., Chakrabarti P., Bahadur R.P. (2007). *Acta Crystallogr Sect D: Biol Crystallogr* 63: 1–8.
14. Jones S., Thornton J.M. (1997). *J Mol Biol* 272: 133–143.
15. Bogan A.A., Thorn K.S. (1998). *J Mol Biol* 280: 1–9.
16. Chakrabarti P., Janin J. (2002). *Proteins* 47: 334–343.
17. Keskin O., Ma B.Y., Nussinov R. (2005). *J Mol Biol* 345: 1,281–1,294.
18. Lo Conte L., Chothia C., Janin J. (1999). *J Mol Biol* 285: 2,177–2,198.
19. Reichmann D., Rahat O., Cohen M., Neuvirth H., Schreiber G. (2007). *Curr Opin Struct Biol* 17: 67–76.
20. Stites W.E. (1997). *Chem Rev* 97: 1,233–1,250.
21. Tsai C.J., Lin S.L., Wolfson H.J., Nussinov R. (1997). *Protein Sci* 6: 53–64.
22. Xu D., Lin S.L., Nussinov R. (1997). *J Mol Biol* 265: 68–84.
23. Rodier F., Bahadur R.P., Chakrabarti P., Janin J. (2005). *Proteins* 60: 36–45.

24. Xu D., Tsai C.J., Nussinov R. (1997). *Protein Eng* 10: 999–1,012.
25. McCoy A.J., Epa V.C., Colman P.M. (1997). *J Mol Biol* 268: 570–584.
26. Clackson T., Wells J.A. (1995). *Science* 267: 383–386.
27. Gasser R.P.H., Richards W.G. (1995). *An Introduction to Statistical Thermodynamics.* World Scientific.
28. Kollman P. (1993). *Chem Rev* 93: 2,395–2,417.
29. Srinivasan J., Cheatham T.E. 3rd, Cieplak P., Kollman P., Case D.A. (1998). *J Am Chem Soc* 120: 9,401–9,409.
30. Chong L.T., Duan Y., Wang L., Massova I., Kollman P.A. (1999). *Proc Natl Acad Sci USA* 96: 14,330–14,335.
31. Massova I., Kollman P. (1999). *J Am Chem Soc* 121: 8,133–8,143.
32. Aqvist J. (1994). *J Phys Chem* 94: 8,021–8,024.
33. Brandsdal B.O., Osterberg F., Almlof M., Feierberg I., Luzhkov V.B., Aqvist J. (2003). *Adv Protein Chem* 66: 123–158.
34. Wang W., Donini O., Reyes C.M., Kollman P.A. (2001). *Annu Rev Biophys Biomol Struct* 30: 211–243.
35. Michielin O., Karplus M. (2002). *J Mol Biol* 324: 547–569.
36. Taranta M., Bizzarri A.R., Cannistraro S. (2009). *J Mol Recognit* 22:215-222.
37. Charlier L., Nespoulous C., Fiorucci S., Antonczak S., Golebiowski J. (2007). *PCCP* 9: 5,761–5,771.
38. Aqvist J., Marelius J. (2001). *Comb Chem High Throughput Screen* 4: 613–626.
39. Zoete V., Meuwly M., Karplus M. (2005). *Proteins* 61: 79–93.
40. Gohlke H., Kiel C., Case D.A. (2003). *J Mol Biol* 330: 891–913.
41. Brandsdal B.O., Aqvist J., Smalas A.O. (2001). *Protein Sci* 10: 1,584–1,595.
42. Almlof M., Aqvist J., Smalas A.O., Brandsdal B.O. (2006). *Biophys J* 90: 433–542.
43. Wang W., Kollman P.A. (2000). *J Mol Biol* 303: 567–582.
44. Brandsdal B.O., Smalas A.O. (2000). *Protein Eng* 13: 239–245.
45. Kortemme T., Baker D.(2002). *Proc Natl Acad Sci USA* 99: 14,116–14,121.
46. Guerois R., Nielsen J.E., Serrano L. (2002). *J Mol Biol* 320: 369–387.
47. Pokala N., Handel T.M. (2005). *J Mol Biol* 347: 203–227.
48. Vajda S., Sippl M., Novotny J. (1997). *Curr Opin Struct Biol* 7: 222–228.
49. Smith G.R., Sternberg M.J. (2002). *Curr Opin Struct Biol* 12: 28–35.
50. Poole A.M., Ranganathan R. (2006). *Curr Opin Struct Biol* 6: 508–513.
51. Jiang L., Gao Y., Mao F., Liu Z., Lai L. (2002). *Proteins* 46: 190–196.
52. Chivian D., Kim D.E., Malmstrom L., Bradley P., Robertson T., Murphy P., Strauss C.E., Bonneau R., Rohl C.A., Baker D. (2003). *Proteins* 53 Suppl 6: 524–533.
53. Guerois R., Serrano L. (2000). *J Mol Biol* 304: 967–982.
54. Hurley J.H., Mason D.A., Matthews B.W. (1992). *Biopolymers* 32: 1,443–1,446.
55. Tokuriki N., Stricher F., Serrano L., Tawfik D.S. (2008). *PLoS Comput Biol* 4: e1000002.
56. Kolsch V., Seher T., Fernandez-Ballester G.J., Serrano L., Leptin M. (2007). *Science* 315: 384–386.

57. Tur V., van der Sloot A.M., Reis C.R., Szegezdi E., Cool R.H., Samali A., Serrano L., Quax W.J. (2008). *J Biol Chem* 283: 20,560–20,568.
58. Simons K.T., Kooperberg C., Huang E., Baker D. (1997). *J Mol Biol* 268: 209–225.
59. Bowers P.M., Strauss C.E., Baker D. (2000). *J Biomol NMR* 18: 311–318.
60. Kuhlman B., Baker D. (2000). *Proc Natl Acad Sci USA* 97: 10,383–10,388.
61. Kumar M., Bava K., Gromiha M., Prabakaran P., Kitajima K., Uedaira H., Sarai A. (2006). *Nucleic Acids Res* 34: D204–206.
62. Joachimiak L.A., Kortemme T., Stoddard B.L., Baker D. (2006). *J Mol Biol* 361: 195–208.
63. Kortemme T., Joachimiak L.A., Bullock A.N., Schuler A.D., Stoddard B.L., Baker D. (2004). *Nat Struct Mol Biol* 11: 371–379.
64. Sammond D.W., Eletr Z.M., Purbeck C., Kimple R.J., Siderovski D.P., Kuhlman B. (2007). *J Mol Biol* 371: 1,392–1,404.
65. Pokala N., Handel T.M. (2001). *J Struct Biol* 134: 269–281.
66. Clark L.A., van Vlijmen H.W. (2008). *Proteins* 70: 1,540–1,550.
67. Chen Y., Kortemme T., Robertson T., Baker D., Varani G. (2004). *Nucleic Acids Res* 32: 5,147–5,162.
68. Zhang C., Liu S., Zhu Q., Zhou Y. (2005). *J Med Chem* 48: 2,325–2,335.

CHAPTER 12

Small-Molecule Inhibitors of Protein–Protein Interactions

Thorsten Berg

Max Planck Institute of Biochemistry,
Department of Molecular Biology,
Am Klopferspitz 18,82152 Martinsried, Germany,
and Center for Integrated Protein Science Munich
(CIPSM)
E-mail: berg@biochem.mpg.de

The majority of current FDA-approved small molecules target enzymes and cell surface receptors. While there is no doubt that these protein classes are excellent drug targets, it is also well-known that they constitute only a minor proportion of all human proteins. Therefore, the development of approaches that allow for the expansion of the number of protein classes that can be targeted by small molecules is a major challenge to academic research. Because most biological processes are performed by protein complexes and the function of a protein can depend on its binding partners, the inhibition of protein–protein interactions bears tremendous potential to render additional protein families accessible to functional modulation by small molecules. This chapter highlights the case stories of selected inhibitors of protein–protein interactions which illustrate the current scope of the field.

12.1 Introduction

Approximately 80% of small organic molecules approved for human use by the U.S. Food and Drug Administration (FDA) target cell-surface receptors and enzymes. Other prominent drug targets include nuclear receptors, ion channels and transporters.[1–3] Overall, the protein classes

318

perceived as 'druggable' have been estimated to be only 10–15% of all human proteins.[1] This indicates the tremendous potential that expansion of the perceived spectrum of druggable proteins could have for drug discovery. The expansion of the target spectrum of small molecules would also be very beneficial for basic research, since small-molecule inhibitors are excellent tools for the analysis of protein functions in genetically unmodified systems.[4]

Most biological processes are performed not by single proteins, but by protein complexes, which often consist of ten or more proteins.[5,6] Since the function of a given protein can depend on its binding partners, cell-permeable inhibitors of protein–protein interactions can potentially influence protein functions.[7] Thus, inhibitors of protein–protein interactions open up novel possibilities for expanding the proportion of proteins which can be targeted by small molecules; moreover, they can provide access to alternative modes of inhibition of established classes of small-molecule targets. In this review, I shall discuss recent progress in the field of small-molecule inhibitors of intracellular protein–protein interactions.[7–12] This review does not cover the development of inhibitors of the protein–protein interactions between integrins and components of the extracellular matrix,[13,14] a field which is particularly advanced and which has recently been reviewed elsewhere.[15,16]

12.2 Inhibitors of Binding between Anti-apoptotic Bcl-2 Proteins and BH3 Domains of Pro-apoptotic Bcl-2 Proteins

Many human tumours overexpress anti-apoptotic Bcl-2 family members, such as Bcl-2 itself, Bcl-x_L, Mcl-1 or Bcl-w. Anti-apoptotic Bcl-2 proteins bind to and thereby inactivate pro-apoptotic Bcl-2 proteins (e.g., Bad, Bak or Bim); the interaction is mediated by a hydrophobic cleft formed by Bcl-2 homology (BH) domains 1–3 of anti-apoptotic Bcl-2 proteins and the helical BH3 domain of pro-apoptotic Bcl-2 proteins.[17] Small-molecule inhibitors of the interaction between pro- and anti-apoptotic Bcl-2 family members can restore the activities of pro-apoptotic Bcl-2 family members in tumour cells and thereby induce apoptosis. The combination of the clearly defined biological effect of restoring the activity of pro-apoptotic Bcl-2 proteins (i.e., the induction of apoptosis in tumour cells), and the presence of a structurally well-characterised and not too extensive binding pocket, has resulted in large research efforts aimed at finding small-molecule inhibitors of this interaction. Because many of them have already been covered by recent reviews,[18,19] this section will focus on a selection of recent developments in the field.

The natural product (*R*)-(-)-gossypol (**1**) (Fig. 12.1) was one of the first small molecules discovered to inhibit binding of pro-apoptotic Bcl-2 family members to anti-apoptotic Bcl-2 family members.[20] It inhibits the anti-apoptotic Bcl-2 family members Bcl-2, Bcl-x_L, and Mcl-1 in the submicromolar concentration range.[21] (*R*)-(-)-gossypol, which has several additional known activities, is currently undergoing clinical trials as an anti-cancer agent.[22] A combination of molecular modelling and synthetic efforts led to the synthesis of compound **2**, which inhibits Bcl-2 (K_i = 17 nM) and Mcl-1 (K_i = 18 nM), and to a lesser extent, Bcl-x_L (K_i = 1.2 µM).[23] The chemical structure of gossypol was, furthermore, used as a starting point for the generation of TM-1206 (**3**), which shows only limited remaining similarity to gossypol. Nevertheless, TM-1206 is a potent inhibitor with more balanced activities against the anti-apoptotic Bcl-2 proteins Bcl-2, Bcl-x_L, and Mcl-1 in the triple-digit nanomolar concentration range.[24] Both TM-1206 and compound **2** inhibited proliferation of a breast cancer cell line and caused significant cell death at concentrations at or below 1 µM.

WL-276 (**4**)[25] is the most active and best characterised derivative[26] of the BH3-I series of Bcl-x_L inhibitors.[27] It displays approximately 20-fold selectivity for Bcl-x_L over Bcl-2 (K_i-values for Bcl-x_L and Bcl-2: 1.2 µM and 22.8 µM, respectively), and induces apoptosis in the prostate cancer cell line PC-3. Importantly, PC-3 cells with resistance to standard cancer chemotherapeutic agents showed higher susceptibility to WL-276 than parental PC-3 cells, suggesting the potential utility of WL-276 for the treatment of drug-resistant prostate cancers.

A team of scientists led by Fesik and Rosenberg at Abbott Laboratories developed the small molecule ABT-737 (**5**) which inhibits Bcl-2, Bcl-x_L and Bcl-w with subnanomolar activities.[28] Although the agent was shown to exert potent and selective effects in xenograft mouse tumour models, its utility as a drug was limited by its poor bioavailability. Fortunately, the introduction of only a few chemical modifications led to the derivative ABT-263 (**6**), which exhibits similar activity against its targets whilst showing up to 50% bioavailability in animal models.[29] ABT-263 is currently undergoing phase 1/2a clinical trials against several human malignancies.[29]

Another inhibitor of an anti-apoptotic Bcl-2 family protein under clinical investigation is Obatoclax (**7**), which is predicted to bind to a hydrophobic pocket within the BH3 binding groove of Bcl-2.[30] Obatoclax was shown to interfere with the interaction between Mcl-1 and Bak in mitochondrial outer membranes and in cells, and overcomes Mcl-1-mediated resistance to apoptosis against ABT-737 and the proteasome inhibitor bortezomib.

12.3 Inhibitors of Binding between XIAP and Caspases

The protein X-linked inhibitor of apoptosis (XIAP) is overexpressed in many human tumours and contributes to resistance of cancer cells to chemotherapy.[31,32] XIAP contains three baculovirus IAP repeat (BIR) domains. The natural inhibitor of XIAP, Smac/DIABLO (second mitochondria-derived activator of caspases / direct IAP binding protein with low pI targets both the BIR2 and the BIR3 domain of XIAP via its N-terminal AVPI motif.[33] The function of the BIR3 domain is well-understood: it binds to and thereby inhibits the initiator caspase-9. In contrast, the mechanisms in which the BIR2 domain is involved in inhibition of two effector caspases, caspase-3 and caspase-7, is still not entirely clear.[34] Regardless of the exact mechanism of caspase-3/-7 inhibition by XIAP, numerous studies have shown that mimics of the AVPI motif have the potential to relieve caspase inhibition and thereby reactivate apoptotic pathways blocked by aberrant XIAP activity.[31]

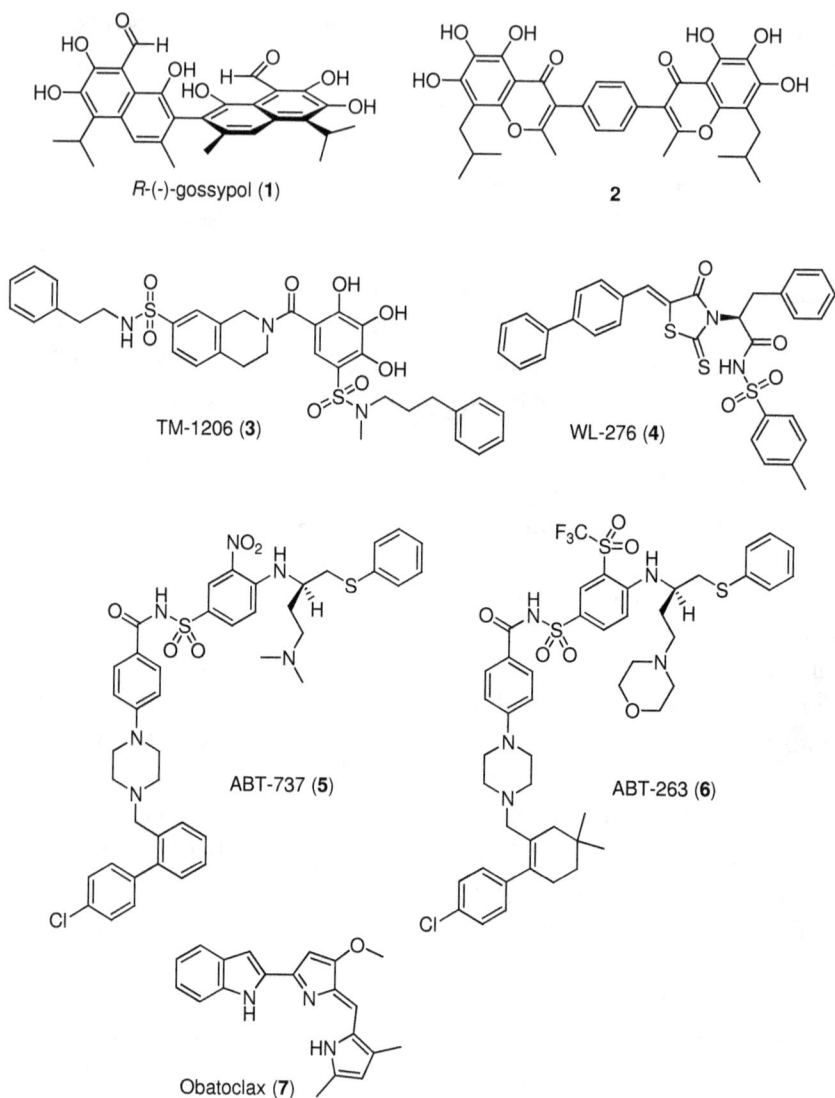

Fig. 12.1. Small-molecule inhibitors of anti-apoptotic Bcl-2 family proteins.

Since nature has designed the XIAP inhibitor Smac/DIABLO as a homodimer, which is believed to target the function of both the BIR3 and the BIR2 domain simultaneously, it seems interesting to mimic this approach with bivalent inhibitors. The first inhibitory agent shown to target the BIR2 and the BIR3 domain simultaneously was the C_2-symmetrical diyne **8**, which was generated in a side reaction of a synthetic manipulation of a monomeric alkyne (Figure 12.2).[35] **8** and the corresponding monomer **9** displayed similar affinities for the purified XIAP BIR3 domain *in vitro*, but **8** was significantly more active in a caspase-3 activation assay using cellular extracts, presumably because it targets both the BIR2 and the BIR3 domain. **8** did not significantly increase the apoptotic rate of HeLa cells on its own, but potently synergised with tumour necrosis factor α (TNFα) and TNF-related apoptosis-inducing ligand (TRAIL) to induce caspase activation and apoptosis in HeLa cells.

Wang and co-workers demonstrated that an [8.5] bycyclic system in compound SM-122 (**10**) could effectively mimic the central VP motif of AVPI. **10** to bind to both the BIR2 and BIR3 domain separately with high affinity.[36] Fortunately, the compound tolerated substitution of one of the phenyl groups for a [1,2,3]-triazole ring, and thereby allowed synthesis of bivalent inhibitors based on SM-157 (**11**) via 'click chemistry'.[37] The most potent bivalent molecule SM-164 (**12**) generated by this approach could be shown to bind to both BIR domains simultaneously, and displayed excellent affinity to XIAP in an *in vitro* competition assay (IC_{50} = 1.4 nM). Furthermore, it inhibited proliferation and induced apoptosis of a human leukemia cell line at concentrations as low as 1 nM.

Very recently, the Wang group demonstrated that cyclic, bivalent peptide mimetics can be designed which are effective in targeting XIAP not only *in vitro* but also in tissue culture.[38] This study was based on previous data showing that the tetrapeptide AKPF methylated on alanine has a higher affinity to XIAP than the native, Smac-derived sequence AVPI. The cyclic peptide **13**, which incorporates two NH(Me)-AKPF motifs, forms a 1:2 complex with the isolated BIR3 domain, but a 1:1 complex with XIAP protein containing both the BIR 2 and the BIR 3 domain. **13** inhibited cell growth and induced apoptosis in the human breast cancer cell line MDA-MB-231 in a caspase-3-dependent manner.

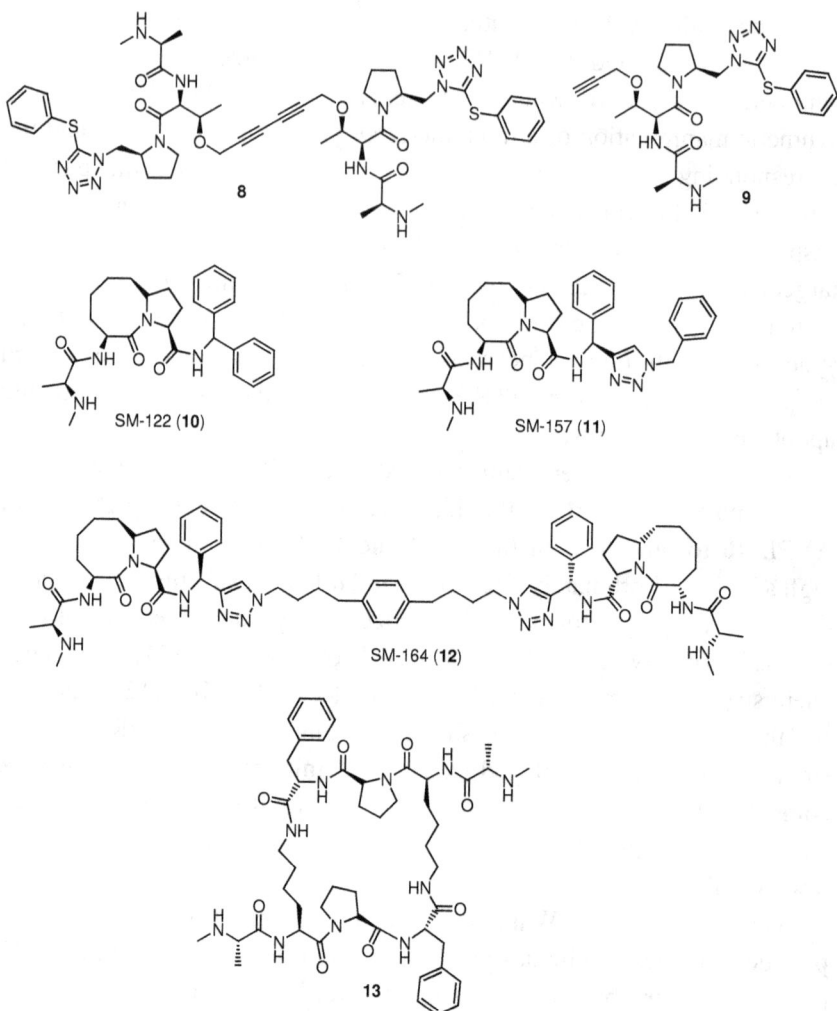

Fig. 12.2. Small-molecule inhibitors of XIAP.

12.4 Inhibitors of the Plk1 PBD Domain

The serine/threonine kinase Polo-like kinase 1 (Plk1) regulates multiple stages of mitosis.[39,40] It is frequently found to be overexpressed in human cancers, and has been established as a negative prognostic marker for tumour therapy.[41] Because inhibition of Plk1 by various approaches has been shown to induce mitotic arrest and apoptosis in tumour cells, large research efforts are being channeled into the development of small-molecule inhibitors of the enzyme's catalytic activity.[41,42] Some small-molecule inhibitors of Plk1's catalytic activity are already being tested for their clinical safety and efficacy against human malignancies.[43,44] Most Plk1 inhibitors target the conserved ATP binding pocket. Unfortunately, the conserved nature of the ATP binding site of protein kinases turns the development of truly mono-specific inhibitors of a single kinase into an enormous challenge, and has stimulated the development of novel methods for the analysis of the activity profiles of ATP-competitive protein kinase inhibitors.[45–47]

Polo-like kinases contain a domain referred to as the polo-box domain (PBD) at their C-terminus.[48] The PBD of Plk1 is believed to serve as an anchor by which the enzyme locates to its intracellular anchoring sites. The Plk1 PBD binds to sequences comprising an S-(pT/pS)-(P/X) motif,[48] but similar sequence motifs are also tolerated. Since correct intracellular localization of Plk1 has been shown to be required for completion of mitosis[49] and the PBD is unique to polo-like kinases, the PBD has been suggested as a target for small organic molecules.[41,50] Screening of chemical libraries in an assay based on fluorescence polarization[51] for functional inhibitors of the Plk1 PBD led to the identification of the thymoquinone derivative Poloxin (**14**) as a selective inhibitor of the function of the Plk1 PBD *in vitro* (Figure 12.3).[52] Thymoquinone (**15**) itself was also identified as a Plk1 PBD inhibitor, but displayed a less desirable specificity profile. Both compounds caused mitotic arrest and chromosome congression defects, the cellular phenotype associated with mislocalization of endogenous Plk1.[49] Immunofluorescence assays demonstrated that the compounds interfered with correct intracellular localization of endogenous Plk1. Since both compounds led to apoptosis in cancer cells, the study validated the Plk1 PBD as an anti-tumour target for small molecules.

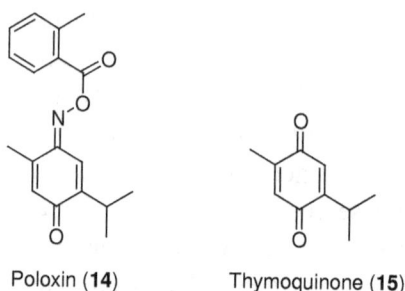

Poloxin (**14**)   Thymoquinone (**15**)

Fig. 12.3. Small-molecule inhibitors of the polo-box domain of Plk1.

12.5 Inhibitors of the MDM2-p53 Interaction

The tumour suppressor p53 is mutated or deleted in approximately 50% of all cancers.[53,54] In a large fraction of the other cases, wild-type p53 is found to be functionally inactivated by binding to overexpressed MDM2 (also referred to as HDM2 in humans). MDM2 is a target gene of p53 and forms an autoregulatory feedback loop by inhibiting p53 by three different mechanisms: (a) it binds to the transactivation domain and thereby inhibits p53's transcriptional activity; (b) it supports nuclear export of p53; and (c) it serves as a ubiquitin ligase and thereby promotes p53's proteasomal degradation. Blocking the p53/MDM2 interaction has been clearly shown to restore p53 levels and activity.[55,56] The nature of the p53/MDM2 interaction as a biologically desirable target for drug development, in combination with the existence of a clearly defined binding pocket characterised by X-ray crystallography and NMR, have stimulated a plethora of drug discovery efforts.[57] For reasons related to space limitation, this section will only cover a selection of prominent and recent inhibitor studies.

The development of a class of *cis*-imidazoles dubbed Nutlins, shown to be selective inhibitors of the p53/MDM2 interaction with activities in the nanomolar concentration range, was a landmark in this research area.[58] The efficacy of Nutlins is based on their structural mimicking of three p53-derived amino acids (Phe19, Trp23 and Leu26), which are crucial for binding to MDM2, by decorating the central imidazole core with suitable chemical groups.[58] The most potent compound of this series, Nutlin-3a (**16**) (the active enantiomer of Nutlin-3), inhibited the p53/MDM2 interaction with an IC_{50} of 90 nM, and was shown to reactivate wild-type p53 in cellular systems as well as in mouse xenograft models. Another inhibitor for which the interaction with MDM2 was studied by X-ray crystallography is the benzodiazepinedione **17**. It was discovered in a high-throughput screen of chemical libraries in an assay that detects MDM2 stabilization by binding of compounds. **17** binds to MDM2 with high affinity ($K_d = 67$ nM) *in vitro*, and showed a p53-dependent effect on the proliferation of cancer cell lines. Crystallographic analysis revealed that similar to Nutlins, **17** occupies the same binding pockets in MDM2 as the side chains of Phe19, Trp23 and Leu26 of p53.

Based on their discovery that oxindole can mimic the side chain of Trp23,[59] Wang and co-workers have reported on a series of spiro-oxindoles as inhibitors of the p53/MDM2 interaction. The spiro-oxindole with the highest affinity for MDM2 described to date is MI-63 (**18**), which possesses a K_i-value of only 3 nM for MDM2.[60] MI-63 selectively inhibits cell growth in cell lines with wild-type p53, and induces apoptosis *ex vivo* in leukemia patient samples.[61] Unfortunately, MI-63 has an unfavourable pharmacokinetics profile, and thus is not suitable as a therapeutic agent. Further compound optimization led to the design of spiro-oxindole MI-219 (**19**), which binds to human MDM2 with almost the same affinity as MI-63 ($K_i = 5$ nM), and is orally bioavailable. MI-219 displays excellent selectivity for MDM2 over the homologous MDMX protein *in vitro* (Fig. 12.4).[62] MI-219 could be shown to activate wild-type p53, leading to cell cycle arrest and apoptosis in tumour cells *in vitro* and in mouse xenograft models. Docking studies predicted that MI-219 occupies the same space of MDM2 as key binding residues of p53.[62]

Using the structures of known p53/MDM2 inhibitors as a starting point, the groups of Holak and Weber searched chemical databases for compounds with structural similarity but different scaffolds. This process led to the prediction that an isoquinolin-1-one could be a p53/MDM2 inhibitor (Fig, 12.4).[63] In order to investigate the prediction, several isoquinolin-1-ones were synthesised. As an example, the isoquinolin-1-one NXN-7 (**20**) was shown to bind to MDM2 with low micromolar affinity by heteronuclear NMR experiments,[64] isothermal titration

calorimetry and Biacore experiments. As expected for an inhibitor of the
p53/MDM2 interaction, NXN-7 increased cellular levels of p53 and its
target genes, and induced apoptosis in an ovarian carcinoma cell line.

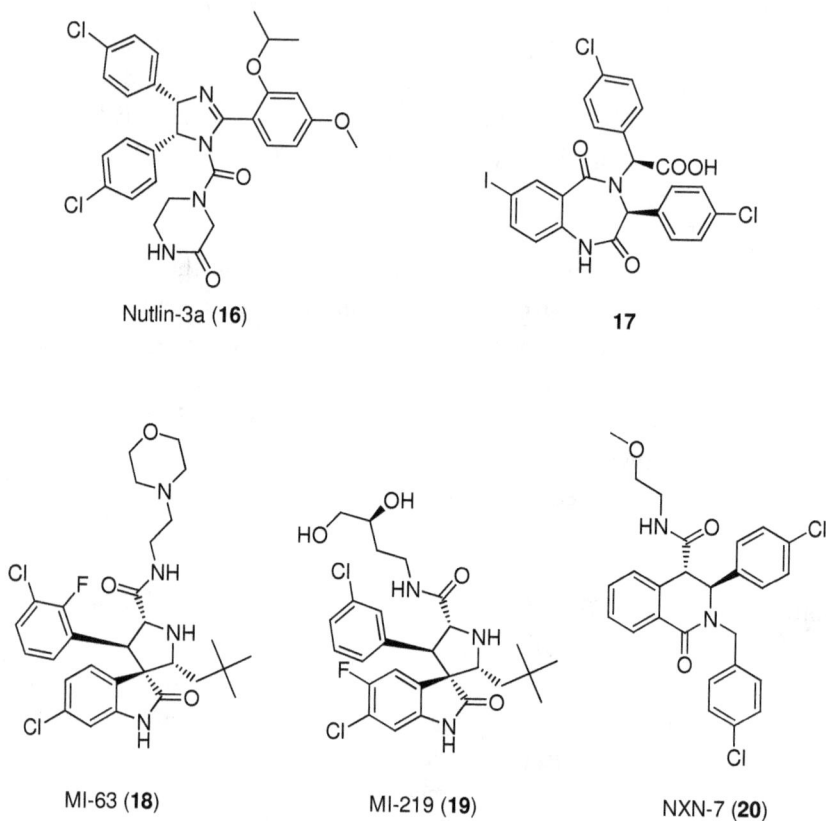

Fig. 12.4. Small-molecule inhibitors of the MDM2/p53 interaction.

12.6 Inhibitors of STAT3 and STAT5

Signal transducers and activators of transcription (STATs) are
transcription factors which transduce signals from the cell surface to the
nucleus.[65,66] Via their SH2 domains, STATs bind to activated cell surface
receptors and non-receptor tyrosine kinases. STATs are subsequently

phosphorylated at a conserved tyrosine residue C-terminal of their SH2 domain by receptor-associated Janus kinases (JAKs), the intrinsic tyrosine kinase activity of growth factor receptors, or other cytoplasmic tyrosine kinases. Tyrosine phosphorylation of STATs induces the formation of STAT dimers formed by reciprocal phosphotyrosine-SH2 domain interactions. These STAT dimers subsequently accumulate in the nucleus and regulate the transcription of their respective target genes.

STAT3 is found constitutively activated in a wide range of human tumours and cancer cell lines.[67] Importantly, inhibition of signalling via STAT3 in cells with constitutive STAT3 activity by a dominant negative mutant,[68,69] antisense approaches,[70] decoy oligonucleotides,[71–73] siRNAs,[74–76] peptide aptamers[77,78] or G-quartet oligonucleotides[79,80] has been uniformly shown to suppress tumour growth and to induce apoptosis. Therefore, STAT3 is regarded as a strong candidate target for cancer therapy.[81–88] Similarly, the STAT5 isoforms STAT5a and STAT5b are overactive in several kinds of human tumours, including leukemias, breast cancer, uterine cancer, prostate cancer, and squamous cell carcinoma of the head and neck (SCCHN).[89] Inhibition of signalling via STAT5, especially STAT5b, has been shown to inhibit tumour growth and to induce apoptosis of tumour cells.[90–92] Therefore, STAT3 and STAT5a/b are considered promising targets for cancer therapy.

Most approaches towards the direct inhibition of STATs target the SH2 domain, as it is required for two steps involved in STAT signalling: tyrosine phosphorylation and dimerization. Functional inhibitors of STAT SH2 domains can therefore be expected to inhibit not only STAT activation, but also dimerization of any STAT molecules which have become activated despite the presence of the inhibitor.[93] The validity of this concept was initially demonstrated using peptides[94,95] and peptidomimetics.[96] The oxazole S3I-M2001 described by Turkson and co-workers (21) is probably the most advanced peptide-derived inhibitor of the STAT3 SH2 domain (Fig. 12.5).[97] S3I-M2001 was designed based on the tripeptide motif A/PpYL known to bind to the STAT3 SH2 domain[94] and the suggested binding mode of a peptide mimetic,[96] but has only minimal remaining peptidic character. Despite the presence of the side chain of the central phosphotyrosine, which should negatively impact cellular uptake and intracellular stability, S3I-M2001 displayed strong STAT3-dependent effect in tissue culture at concentrations of 30–100 μM. The compound was shown to disrupt dimeric, tyrosine phosphorylated STAT3 and to inhibit STAT3-mediated gene transcription, malignant transformation, survival and migration. Importantly, the compound inhibited proliferation of a breast cancer cell line with constitutive STAT3 activation in a mouse xenograft model.

Two independent studies performed virtual screening of chemical databases for inhibitors of the STAT3 SH2 domain.[98,99] These efforts led to the discovery of two structurally unrelated STAT3 inhibitors. STA-21 (NSC 628869) (**22**) inhibits DNA-binding of pre-phosphorylated STAT3, and was shown to inhibit STAT3-mediated gene transcription and to induce apoptosis in a STAT3-dependent manner.[98] S3I-201 (NSC 74859) (**23**) also displayed such biological activity and was additionally shown to inhibit tumour growth in a mouse xenograft model.[99]

Biochemical screening[100] of chemical libraries for compounds with an inhibitory effect on the function of the STAT3 SH2 domain enabled our group to identify Stattic (**24**).[101] Stattic inhibits the function of the STAT3 SH2 domain regardless of the STAT3 activation state *in vitro*. It could be shown to selectively inhibit nuclear translocation of STAT3, and to increase the apoptotic rate of breast cancer cell lines in a STAT3-dependent manner. A similar screening approach as was used for the discovery of Stattic, led to the identification of chromone-based acyl hydrazones as the first non-peptidic inhibitors of the STAT5 SH2 domain.[102,103] Compound **25** and others selectively inhibited the STAT5b SH2 domain *in vitro*, and blocked IFN-α-induced activation of STAT5 in a lymphoma cell line.

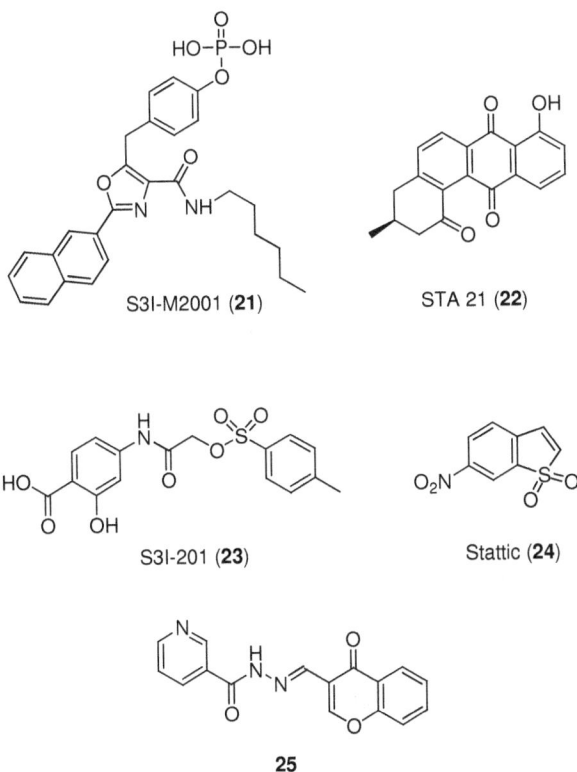

Fig. 12.5. Inhibitors of the SH2 domain of STAT3 and STAT5.

12.7 Inhibitors of c-Myc/Max Dimerization

c-Myc is a member of the basic helix-loop-helix (bHLH-Zip) transcription factor family. It plays a critical role not only in cell cycle progression, growth and oncogenic transformation but also in regulation of apoptosis.[104,105] c-Myc is overexpressed in many human cancers and overexpression of c-Myc in genetic model systems leads to tumourigenesis. Therefore, inhibition of c-Myc by cell permeable agents represents a promising strategy to both elucidate the transcription factor's biological functions, and to interfere with human cancers with increased levels and/or activities of c-Myc.[106] Since all known biological functions of c-Myc require binding to its activation partner Max, which is also a bHLH-Zip protein, inhibition of c-Myc/Max dimerization appears to be the most direct and comprehensive approach towards c-Myc inhibition.[107] The discovery of inhibitors of c-Myc/Max heterodimers seems particularly difficult owing to the large α-helical interface which does not include any obvious binding sites for small molecules.[108] In some cases, c-Myc-induced tumourigenesis in mouse models can be reverted by inactivation of the c-Myc transgene.[109]

The initial report on the feasibility of targeting the c-Myc/Max interaction by small molecules by Vogt and co-workers[110] – which simultaneously represented the first report of small-molecule inhibitors of a transcription factor dimerization – was followed by a number of studies.[111] Prochownik's group screened members of a chemical library for those which could prevent dimerization of c-Myc and Max in a yeast two-hybrid assay.[112] The structure of one inhibitor, 10058-F4 (**26**), was used to search for more potent derivatives (Fig. 12.6). Structural variation of the substituents on the aromatic ring and the rhodanine moiety as found in compound 28RH-NCN-1 (**27**) did not significantly affect the activity, as compared to the parent compound 10058-F4 in a c-Myc/Max DNA binding assay, but inhibited growth of a leukemia cell line almost two-fold more potently (IC_{50} = 29 µM against HL-60 cells).[113]

Janda and Vogt identified four structurally related inhibitors of Myc/Max dimerization and DNA binding, which share planar, aromatic scaffolds.[114] As an example, compound NY2267 (**28**) strongly inhibited c-Myc-dependent oncogenic transformation with very good selectivity over transformation mediated by v-Src or v-Jun, but did not discriminate between transcription mediated by c-Jun and c-Myc in luciferase assays.

Screening of diverse chemical libraries for compounds which inhibited DNA binding of c-Myc by our group led to the discovery of the pyrazolo[1,5-*a*]pyrimidine Mycro1 (**29**)[115] and its close derivative Mycro2. Subsequent screening of 1,439 pyrazolo[1,5-*a*]pyrimidines resulted in the identification of the most selective c-Myc/Max inhibitor of this series. Mycro3 (**30**) (referred to as compound 1 in the original publication) inhibited c-Myc/Max dimerization and DNA binding with very good selectivity *in vitro*, and also showed good potency and selectivity at concentrations of 10–40 µM against c-Myc in cellular assays.[116]

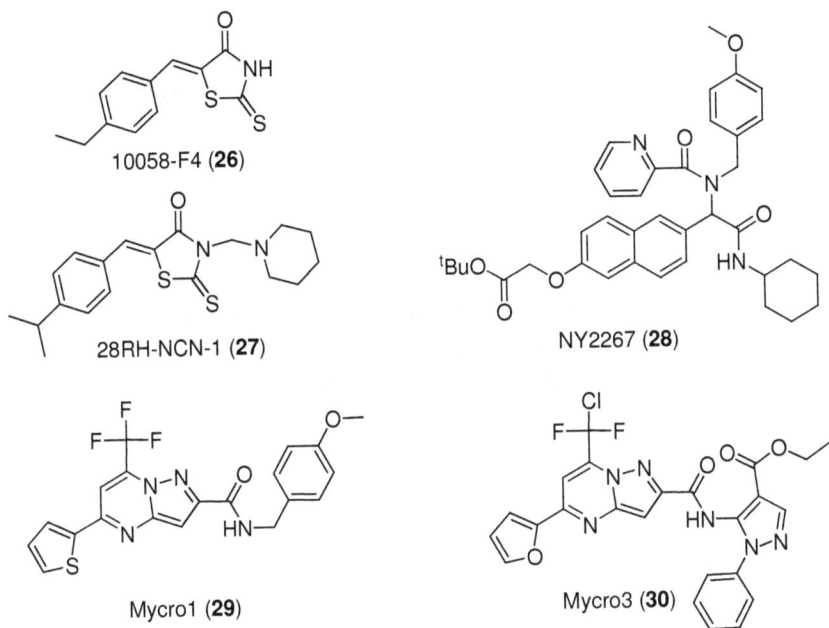

Fig. 12.6. Inhibitors of c-Myc/Max dimerization.

12.8 Conclusions and Outlook

The examples described in this review demonstrate that the inhibition of intracellular protein–protein interactions with cell-permeable molecules has already been achieved in many cases. Some inhibitors of protein–protein interactions have even progressed to clinical trials.[13,21,29,30,117] Small-molecule inhibitors of protein–protein interactions have already altered our view on the nature of druggable proteins. As described in this chapter, they have demonstrated their suitability to the targeted inactivation of anti-apoptotic proteins (Bcl-2 family proteins or XIAP), oncogenic transcription factors (c-Myc, STAT3/5) and to the reactivation of transcription factors with tumour suppressor function (p53). Moreover, the inhibition of protein–protein interactions was demonstrated to provide an alternative entry point to the inhibition of enzymes (exemplified by Plk1). The selected examples discussed here suggest that the modulation of protein–protein interactions by small molecules is likely to play an important role in achieving one of the ultimate goals of chemical biology, that of devising a selective small-molecule probe for every function of every human protein.[118]

Acknowledgements

Work in my research group is generously supported by the Department of Molecular Biology (director: Axel Ullrich) at the Max Planck Institute of Biochemistry, the Bundesministerium für Bildung und Forschung (NGFN-2, grant 01GS0451), and the Deutsche Krebshilfe.

References

1. Hopkins A.L., Groom C.R. (2002). *Nat Rev Drug Discov* 1: 727–730.
2. Overington J.P., Al-Lazikani B., Hopkins A.L. (2006). *Nat Rev Drug Discov* 5: 993–996.
3. Imming P., Sinning C., Meyer A.(2006). *Nat Rev Drug Discov* 5: 821–834.
4. Mayer T.U. (2003). *Trends Cell Biol* 13: 270–277.
5. Alberts B. (1998). *Cell* 92: 291–294.
6. Gavin A.C., Bosche M., Krause R., Grandi P., Marzioch M., Bauer A., Schultz J., Rick J.M., Michon A.M., Cruciat C.M., Remor M., Hofert C., Schelder M., Brajenovic M., Ruffner H., Merino A., Klein K., Hudak M., Dickson D., Rudi T., Gnau V., Bauch A., Bastuck S., Huhse B., Leutwein C., Heurtier M.A., Copley R.R., Edelmann A., Querfurth E., Rybin V., Drewes G., Raida M., Bouwmeester T., Bork P., Seraphin B., Kuster B., Neubauer G., Superti-Furga G. (2002). *Nature* 415: 141–147.
7. Berg T. (2003). *Angew Chem Int Ed Engl* 42: 2,462–2,481.
8. Arkin M.R., Wells J.A. (2004). *Nat Rev Drug Discov* 3: 301–317.
9. Yin H., Hamilton A.D. (2005). *Angew Chem Int Ed Engl* 44: 4,130–4,163.
10. Hershberger S.J., Lee S.G., Chmielewski J. (2007). *Curr Top Med Chem* 7: 928–942.
11. Wells J.A., McClendon C.L. (2007). *Nature* 450: 1,001–1,009.
12. Berg T. (2008). *Curr Opin Drug Disc Devel* 11: 666-674..
13. Dechantsreiter M.A., Planker E., Matha B., Lohof E., Holzemann G., Jonczyk A., Goodman S.L., Kessler H. (1999). *J Med Chem* 42: 3,033–3,040.
14. Heckmann D., Meyer A., Laufer B., Zahn G., Stragies R., Kessler H. (2008). *ChemBioChem* 9: 1,397–1,407.
15. Heckmann D., Kessler H. (2007). *Methods Enzymol* 426: 463–503.

16. Meyer A., Auernheimer J., Modlinger A., Kessler H. (2006). *Curr Pharm Des* 12: 2,723–2,747.

17. Petros A.M., Olejniczak E.T., Fesik S.W. (2004). *Biochim Biophys Acta* 1644: 83–94.

18. Arkin M. (2005). *Curr Opin Chem Biol* 9: 317–324.

19. Zhang L., Ming L., Yu J. (2007). *Drug Resist Updat* 10: 207–217.

20. Wang S., Yang D. (2002). *US-patent application series no. 20030008924.*

21. Wang G., Nikolovska-Coleska Z., Yang C.Y., Wang R., Tang G., Guo J., Shangary S., Qiu S., Gao W., Yang D., Meagher J., Stuckey J., Krajewski K., Jiang S., Roller P.P., Abaan H.O., Tomita Y., Wang S. (2006). *J Med Chem* 49: 6,139–6,142.

22. Kitada S., Leone M., Sareth S., Zhai D., Reed J.C., Pellecchia M. (2003). *J Med Chem* 46: 4,259–4,264.

23. Tang G., Ding K., Nikolovska-Coleska Z., Yang C.Y., Qiu S., Shangary S., Wang R., Guo J., Gao W., Meagher J., Stuckey J., Krajewski K., Jiang S., Roller P.P., Wang S. (2007). *J Med Chem* 50: 3,163–3,166.

24. Tang G., Yang C.Y., Nikolovska-Coleska Z., Guo J., Qiu S., Wang R., Gao W., Wang G., Stuckey J., Krajewski K., Jiang S., Roller P.P., Wang S. (2007). *J Med Chem* 50: 1,723–1,726.

25. Wang L., Sloper D.T., Addo S.N., Tian D., Slaton J.W., Xing C. (2008). *Cancer Res* 68: 4,377–4,383.

26. Xing C., Wang L., Tang X., Sham Y.Y. (2007). *Bioorg Med Chem* 15: 2,167–2,176.

27. Degterev A., Lugovskoy A., Cardone M., Mulley B., Wagner G., Mitchison T., Yuan J. (2001). *Nat Cell Biol* 3: 173–182.

28. Oltersdorf T., Elmore S.W., Shoemaker A.R., Armstrong R.C., Augeri D.J., Belli B.A., Bruncko M., Deckwerth T.L., Dinges J., Hajduk P.J., Joseph M.K., Kitada S., Korsmeyer S.J., Kunzer A.R., Letai A., Li C., Mitten M.J., Nettesheim D.G., Ng S., Nimmer P.M., O'Connor J.M., Oleksijew A., Petros A.M., Reed J.C., Shen W., Tahir S.K., Thompson C.B., Tomaselli K.J., Wang B., Wendt M.D., Zhang H., Fesik S.W., Rosenberg S.H. (2005). *Nature* 435: 677–681.

29. Tse C., Shoemaker A.R., Adickes J., Anderson M.G., Chen J., Jin S., Johnson E.F., Marsh K.C., Mitten M.J., Nimmer P., Roberts L., Tahir S.K., Xiao Y., Yang X., Zhang H., Fesik S., Rosenberg S.H., Elmore S.W. (2008). *Cancer Res* 68: 3,421–3,428.

30. Nguyen M., Marcellus R.C., Roulston A., Watson M., Serfass L., Murthy-Madiraju S.R., Goulet D., Viallet J., Belec L., Billot X., Acoca S., Purisima E., Wiegmans A., Cluse L., Johnstone R.W., Beauparlant P., Shore G.C. (2007). *Proc Natl Acad Sci USA* 104: 19,512–19,517.

31. Rajapakse H.A. (2007). *Curr Top Med Chem* 7: 966–971.

32. Vucic D. (2008). *Curr Cancer Drug Targets* 8: 110–117.

33. Shiozaki E.N., Shi Y. (2004). *Trends Biochem Sci* 29: 486–494.

34. Eckelman B.P., Salvesen G.S., Scott F.L. (2006). *EMBO Rep* 7: 988–994.

336 T. Berg

35. Li L., Thomas R.M., Suzuki H., De Brabander J.K., Wang X., Harran P.G. (2004). *Science* 305: 1,471–1,474.
36. Sun H., Nikolovska-Coleska Z., Lu J., Meagher J.L., Yang C.Y., Qiu S., Tomita Y., Ueda Y., Jiang S., Krajewski K., Roller P.P., Stuckey J.A., Wang S. (2007). *J Am Chem Soc* 129: 15,279–15,294.
37. Moses J.E., Moorhouse A.D. (2007). *Chem Soc Rev* 36: 1,249–1,262.
38. Nikolovska-Coleska Z., Meagher J.L., Jiang S., Yang C.Y., Qiu S., Roller P.P., Stuckey J.A., Wang S. (2008). *Biochemistry* 47: 9811-9824.
39. Petronczki M., Lénárt P., Peters J.M. (2008). *Dev Cell* 14 : 646–659.
40. Barr F.A., Sillje H.H., Nigg E.A. (2004). *Nat Rev Mol Cell Biol* 5: 429–440.
41. Strebhardt K., Ullrich A. (2006). *Nat Rev Cancer* 6: 321–330.
42. McInnes C., Mezna M., Fischer P.M. (2005). *Curr Top Med Chem* 5: 181–197.
43. Steegmaier M., Hoffmann M., Baum A., Lenart P., Petronczki M., Krssak M., Gurtler U., Garin-Chesa P., Lieb S., Quant J., Grauert M., Adolf G.R., Kraut N., Peters J.M., Rettig W.J. (2007). *Curr Biol* 17: 316–322.
44. Gumireddy K., Reddy M.V., Cosenza S.C., Boominathan R., Baker S.J., Papathi N., Jiang J., Holland J., Reddy E.P. (2005). *Cancer Cell* 7: 275–286.
45. Daub H. (2005). *Biochim Biophys Acta* 1754: 183–190.
46. Daub H., Specht K., Ullrich A. (2004). *Nat Rev Drug Discov* 3: 1,001–1,010.
47. Bantscheff M., Eberhard D., Abraham Y., Bastuck S., Boesche M., Hobson S., Mathieson T., Perrin J., Raida M., Rau C., Reader V., Sweetman G., Bauer A., Bouwmeester T., Hopf C., Kruse U., Neubauer G., Ramsden N., Rick J., Kuster B., Drewes G. (2007). *Nat Biotechnol* 25: 1,035–1,044.
48. Elia A.E., Cantley L.C., Yaffe M.B. (2003). *Science* 299: 1,228–1,231.
49. Hanisch A., Wehner A., Nigg E.A., Sillje H.H. (2006). *Mol Biol Cell* 17: 448–459.
50. Elia A.E., Rellos P., Haire L.F., Chao J.W., Ivins F.J., Hoepker K., Mohammad D., Cantley L.C., Smerdon S.J., Yaffe M.B. (2003). *Cell* 115: 83–95.
51. Reindl W., Strebhardt K., Berg T. (2008). *Anal Biochem* 383: 205-209.
52. Reindl W., Yuan J., Krämer A., Strebhardt K., Berg T. (2008). *Chem Biol* 15: 459-466.
53. Feki A., Irminger-Finger I. (2004). *Crit Rev Oncol Hematol* 52: 103–116.
54. Hainaut P., Hollstein M. (2000). *Adv Cancer Res* 77: 81–137.
55. Dudkina A.S., Lindsley C.W. (2007). *Curr Top Med Chem* 7: 952–960.
56. Vassilev L.T. (2007). *Trends Mol Med* 13: 23–31.
57. Dömling A. (2008). *Curr Opin Chem Biol* 12: 281–291.
58. Vassilev L.T., Vu B.T., Graves B., Carvajal D., Podlaski F., Filipovic Z., Kong N., Kammlott U., Lukacs C., Klein C., Fotouhi N., Liu E.A. (2004). *Science* 303: 844–848.
59. Ding K., Lu Y., Nikolovska-Coleska Z., Qiu S., Ding Y., Gao W., Stuckey J., Krajewski K., Roller P.P., Tomita Y., Parrish D.A., Deschamps J.R., Wang S. (2005). *J Am Chem Soc* 127: 10,130–10,131.

60. Ding K., Lu Y., Nikolovska-Coleska Z., Wang G., Qiu S., Shangary S., Gao W., Qin D., Stuckey J., Krajewski K., Roller P.P., Wang S. (2006). *J Med Chem* 49: 3,432–3,435.
61. Saddler C., Ouillette P., Kujawski L., Shangary S., Talpaz M., Kaminski M., Erba H., Shedden K., Wang S., Malek S.N. (2008). *Blood* 111: 1,584–1,593.
62. Shangary S., Qin D., McEachern D., Liu M., Miller R.S., Qiu S., Nikolovska-Coleska Z., Ding K., Wang G., Chen J., Bernard D., Zhang J., Lu Y., Gu Q., Shah R.B., Pienta K.J., Ling X., Kang S., Guo M., Sun Y., Yang D., Wang S. (2008). *Proc Natl Acad Sci USA* 105: 3,933–3,938.
63. Rothweiler U., Czarna A., Krajewski M., Ciombor J., Kalinski C., Khazak V., Ross G., Skobeleva N., Weber L., Holak T.A. (2008). *Chem Med Chem* 51: 5035–5042.
64. Krajewski M., Rothweiler U., D'Silva L., Majumdar S., Klein C., Holak T.A. (2007). *J Med Chem* 50: 4,382–4,387.
65. Darnell J.E. Jr (2002). *Nat Rev Cancer* 2: 740–749.
66. Schindler C., Levy D.E., Decker T. (2007). *J Biol Chem* 282: 20,059–20,063.
67. Buettner R., Mora L.B., Jove R. (2002). *Clin Cancer Res* 8: 945–954.
68. Niu G., Heller R., Catlett-Falcone R., Coppola D., Jaroszeski M., Dalton W., Jove R., Yu H. (1999). *Cancer Res* 59: 5,059–5,063.
69. Catlett-Falcone R., Landowski T.H., Oshiro M.M., Turkson J., Levitzki A., Savino R., Ciliberto G., Moscinski L., Fernandez-Luna J.L., Nunez G., Dalton W.S., Jove R. (1999). *Immunity* 10: 105–115.
70. Redell M.S., Tweardy D.J. (2005). *Curr Pharm Des* 11: 2,873–2,887.
71. Xi S., Gooding W.E., Grandis J.R. (2005). *Oncogene* 24: 970–979.
72. Leong P.L., Andrews G.A., Johnson D.E., Dyer K.F., Xi S., Mai J.C., Robbins P.D., Gadiparthi S., Burke N.A., Watkins S.F., Grandis J.R. (2003). *Proc Natl Acad Sci USA* 100: 4,138–4,143.
73. Barton B.E., Murphy T.F., Shu P., Huang H.F., Meyenhofer M., Barton A. (2004). *Mol Cancer Ther* 3: 1,183–1,191.
74. Konnikova L., Kotecki M., Kruger M.M., Cochran B.H. (2003). *BMC Cancer* 3: 23.
75. Ling X., Arlinghaus R.B. (2005). *Cancer Res* 65: 2,532–2,536.
76. Gao L., Zhang L., Hu J., Li F., Shao Y., Zhao D., Kalvakolanu D.V., Kopecko D.J., Zhao X., Xu D.Q. (2005). *Clin Cancer Res* 11: 6,333–6,341.
77. Borghouts C., Kunz C., Delis N., Groner B. (2008). *Mol Cancer Res* 6: 267–281.
78. Nagel-Wolfrum K., Buerger C., Wittig I., Butz K., Hoppe-Seyler F., Groner B. (2004). *Mol Cancer Res* 2: 170–182.
79. Jing N., Li Y., Xiong W., Sha W., Jing L., Tweardy D.J. (2004). *Cancer Res* 64: 6,603–6,609.
80. Jing N., Zhu Q., Yuan P., Li Y., Mao L., Tweardy D.J. (2006). *Mol Cancer Ther* 5: 279–286.
81. Fletcher S., Turkson J., Gunning P.T. (2008). *ChemMedChem* 3: 1,159–1,168.

82. Aggarwal B.B., Sethi G., Ahn K.S., Sandur S.K., Pandey M.K., Kunnumakkara A.B., Sung B., Ichikawa H. (2006). *Ann N Y Acad Sci* 1091: 151–169.
83. Desrivieres S., Kunz C., Barash I., Vafaizadeh V., Borghouts C., Groner B. (2006). *J Mammary Gland Biol Neoplasia* 11: 75–87.
84. Darnell J.E. (2005). *Nat Med* 11: 595–596.
85. Klampfer L. (2006). *Curr Cancer Drug Targets* 6: 107–121.
86. Leeman R.J., Lui V.W., Grandis J.R. (2006). *Expert Opin Biol Ther* 6: 231–241.
87. Deng J., Grande F., Neamati N. (2007). *Curr Cancer Drug Targets* 7: 91–107.
88. Al Zaid Siddiquee K., Turkson J. (2008). *Cell Res* 18: 254–267.
89. Wittig I., Groner B. (2005). *Curr Drug Targets Immune Endocr Metabol Disord* 5: 449–463.
90. Xi S., Zhang Q., Gooding W.E., Smithgall T.E., Grandis J.R. (2003). *Cancer Res* 63: 6,763–6,771.
91. Mohapatra S., Chu B., Wei S., Djeu J., Epling-Burnette P.K., Loughran T., Jove R., Pledger W.J. (2003). *Cancer Res* 63: 8,523–8,530.
92. Demoulin J.B., Uyttenhove C., Lejeune D., Mui A., Groner B., Renauld J.C. (2000). *Cancer Res* 60: 3,971–3,977.
93. Berg T. (2008). *ChemBioChem* 9: 2,039–2,044.
94. Turkson J., Ryan D., Kim J.S., Zhang Y., Chen Z., Haura E., Laudano A., Sebti S., Hamilton A.D., Jove R. (2001). *J Biol Chem* 276: 45,443–45,455.
95. Coleman D.R., Ren Z., Mandal P.K., Cameron A.G., Dyer G.A., Muranjan S., Campbell M., Chen X., McMurray J.S. (2005). *J Med Chem* 48: 6,661–6,670.
96. Turkson J., Kim J.S., Zhang S., Yuan J., Huang M., Glenn M., Haura E., Sebti S., Hamilton A.D., Jove R. (2004). *Mol Cancer Ther* 3: 261–269.
97. Siddiquee K.A., Gunning P.T., Glenn M., Katt W.P., Zhang S., Schroeck C., Sebti S.M., Jove R., Hamilton A.D., Turkson J. (2007). *ACS Chem Biol* 2: 787–798.
98. Song H., Wang R., Wang S., Lin J. (2005). *Proc Natl Acad Sci USA* 102: 4,700–4,705.
99. Siddiquee K., Zhang S., Guida W.C., Blaskovich M.A., Greedy B., Lawrence H.R., Yip M.L., Jove R., McLaughlin M.M., Lawrence N.J., Sebti S.M., Turkson J. (2007). *Proc Natl Acad Sci USA* 104: 7,391–7,396.
100. Schust J., Berg T. (2004). *Anal Biochem* 330: 114–118.
101. Schust J., Sperl B., Hollis A., Mayer T.U., Berg T. (2006). *Chem Biol* 13: 1,235–1,242.
102. Muller J., Schust J., Berg T. (2008). *Anal Biochem* 375: 249–254.
103. Muller J., Sperl B., Reindl W., Kiessling A., Berg T. (2008). *Chembiochem* 9: 723–727.
104. Adhikary S., Eilers M. (2005). *Nat Rev Mol Cell Biol* 6: 635–645.
105. Cowling V.H., Cole M.D. (2006). *Semin Cancer Biol* 16: 242–252.
106. Vita M., Henriksson M. (2006). *Semin Cancer Biol* 16: 318–330.
107. Berg T. (2008). *Curr Opin Chem Biol* 12: 464–471.

108. Nair S.K., Burley S.K. (2003). *Cell* 112: 193–205.
109. Felsher D.W. (2006). *Cell Cycle* 5: 1,808–1,811.
110. Berg T., Cohen S.B., Desharnais J., Sonderegger C., Maslyar D.J., Goldberg J., Boger D.L., Vogt P.K. (2002). *Proc Natl Acad Sci USA* 99: 3,830–3,835.
111. Lu X., Vogt P.K., Boger D.L., Lunec J. (2008). *Oncol Rep* 19: 825–830.
112. Yin X., Giap C., Lazo J.S., Prochownik E.V. (2003). *Oncogene* 22: 6,151–6,159.
113. Wang H., Hammoudeh D.I., Follis A.V., Reese B.E., Lazo J.S., Metallo S.J., Prochownik E.V. (2007). *Mol Cancer Ther* 6: 2,399–2,408.
114. Xu Y., Shi J., Yamamoto N., Moss J.A., Vogt P.K., Janda K.D. (2006). *Bioorg Med Chem* 14: 2,660–2,673.
115. Kiessling A., Sperl B., Hollis A., Eick D., Berg T. (2006). *Chem Biol* 13: 745–751.
116. Kiessling A., Wiesinger R., Sperl B., Berg T. (2007). *Chem Med Chem* 2: 627–630.
117. Nabors L.B., Mikkelsen T., Rosenfeld S.S., Hochberg F., Akella N.S., Fisher J.D., Cloud G.A., Zhang Y., Carson K., Wittemer S.M., Colevas A.D., Grossman S.A. (2007). *J Clin Oncol* 25: 1,651–1,657.
118. Schreiber S. L. (2005). *Nat Chem Biol* 1: 64–66.

CHAPTER 13

Protein Dynamics and Drug Design: The Role of Molecular Simulations

Giulia Morra, Alessandro Genoni, Giorgio Colombo

Istituto di Chimica del Riconoscimento Molecolare, CNR,
Via Mario Bianco 9, 20131 Milano, Italy
E-mail: giorgio.colombo@icrm.cnr.it

The motions of proteins underlie all processes in cells, ranging from substrate transport to signal transmission, trafficking, formation of complexes and catalysis. Taking dynamics into account in molecular recognition may hold great promise in understanding the determinants of complex formation, in the identification of new binding sites and in the discovery of new drugs. Several groups have started tackling these problems with the use of simulation methods. The study of ligand-induced dynamic variations has also been exploited to review the concept of allosteric changes. The dynamics of proteins and complexes has also been used to develop pharmacophore models based on ensembles of protein conformations. These models, taking flexibility explicitly into account, are able to distinguish active inhibitors vs non-active drug-like compounds, to define new molecular motifs and to preferentially identify specific ligands for a certain protein target. In this chapter, examples illustrating how simulations can be used to understand dynamics in relation to ligand binding and eventually to drug design will be presented. Finally, we will present two examples illustrating the utility of including dynamics in the design process of inhibitors against a well-defined protein receptor and against the formation of self-aggregated peptide oligomers.

13.1 Introduction

Proteins control most of the fundamental biochemical pathways in the cell. They are not static entities. On the contrary, they are subject to constant motions and interactions that determine their recognition and functional properties.

Protein conformational dynamics is fundamental in understanding their functional mechanisms. Hemoglobin represents a paradigmatic example of this: since the early days of protein crystallography and biochemical studies, it has been clear that the protein can be stabilised in two considerably different structures depending on its degree of oxygenation.

Conformational dynamics can also be exploited by enzymes to define and control reactivity in catalytic cycles. Dihydropholate reductase (DHFR) for instance, catalyses the reduction of dihydropholate to tetrahydropholate with the concurrent oxidation of the NADPH cofactor. In order to carry out the reaction, the catalytic residues of the enzyme, the substrate and the cofactor must be optimally arranged in space.[1,2] DHFR is characterised by the presence of a flexible loop, the M20 loop, that can access three different conformations: closed, open and occluded.[3] When both the substrate and cofactor are bound, the closed conformation is favoured over the others and the M20 loop is packed against the nicotinamide ring of NADPH. The closed form is the only conformation of the enzyme in which the substrate and the cofactor are optimally aligned for the reaction.

Flexibility and dynamics also play a primary role in protein–protein interactions. Protein–protein complex formation often involves structural changes that may extend well beyond local scale re-arrangements and that cannot be understood in terms of the lock-and-key model,[4] or be described by simple models based only on surface accessibility. In forming a complex with a given partner, they must adopt specific structures. These may be pre-existing in the accessible conformational space of the protein and the conformational equilibrium may be shifted towards them upon binding.[5] Otherwise specific structures are induced after the formation of a complex with the binding partner. In this context, it is important to underline that proteins very seldom function in isolation. Rather, they typically work within an ensemble, so that they must be sufficiently flexible to interact with multiple partners and carry out diverse tasks.

The importance of taking the highly dynamic nature of protein and peptide complexes into account has emerged also in the field of aggregation and self-organisation. Aggregation of peptides and proteins into cytotoxic oligomers and formation of insoluble amyloid fibrils have been recognised as the central molecular event in more than 20 human pathogenic conditions.[6] Drugs able to interfere with these processes may cure relevant diseases and are thus urgently needed. Recently, soluble beta-sheet oligomeric species have been identified as the main cause of cytotoxicity,[7–12] making them clear targets for the design of antiamyloidogenic drugs. However, the transient and highly dynamic nature of these early oligomers has hampered the characterization of their structural-dynamical properties at the atomic level of resolution. The lack of detailed structural models for the different species on- and off- the amyloid pathway that might represent potential drug targets has seriously limited the potential of rational drug-design and hindered efforts to rationally improve the efficacy of lead molecules. Detailed analysis of the structure and dynamics of atomically detailed structural models of peptide oligomers may help unveil the molecular determinants of aggregation, opening up the possibility to design new molecules mimicking fundamental interactions and with the ability to derail peptide oligomers from toxic pathways.

Finally, the possibility to alternate and to trap different specific states through ligand binding at sites distal from the active site represents one of the most common and powerful means of regulating protein function (allosteric regulation) in cellular metabolism. Allosteric regulation does not necessarily imply major conformational changes. Recent thermodynamic and conformational analyses have shown that allosteric communications can be transmitted simply through variations in protein dynamics that do not determine a macroscopic change of shape. The identification and targeting of allosteric sites have thus become the focus of both basic research and several drug design efforts. From the basic point of view, the study of allosteric sites and signal transduction mechanisms will provide precious insights into how nature has evolved control strategies in protein architectures. From the pharmaceutical point of view, these sites offer unique opportunities to discover new chemical entities for enzymes and other targets for which the discovery of active-site inhibitors has proven challenging.

Based on the simple examples and considerations reported, it appears clear that the study of the influence of dynamics on different aspects of protein properties (form function to molecular recognition) may be of great help in the development of new molecules with specific activities. Computational and theoretical approaches offer a unique means to investigate these aspects at different levels of resolution, and to include them into the processes of rational design and discovery of new drug-like molecules. Including target dynamics appears to hold a great deal of promise in advancing structure-based molecular design and medicinal chemistry in general, allowing to optimise molecules targeting active sites and to determine the presence of allosteric sites for which inhibitors and/or modulators can be developed. The application of these methods will broaden the chemical space of available active molecules providing access to new chemotypes that bind known receptors and possible new proteins.

In this chapter, we will present examples illustrating the extent to which simulations can be used to understand dynamic properties of proteins along with examples of how to include flexibility in the rational design of new molecules. In the final part of this chapter, in particular, we will illustrate two examples in which the description of protein dynamics is instrumental to the identification of new molecules with specific inhibitory properties. In the first case, we will discuss the targeting of a well-defined active site in a chaperone protein fundamental for cancer development. In the second example, we will discuss the use of molecular dynamics simulations to identify the molecular determinants of the inhibitory core of a series of small peptides able to interfere with amyloidogenesis. We will show how these inhibitory determinants can be used as structural templates in the development of pharmacophore models for the identification of novel non-peptidic, small molecule inhibitors of aggregation.

13.2 Dynamic Regulation of Protein Function

Allostery indicates the coupling of conformational changes between two different sites of a protein. The term 'allosteric inhibition' was introduced in 1961 by Monod and Jacob[13] to describe inhibitory interaction with a site distinct from the active site of an enzyme. Several enzyme systems can exhibit heterotrophic regulation (induced by small molecules that are distinct from the primary ligand) or positive homotropic regulation, where the binding of the primary ligand at one site cooperatively increases the affinity for the same ligand at another site. The latter is the case for haemoglobin where oxygen is both the primary ligand and the allosteric effector.

The Monod–Wyman–Changeux (MWC)[14] and Koshland–Nemethy–Filmer (KNF)[15] models were both proposed in the 1960s to explain the allosteric regulation seen in hemoglobin and in several metabolic regulatory enzymes. The concerted (MWC) model suggests that allosteric proteins are symmetric oligomers with identical protomers. Each protomer can exist in two conformational states (T and R) and the protein interconverts between the two conformations in a concerted manner. The conformational transition is therefore seen as a concerted action between two co-existing discrete states (T and R), 'with the ligand stabilizing the conformation to which it binds with higher affinity'.[14] On the other hand, in the sequential model, subunits change conformation one at time. Here, the binding of a ligand changes the conformation of one protomer without affecting the other subunits.

Although originally introduced for oligomeric proteins, allostery does not require the protein to be multimeric and the modulation of protein function by means of interactions located at long distance from the active site is now recognised as a general feature of many monomeric proteins, as pointed out in Ref. 16. For instance, myoglobin can be allosterically regulated.[16]

From the very beginning, it was suggested that allostery might act as a general form of protein regulation, as it constitutes a direct and efficient mechanism for the modulation of cellular function in response to changes in concentration of small molecules. This led to the suggestion that allosteric sites might be even more useful as drug targets than active sites (see Ref. 17 and references therein). The main advantage of exploiting allosteric effects in drug design is the specificity of the targets. While active sites are overall well conserved within whole protein families, such that it is difficult to target selected members of the family, allosteric sites are not evolutionarily constrained and may be much more specific to single proteins.[18]

However, the prediction of allosteric phenomena in proteins and the identification of the involved sites and effectors have not been subject of as intense research as other problems in molecular biology, like for example protein structure and function determination. Due to the heterogeneous chemical nature of allosteric modulators and the lack of systematic techniques to discover them,[19] a methodological development is required.

Herein, we will present theoretical and computational approaches to the study of protein dynamics and allostery, with an eye on drug-design.

13.2.1 *Theory*

Protein motions in the conformational space, such as local folding/unfolding re-arrangements and also allosteric changes, have a cooperative character. The protein behaves like a cohesive unit, where any change results from the concerted contribution of all parts, and is not the sum of their properties. In the case of allostery, cooperativity explains why local phenomena such as phosphorylation, ATP binding or hydrolysis can switch-on large-scale conformational transitions.

Any conformational change involves a free energy variation, resulting from a net combination of enthalpy and entropy changes. For instance, a favourable enthalpy change upon ligand binding is due to the formation of tighter attractive interactions, which in turn may reduce the protein entropy due to a loss of flexibility of the molecule, or increase the solvent entropy by means of the hydrophobic effect. The balance between all these contributions yields the net free energy change that makes the binding favourable or unfavourable. In the case of allostery, at least two such binding events occur showing cooperativity. This cooperativity can be positive, when the second binding event becomes more favourable in the presence of the first ligand (allosteric activator), or negative, when the second binding event is more costly due to the previous binding of the first ligand (allosteric inhibitor).

The current view of protein allostery is connected to the energy landscape model of protein folding, defining the native state of a protein as a conformational ensemble. Since proteins are not rigid, the native state exists as a statistical ensemble with a number of local minima, separated by locally unfolded regions given by transiently populated higher free energy states. The rough profile at the bottom of the protein folding funnel constitutes the range of the allosterically accessible folded substates. Binding the allosteric effector induces a shift in the population of states, by stabilising one conformational state at the expense of others, thus reducing the heterogeneity of the native ensemble. The stabilised conformation will be in turn more or less prone to binding the second ligand.

This view is related to the conformational selection model,[20] which re-interprets the relationship between ligand binding and conformational change in an enzyme (traditionally referred to as the 'induced fit' mechanism).[21] The conformational selection model is based on the assumption that structural changes observed in the presence of different substrates are already accessible in the absence of the ligand (see Ref. 22 and references therein).[23] Evidence of a pre-existing equilibrium between conformational states was provided by several experiments and also by computational studies.[24] An early indication of 'pre-sampling' of the conformational change came from molecular dynamics studies on myoglobin, which showed that the largest fluctuations are located in the same protein regions that experience the largest conformational change between the limiting static X-ray structures.[25] More importantly, not only structural changes, but also motions involved in biophysical (e.g. substrate recognition and binding), biochemical (e.g. catalysis) or biological (e.g. signalling, transcription) activities are observed to be pre-sampled before protein activity. This broadens the view of conformational changes upon binding, and also of allosteric changes, to include different dynamical states. The pre-existence of motions characteristic of catalytic function was detected by the Kern group in the case of cyclophilin.[26] Computational studies based on molecular dynamics simulations on the N-terminal domain of human Hsp90 pointed out, in agreement with the conformational selection model, that different functional motions, – as detected by essential dynamics analysis, and elicited by different ligands – are basically all present in the free state of the molecule.[27]

The view of allostery in terms of conformational and dynamic selection offers a generalization in the same spirit of the MWC model to a wider set of conformations and dynamical states than two single discrete structures, like the T and R conformations of hemoglobin.

The role of entropy and dynamics in allosteric communication between distant binding sites has been recently reviewed by Kern[28] and Nussinov.[5] Given the thermodynamic nature of allostery, communication across the protein can be mediated not only by enthalpy, by means of protein backbone re-arrangements, but also by entropy, via changes in the dynamic fluctuations around the mean structure.

An entropic effect in positive cooperativity might generally be present, as recognised by Cooper and Dryden in the 1980s.[29] They formalised an entropic component to the allosteric interaction free energy, which can be estimated in some kcal/mole as a function of tiny rms changes of atoms and is due to the propagation of protein rigidity.[29] Namely, binding increases packing and locally rigidifies the protein structure. In an allosteric protein containing two or more binding sites, each site may cause partial rigidification upon individual binding. The global rigidification of the protein due to the binding of both ligands might partly be accomplished by either of the ligand binding steps, inducing positive cooperativity for the binding of the second ligand.

13.2.2 *Molecular Basis of Allostery: Insights from Theory and Simulations*

Signal transduction along well-defined paths connecting distant residues has been recognised as a common property of allosteric proteins. In the understanding of allosteric effects, theory aims at describing the site to site communication at the molecular or atomic level. A number of attempts have been carried out by means of sequence or structure-based computational methods, in order to predict allosteric sites and pathways in proteins.

Based on the sequence, the statistical site coupling approach (SCA) of Ranganathan allows the identification of evolutionarily conserved couplings between residue pairs.[30,31] The method is based on the assumption that if the interaction between a pair of residues is relevant for protein function or folding, then the two residues should co-evolve, showing a relevant covariation in multiple sequence alignments (MSA) of a protein family. A statistical energy term can be derived from this covariance. This method was originally applied to the human tyrosine phosphatase PDZ domain family and led to the identification of residue patterns and networks linking active sites to distant regions, predicting a signalling pathway subsequently confirmed by NMR experimnts.[32]

In the analysis of allosteric properties of PDZ domain, a structure-based approach making use of non-equilibrium MD simulations was introduced by Ota and Agard.[33] The method, called anisotropic thermal diffusion, relies on the assumption that the communication properties of residues depend on the energetic couplings among them and signals are driven by the same pathways on which the kinetic energy, originated when perturbing a site, dissipates through the structure. Therefore, a non-equilibrium MD simulation is set up, where a significant perturbation is created by artificially heating a single residue with a temperature bath of 300K, while the rest of the protein is thermalised at low temperature. The dissipation of the thermal fluctuation, tracked along the structure during dynamics, turns out to be highly anisotropic and the presence of preferred signalling pathways can be detected, in agreement with experimental data and the sequence-based computational methods of Ranganathan.

Another application of the statistical coupling analysis (SCA), in combination with a coarse-grained Brownian dynamic simulation, was introduced by Chien *et al.*[34] to analyse the signalling network of *E. coli dihydrofolate reductase.* This enzyme undergoes an allosteric conformational transition involving the Met20 loop during the catalytic cycle. By means of the SCA, a number of highly covarying residues is hypothesised to be the minimal network signalling the kinetics of the conformational transition. A long term simulation of the protein using the Self-Organizing Polymer (SOP) coarse-grained description is then produced to generate the covariance map, showing that correlated and anticorrelated motions on the microsecond time scale involve the same conserved residues, even if they are spatially separated.

The role of low frequency, collective modes in mediating the signal propagation along given pathways is emerging as a plausible mechanism to explain allosteric effects. Recently, Bahar *et al.* introduced a novel information-theoretic approach called MAPS (MArkovian Propagation of Signals) to study the transition of GroEL from the T (apo) state into the R (ADP and ATP bound, respectively to cis and trans rings) state.[35] The protein structure is modelled as a network of interactions, where the interaction strength between two residues depends on the number of atom–atom contacts between the two. When a site is perturbed, the perturbation propagates through the network, being dissipated by means of coupled fluctuations in residue positions (fluctuation–dissipation) and eventually relaxing to a new equilibrium. It has been shown that signal propagation time is shorter when pairs of residues are subject to smaller fluctuations in their inter residue distance, which is the case for instance for tightly interacting residues. By treating the network with a clustering hierarchical reduction, sites with a 'high allosteric potential' can be identified while communication pathways can be defined by tracking the maximum-likelihood paths along the network. The analysis of the ADP-bound GroEL–GroES structure reveals strong inter-subunit couplings between the cis and the trans ring and with the co-chaperonin, and a number of residues or chain regions are correctly identified as important for allosteric communication. Moreover, the global motions of the complex, described by the lowest eigenmode in an ANM representation, are responsible for modulating the interactions of the allosterically relevant residues at the interface of the subunits.

The analysis of the inter-residue distance fluctuations has been applied also in the framework of molecular dynamics simulations to investigate the large-scale conformational re-arrangements of the yeast Hsp90 dimer upon ATP binding and hydrolysis. The perspective of MD is useful in capturing the global changes occurring upon ligand binding, without neglecting the details of the side chain orientation at the local level. In fact, the subtle changes in side chain arrangement constitute the first step in response to a binding event and then accumulate to produce a global change in the structure. The communication propensity between the ATP binding site and the distant C-terminal interface (more than 80 Angstrom apart), measured in terms of inter-residue distance fluctuations, is increased in the presence of ATP or ADP and is switched off in the unbound complex. Moreover, ATP and ADP appear to communicate with distinct regions of the C-terminal area, indicating a specific allosteric effect of each ligand.[36]

Another network approach based on the topology of the protein, which also makes use of MD simulations to take side chain details into account, is the protein structure network (PSN) approach of Vishveshwara et al.[37,38] The method of Protein Structure Network analysis has been applied to an allosteric protein, member of the class of aminoacyl tRNA synthetases and responsible for picking up the cognate amino acid in the active site and cognate tRNA in the anticodon binding site. The method consists of carrying out molecular dynamics simulations of the protein in different liganded states, and then for each dynamic snapshot constructing a network of interactions between the residues where each residue is a node, and an edge between two residues represents the interaction defined by the fraction of atom pairs that are closer than 4.5 Angstroms. These structure networks evolve during dynamics and can be used to find the communication pathways between two endpoints[37] (selecting the shortest and most dynamically correlated path). The paths turn out to be inherently present in the enzyme. They establish efficient communication at distance of 70 Angstroms between the anticodon region and the activation site, in the presence of both the tRNA and MetAMP that drive specific residue–residue contacts. The analysis of the graph in terms of cliques (sets of mutually connected nodes), communities of cliques (non-identical cliques sharing nodes) and hubs (nodes making more than four edges) also reveals a different dynamic behaviour for communication paths, which are flexible and robust. The network parameters describing the non-covalent interaction of side chains are able to discriminate subtle differences in the side chain interactions, taking place in response to different ligands.[38]

The knowledge of the allosteric pathways transmitting signals in a protein allows in principle to regulate its function by means of suitable small molecules that can bind to the allosteric sites, but also by designing ad hoc mutations able to interfere with the protein (mis)function. A suggestion in this sense comes from a recent study by Liu and Nussinov on the pVHL protein, a tumour suppressor protein. A naturally occurring mutant of this protein with reduced thermodynamic stability is associated to a number of cancer diseases.[39] By means of MD simulations the region mainly responsible for the low stability of the protein is identified. In order to avoid mutations directly on this area, which would perturb the protein function, regions dynamically correlated with it are found by analysing the covariance matrix and mutations are inserted there. The simulation of the new mutants suggests a repairing effect of the mutations that appear to be able to restore the wild type stability.

The next section will be based on the description of computational approaches exploiting flexibility for drug design.

13.3 Protein Flexibility in Drug Design and Discovery

Nowadays, the most sophisticated and rational procedures to discover novel drugs consist in computer-based approaches. All these strategies can be generally classified into two main categories: *de novo* drug-design methods and database virtual screening techniques. The former look for new possible ligands of therapeutic relevance that fit in a complementary way the electrostatic and hydrophobic properties of the target receptor, whereas the latter apply one or more filters to decide which compounds collected in given databases can be considered as potential drug leads.

In spite of the subdivision into the two main groups described above, most of the computational drug-design strategies use a rigid structure for the target protein, although the importance of the receptor flexibility has been very well known for a long time. In fact, proteins are such flexible systems that they can undergo quite large conformational changes and it is worthwhile to observe that these re-arrangements are often strongly associated with fundamental protein functions. Accordingly, proteins can be subdivided into three categories: (a) 'rigid' proteins, which show only very small side chain movements upon ligand binding; (b) flexible proteins, whose ligand-induced changes are relatively large re-arrangements in the neighbourhood of hinge points or at the active site in conjunction with side chain motions; (c) unstable proteins, which are characterised by a dynamic equilibrium between several conformations often structurally similar and at the same time energetically different.[40] This classification is strongly connected with the three different models that have been proposed in order to explain the ligand–protein interactions. The simplest one is the well-known 'Lock and Key' model introduced by Fisher.[4] It describes the chemical complementarity between a receptor and the corresponding ligand as the complementarity between a lock and a key and, of course, it works when the protein is strictly rigid. Unfortunately, when the receptor flexibility plays a fundamental role in the molecular recognition process, it is necessary to consider the more advanced 'Induced Fit' description proposed by Koshland[41] according to which both the protein and ligand structures change during the binding process as a result of the introduction of the substrate in the chemical and structural environment of the receptor. Finally, for really unstable proteins, we have to take into account a more modern model[42] that describes the molecular recognition as a process in which the substrate selects the most suited receptor conformation among an ensemble of metastable states, shifting the dynamic population equilibrium towards the configuration adopted by the bound protein. These models clearly point out that both experimentally and theoretically derived rigid structures are not suitable when dealing with very flexible receptors, whose conformational changes are driven by ligand binding and are crucial to express protein functions. The reason why most of the present drug-discovery strategies do not take into account this aspect could be ascribed to the ill-founded belief that the crystal structure is always the correct structure and to the conceptual and technical problems of working with non-defined targets. Therefore, the need to devise new computational methods able to take into account and evaluate dynamic re-arrangements in molecular recognition processes is really compelling and, in this review, we will show some of the current efforts made by the drug-design community to accomplish this task. It is important to note that, among them, there are also noteworthy techniques that efficiently couple Molecular Dynamics (MD) or Monte Carlo (MC) simulations with docking experiments. We believe that computer simulation strategies are probably one of the best tools to get a more complete set of receptor conformations that can be afterwards used in docking or virtual screening methods. This is especially so if we also want to consider high-energy configurations that cannot be detected by current experimental techniques, such as X-ray crystallography and Nuclear Magnetic Resonance. Nevertheless, it is extremely worth noting that, due to some non-negligible drawbacks such as the stochastic nature of Molecular Dynamics and the rather limited length of simulations, we often sample only a small subset of the protein conformational space.

According to the classification introduced by Teodoro and Kavraki,[43,44] the strategies that consider receptor flexibility in modelling of protein–ligand interactions can be subdivided into five categories, which will be discussed with more details in the following sections: (a) soft potential methods; (b) techniques that take into account a reduced number of degrees of freedom; (c) strategies which exploit multiple receptor structures; (d) modified computer simulations (MD or MC) methods; and (e) strategies that use collective degrees of freedom in order to represent very large protein conformational re-arrangements. In general, the accuracy provided by the different approaches is directly proportional to the associated computational complexity and, for the sake of completeness, it is also important to stress that the proposed subdivision is not so strict because many of the strategies that we will describe may belong to more than one category.

Soft potential methods are probably the simplest techniques that take into account a part of receptor flexibility and they essentially consist in a reduction of the van der Waals contributions to the total energy score. This allows relaxing the high-energy penalty associated with clashes between ligand and receptor atoms, and this is the reason why larger ligands are able to fit in binding sites suited for smaller molecules. It is worth to note that the 'Induced Fit' model underlies this kind of approaches because the receptor is allowed to undergo some changes, both in backbone and side chains, in order to respond to the presence of the new ligand. These strategies are very useful for their easy implementation and computational speed, but great attention has to be paid in the definition of the 'soft region' because if it is too large (and this is usually the case when we want to account for large protein re-arrangements) errors can arise. Among this group of methods, we believe that the approach developed by Apostolakis and co-workers[45] is particularly interesting. At the initial stage, they apply the softening of the van der Waals terms in order to avoid high-energy gradients due to steric clashes and then gradually restoring the standard values for the non-bonded energy contributions, a minimization process is performed to slowly adapt each other the receptor and the ligand.

As mentioned above, another possibility in dealing with protein flexibility consists in choosing those fundamental degrees of freedom, usually torsions in active-site side chains, that we want to model explicitly. Selecting is probably the trickiest part of these techniques, and to accomplish this task both experimental and theoretical (e.g. from MD or MC simulations) *a priori* knowledge about alternative binding modes for the receptor in exam is usually necessary. To solve this problem, Anderson *et al.*[46] have developed an interesting algorithm that at first, starting from only a single protein structure, identifies the most flexible zones of the receptor (namely, those that are the most involved in the binding process) and then afterwards, using rotamer libraries, predicts the most likely conformations for the previously detected regions. In this way the bias due to structures of proteins co-crystallized with inhibitors is sensitively reduced in following docking experiments.

The first attempt of modelling the receptor flexibility using a well determined subset of degrees of freedom was made by Leach.[47] Exploring the protein and ligand conformational spaces by means of the Dead End Elimination (DDE)[48] and the A* algorithm,[49] he was able to find for each orientation of the ligand relative to the receptor, potential binding structures corresponding to combinations of residue side chains and ligand conformations whose energy is lower than a predetermined threshold. In this context, Schaffer and Verkhivker[50] predicted the structures of complexes between two HIV-1 protease inhibitors and two mutants of this protein. In this case they performed flexible docking computations that combine the MC simulated annealing technique to determine the ligand bound conformation and the DDE for the side chain optimization of the receptor binding site residues.

Moreover, it is worthwhile to note that some popular docking programmes such as AutoDock[51,52] and GOLD[53] have introduced the possibility to explicitly deal with the side chains flexibility. In particular, GOLD uses a genetic algorithm that allows describing not only ligand positions and conformations, but also hydrogen-bonding networks in the active site. Another GOLD-based strategy is the one devised by Verdonk *et al.*, [54] where only a small subset of protein side chains and terminal rotating hydrogen atoms are explicitly modelled in order to optimise the hydrogen bond interactions.

In this group of techniques we can also consider SLIDE.[55] This method, after looking for the best matching between ligand and receptor using as criteria the best steric complementarity and the best matching between hydrophobic and hydrogen-bonding sites, resolves 'van der Waals collisions' between atoms of the two molecules performing a minimal number of side chain rotations, with the cost of a side chain movement evaluated as the product of the rotation angle and the number of atoms moved. A similar approach is SPECITOPE[56] that uses minimal angle side chain rotations at the end of the docking procedure to remove steric clashes between the ligand and the protein. Unfortunately, although this algorithm is able to resolve many cases of overlap, it does not sample the conformational space completely and it is not able to find the minimum energy conformation.

Finally, among this group of strategies, the SCARE approach recently devised by Abagyan and co-workers[57] is also very interesting. The algorithm scans every pair of neighbouring side chains in the receptor binding site and replaces them with alanine residues. Afterwards, the ligands under study are docked to the previously determined 'gapped' binding pockets and all the poses are scored. In the final stage, the original side chains are reintroduced and, after a proper optimization, each pose is scored again. Although only recently developed, the method is promising due to the fact that the number of corrected predictions greatly increased with respect to highly optimised docking strategies that use a single pocket conformation. These results are even more noteworthy considering that the technique proposed by Abagyan is fully automated and that it does not need to know *a priori* the ligand binding position or any information about the sites of potential variability of the receptor pocket.

Although the strategies described above can be considered as a step forward in dealing with receptor flexibility, the choice of selecting only a subset of degrees of freedom is not always the best one, and of course it depends on the system under exam. For instance, the HIV-1 protease conformational space can be sufficiently sampled by modelling some side chain movements and a water molecule,[52] but on the other hand, there are many systems that undergo remarkable re-arrangements during the ligand binding process. In these situations the 'Induced Fit' model[41] is no longer a good description for molecular recognition and it is necessary to turn to the more advanced representation already mentioned above, according to which the ligand selects the optimal protein conformation among a conformational ensemble.[42] The simplest way to realise it computationally, consists in developing strategies that simultaneously consider multiple receptor conformations (MRC) obtained from experimental (i.e., from X-ray or NMR techniques) or computational (i.e., from MD or MC simulations) sampling procedures.

The main advantage of this group of methods is that it is possible to model protein flexibility in more than just a specific small region. This also means that the MRC strategies allow describing the full receptor flexibility at a much lower computational cost compared to the one associated with possible techniques that include all the receptor degrees of freedom. Nevertheless, flexibility is only modelled implicitly, which means that only a small fraction of protein conformational space is really taken into account. The effect of this poor sampling, particularly in the case of experimental sampling, is that MRC docking results are mainly successful when the available receptor conformers derive from complexes with ligands similar to the one in exam. Finally, another important problem is the need for a reliable energy function that enables to distinguish the small group of real low-energy structures among the hypothetical conformations generated during the sampling. This is a crucial aspect because an optimal energy function allows avoiding unpleasant drawbacks as incorrect pose predictions and false positives in virtual screening outputs.

Among the MRC techniques, FLEXE[58] is probably one of the most remarkable examples. After superimposing different target structures, it merges the similar parts and considers the dissimilar ones as possible alternatives. Therefore, other than the receptor conformations in the starting set, new structures are generated and a more complete 'ensemble' is taken into account during the docking experiments. Nevertheless, a recent study has pointed out that the method is not able to deal with large loop re-arrangements and that furthermore, it shows a worse performance in terms of the enrichment factor than running multiple docking calculations for each receptor structure.[59] The FLEXE philosophy also underlies the FITTED technique[60] in the representation of receptor flexibility. In particular, when flexibility is fully considered, a genetic algorithm is used to combine different side chain rotamers and backbone conformations included in the starting set of structures. In this context, Zavodszky *et al.* proposed an alternative approach.[61] At first, in order to form the ensemble of receptor conformations, they analysed protein flexibility using a graph-theory based algorithm and performed a random walk sampling. Afterwards, they applied the SLIDE method[55] to the set of structures previously obtained. FLIPDock[62] is another strategy recently developed to simultaneously take into account a great number of protein conformations. In order to accomplish this task, it exploits the sophisticated 'Flexibility Tree' algorithm, which in describing protein flexibility in terms of a hierarchical classification of movements, greatly reduces the computational cost associated with the description of receptor flexibility in docking computations. Shoichet and co-workers have also devised an interesting technique[63] that is based on the DOCK programme.[64–67] This method initially considers an ensemble of predetermined protein structures and then, moving independently the flexible regions of the receptor and recombining their conformations, it implicitly takes into account a much greater number of structures. However, the most relevant feature is the fact that, for a given pose of the ligand, the algorithm chooses the best conformations of each part of the protein and this allows a sensitive reduction of the computational effort. Another strategy to examine is the one proposed by Sherman *et al.*[68] which performed an iterative combination of rigid receptor docking and protein structure prediction using Glide[69] and Prime,[70,71] respectively. In particular, the method essentially consists of four main steps: (a) docking the ligands into the rigid receptor structure using a softened potential energy; (b) sampling and minimising the best protein–ligand complexes obtained at the previous step; (c) docking the ligands into the best refined protein structures from step (b) using a hard energy function; and (d) ranking the complexes using a composite scoring function that accounts for both the receptor–ligand interaction and for strain and solvation.

In the framework of the MRC methods, grid-based approaches are particularly noteworthy. As it is well known, to efficiently evaluate the interaction energy between ligands and receptor, many docking strategies use interaction energy grids, which are calculated placing probe atoms at discrete points in the space around a target protein and assigning to these grid points the value of the interaction energy between the probes and the protein. The peculiarity of grid-based MRC techniques is that instead of considering one grid for each conformation in the ensemble, only one grid that represents all the structures is determined and used. The '*in situ* cross docking' approach[72,73] is probably the simplest example and it consists of joining together the grids computed for each protein binding-site. Despite its simplicity, this method is characterised by a sensitive speedup in terms of computation time and it can simultaneously deal with a wide range of protein conformations, although the number of structures that can be examined in each computation is limited. More advanced grid-based MRC strategies exploit grids averaging techniques. As for the *in situ* docking, the main advantage is the improved computational speed, whereas the main drawback is that the average structure represented by the grid may not correspond to a real target and, therefore, the resulting ligand poses may be significant only for the average representation.

According to Osterberg *et al.*,[52] the grid averaging methods can be classified into four main categories. The first two are very simple and consist in creating (a) a new grid of mean values calculated all over the grids and (b) a new grid that considers only the minimum values all over the grids. Of course, as one should expect, these techniques perform poorly. More advanced averaging strategies are (c) the one proposed by Knegtel and co-workers,[74] which either consists of an energy-weighted average or a geometry-weighted average, and (d) a simple Boltzmann-weighted averaging of the interaction energies. The last two techniques provide better results in docking calculations, even if they may sometimes introduce potential dangerous artefacts.

As already mentioned, it is extremely important to note that almost all the MRC methods presented above can be applied to sets of conformations that derive both from experiments and from computer simulations. In this context, it is necessary to observe that experimental techniques, such as X-ray crystallography and NMR, can help in describing the structures and flexibility of proteins, but they are not able to provide complete insights of all molecular re-arrangements. This is because only lower energy states will be detected, while higher energy conformations, which can be eventually stabilised by ligand binding, are usually neglected. Furthermore, other aspects, such as crystallization difficulties and protein molecular weight, can sometimes prevent to apply experimental techniques to the bio-molecule that we want to study. These are the main reasons why computational samplings, in particular the ones that use Molecular Dynamics, are becoming more and more essential to generate large ensembles of protein structures to be used in MRC strategies. Nevertheless, although nowadays computer power is rapidly increasing (and the parallelization of many MD codes has greatly enhanced this aspect), the critical issue associated with Molecular Dynamics simulations is their length. In fact, due to time scale limitations, the MD trajectories are also often not able to sample all the protein structures relevant from a biological point of view and, therefore, large conformational re-arrangements are not always reproduced. However, these drawbacks can be partially overcome by applying replica exchange methods, which allow a better exploration of conformational space, or exploiting coarse-grained samplings[75] that describe large receptor motions and can be combined with a fine-scaled MD sampling to account for local transitions.

Pang and Kozikowski were probably the first to use structures generated by Molecular Dynamics in an MRC strategy.[76] In particular, by studying the binding of huperzine A (HA) to acetylcholinesterase (AChE), they carried out a short MD simulation of AChE and extracted a group of conformations which were afterwards used to rigidly dock the HA ligand. McCammon and co-workers proposed developments of this pioneer approach introducing the 'Relaxed Complex Method'.[77-79] The technique is based on longer and more accurate Molecular Dynamics samplings and it takes into account many more receptor structures which can be selected at regular time intervals, or by considering their conformational variability in order to form a complete ensemble. Another interesting example of an MRC method that uses snapshots from a MD simulation is the one devised by Broughton:[80] by using a weighted average method, it combined the interaction energy grids associated with the collected structures into a single grid.

However, one of the most interesting and outstanding approaches that consider multiple receptor conformations performing computer simulations has been proposed by Carlson and co-workers.[81-94] In this strategy the active site of each structure is flooded with hundreds of small molecular probes, namely benzene molecules to identify hydrophobic and aromatic regions, ethane molecules to distinguish hydrophobic interactions from the aromatic ones and methanol molecules to define hydrogen-bonding sites. These probes are simultaneously optimised by means of a low temperature Monte Carlo minimization, during which the protein is held fixed and probe–probe interactions are completely ignored. After the minimization step, significant clusters of probes (namely, clusters constituted by eight or more probes) can be easily detected and each of them is represented by a 'parent', which is the lowest-energy element in the group. At this point, the results obtained for the different conformations are superimposed and, if a 'cluster parent' is conserved over many of the receptor structures (namely, for more than 40% or 50%), it is possible to define a 'consensus cluster' that is associated with a spherical pharmacophore element. Therefore, the Carlson method allows obtaining pharmacophore models that implicitly account for receptor flexibility, models that can be subsequently used to design new ligands/inhibitors or virtual screenings for databases. Of course, it is worth observing that this technique can also be applied to sets of protein conformations that are derived experimentally.

The philosophy underling the strategy described above is also the basis for the approach recently devised by Colombo and co-workers.[95–97] As in the Carlson method, the final aim is to construct a new pharmacophore, but instead of optimising the placement of small probe molecules, they performed MD simulations of receptors bounded with already known ligands. Analysing these simulations, it is possible to determine those ligand–protein interactions that conserve over the simulations and that, for this reason, also define pharmacophore elements. Although this strategy is conceptually very similar to the previous one, it is worthwhile to observe that using known ligands instead of small molecular probes allows to account for useful experimental information, even if the ligands considered in the simulations are often small peptides, which unfortunately do not completely cover all the features of real drugs.

As just described above, MD simulations play a very important role in improving the MRC techniques. Nevertheless, in order to better describe the molecular recognition process, it is necessary to simulate the interaction between ligand and receptor considering all the possible degrees of freedom and going beyond the limitations associated with the flexibility models previously presented (e.g. MRC models). To accomplish this task, we should perform full Molecular Dynamics or Monte Carlo simulations, which are characterised by a high accuracy, but also by a large computational cost due to the fact that the binding processes between ligands and proteins are usually very slow. Therefore, despite the rapidly increasing computer power, nowadays it is not possible to carry out standard MD or MC simulations to screen large databases of compounds. Nevertheless, introducing some reasonable approximations, which make the calculations less expensive and at the same time less accurate, we can obtain fundamental information that would be unfortunately lost using less flexible receptor descriptions.

The easiest way to reduce the computational cost consists in restricting the full simulations only to the active site, to the ligand, and if necessary, to regions near them too. Mangoni *et al.*[98] have proposed an interesting improvement of this technique where they limit again the simulation to a well-defined region, but allowing a faster exploration of the receptor conformational space. In particular, they simulated the ligand internal motions, the solvent and the receptor at room temperature, whereas they set a much higher temperature for the ligand translational modes of motion. Other interesting ways to speed up the sampling is the strategy devised by Wang and Pack,[99] which applied a scaling function to the equations of motion in order to promote the crossing of barriers, and the multicanonical algorithms, which have been developed in the framework of both the Monte Carlo and Molecular Dynamics techniques,[100–102] and consist of a smoothing of the potential energy surfaces. Finally, in the context of the Monte Carlo methods, we can also consider Prodock[103,104] which is a modified MC strategy, mainly characterised by local gradient-based minimizations after each random move and by the scaling of the potential energy terms during the docking.

Unfortunately, despite the development of modified MD or MC techniques as those just described, to represent large-scale receptor flexibility as large conformational re-arrangements of protein domains it is necessary to take into account strategies that use collective degrees of freedom. These methods are less computationally expensive than the traditional computer simulations because the number of independent degrees of freedom is sensitively reduced, but on the other hand, these degrees of freedom are not the native ones and this may affect the final results.

Typical methods in this framework are those that consider the receptor harmonic normal modes of vibration and, in particular, the low-frequency modes that are associated with very delocalised motions. Two interesting applications of this kind of technique to the problem of molecular recognition are the study by Zacharias and Sklenar[105] which deals with DNA flexibility for the binding of small molecules, and the study by Keserû and Kolossváry[106,107] which investigates the inhibitors binding to HIV integrase. Another group of methods that exploit the receptor collective degrees of freedom relies on techniques that reduce the dimensions of the system under study and, in particular, on the Principal Component Analysis (PCA). Both the low-frequency normal modes in the computation of harmonic normal modes and also the most significant principal components in PCA (namely, the ones associated with the highest eigenvalues) describe very large re-arrangements and most of the conformational variations in large bio-molecules. In this context, Teodoro *et al.*[43,44] developed a strategy that enables to sensitively simplify the description of the protein–ligand interaction, significantly reducing the high-dimensional representation of the protein flexibility. Finally, the approach proposed by Tatsuni *et al.*[108] is particularly interesting. At first, they performed a Principal Component Analysis to determine the receptor global conformational changes and afterwards, they coupled harmonic dynamics with molecular dynamics to take into account protein large re-arrangements and local side chain flexibility, respectively. From preliminary studies on the HIV-1 protease and its ligands, it seems that the new strategy is able to efficiently reproduce the formation of molecular complexes.

Considering all the techniques presented so far, it is obvious that the ones that heavily use MD or MC simulations are the most accurate. Nevertheless, they are also the most computationally demanding and therefore, as already stressed above, they cannot be used when we have to perform screenings of very large databases. Hence, to overcome this problem and to preserve some accuracy, several research groups have recently developed hierarchical methods that combine different strategies, using less accurate and faster approaches for preliminary screenings and then higher-level of theory and expansive strategies for later stages. Machicado et al.[109] proposed a three-step virtual screening strategy that enables to identify active ligands for buried protein cavities. It consists of applying some physicochemical filters, a fast docking procedure and at the final stage, a finer flexible docking algorithm that exploits a Monte Carlo search technique and which uses a purposely defined binding free energy function to properly score the docked complexes. Another example is Extra Precision GLIDE[69,110] which, at first, carries out docking calculations using relaxed criteria and then allows refining the results by means of a more sophisticated computation of the binding free energies. More recently, Lee and Sun have introduced another method[111] that actually consists of four different protocols that optimise the poses obtained from the classical docking programme DOCK 4.0[112] by means of proper successions of minimizations and molecular dynamics simulations followed by a final MM/GBSA (Molecular Mechanics/Generalized Born Surface Area) scoring.

Finally, in this category of strategies, it is worthwhile to also consider the approaches devised by Wang et al.[113] and by Graves et al.[114] The former consists of the following four hierarchical filters to screen ligand databases: (a) pharmacophore model, (b) rigid docking, (c) solvation docking and (d) molecular dynamics simulation combined with the MM/PBSA (Molecular Mechanics/Poisson–Boltzmann Surface Area) method, which is exploited to compute the binding free energies of the 30 most promising hits that survive from the previous steps. It is extremely important to note that, while the first three filters take into account only the ligand flexibility, the last one consists of a MD simulation that samples a part of the conformational space of both the inhibitors and the receptor. The latter approach is based on the same philosophy, but the initial docking calculations are combined with MM/GBSA energy evaluations to re-score the best poses. Also in this case it is worth to stress that the MM/GBSA technique introduces a dynamic sampling of the protein–ligand complexes, although this is limited to the configurational space in the neighbourhood of the starting docking poses. Analysing the results obtained from preliminary tests performed on model cavity sites, it seems that this new strategy allows rescuing many docking false negatives, improves the binding geometry of most of the predicted structures and increases the diversity of the hit lists. Nevertheless, the rescoring technique introduces a great number of false positives, especially among the very top ranking ligands, and unfortunately this is probably due to the introduction of the protein flexibility by means of the MM/GBSA method.

13.4 Examples of the Use of MD Simulations for the Discovery of Small-Molecule Inhibitors

13.4.1 *Design of Inhibitors of the Molecular Chaperone Hsp90*

In this paragraph, we will describe an example of the use of all-atom Molecular Dynamics simulations in understanding the main molecular determinants of peptide–protein binding and how to translate this information into pharmacophore models useful for the screening of virtual libraries. The application described here refers to the inhibition of a protein whose role is fundamental in cancer development. The results presented here are based on Refs 95, 96 and 115.

Cancer therapy now aims at disabling oncogenic pathways that are selectively operative in tumour cells, so to spare normal tissues and limit side effects in humans. This 'targeted therapy' relies on a better understanding of cancer genes, particularly those implicated in tumour cell proliferation and survival.[116] Accordingly, targeted inhibition of the Bcr-Abl kinase with small molecule antagonists has produced dramatic clinical responses in malignancies driven by this oncogene.[117] However, such approach may not be immediately available for the majority of tumours where multiple molecular abnormalities and genetic instabilities may elude the identification of one single, disease-driving oncogene.[116] Conversely, pathways that intersect multiple essential functions of tumour cells may provide wider therapeutic opportunities. A prime target for this strategy is Heat Shock Protein 90 (Hsp90), a molecular chaperone that oversees the correct conformational development of polypeptides and protein refolding through sequential ATPase cycles and stepwise recruitment of cochaperones. This adaptive pathway contributes to the cellular stress response to environmental threats, including heat, heavy metal poisoning, hypoxia etc., and is dramatically exploited in cancer, where Hsp90 ATPase activity is upregulated by ~100-fold.[118] The repertoire of Hsp90 client proteins is restricted mainly to growth-regulatory and signalling molecules, especially kinases and transcription factors, which may contribute to tumour cell maintenance.[118] Therefore, targeted suppression of Hsp90 ATPase activity with a small molecule inhibitor, the benzoquinone ansamycin antibiotic 17-allylamino-17-demethoxygeldanamycin (17-AAG) showed promising anti-cancer activity in pre-clinical models, and recently completed safety evaluation in humans.[119] One Hsp90 client protein with critical roles in tumour cell proliferation and cell viability is Survivin, an Inhibitor of Apoptosis (IAP) protein selectively over-expressed in cancer.[120,121] Accordingly, targeting the Survivin-Hsp90 complex may provide a strategy to simultaneously disable multiple signalling pathways in tumours, and a peptidomimetic antagonist of this interaction structurally different from 17-AAG, Shepherdin, inhibited the chaperone activity and exhibited potent and selective anti-cancer activity in pre-clinical models.[95]

In this paragraph, we report the computational/theoretical structure-based design and characterization of *Shepherdin*, a novel peptidomimetic antagonist of the complex between Hsp90 and Survivin.[95] For its potent and broad anti-tumour activity, selectivity of action in tumour cells vs normal tissues, and inhibition of tumour growth in vivo without toxicity, Shepherdin (K79-L87, KHSSGCAFL) and its retroinverso-analog Shepherdin-RV may offer a promising approach for rational cancer therapy. We will then show how we could use MD to identify the minimum sequence of Shepherdin, labelled Shepherdin-min (K79-G83, KHSSG) required for activity in acute leukemia cancer cells. The structures of these peptides are studied by means of long time-scale MD simulations in explicit water. Subsequently, the dominant structures are docked to Hsp90, and the resulting complexes are also relaxed by means of long time-scale MD simulations to identify at equilibrium the dominant interactions responsible for binding. Finally, we will describe the use of the information developed in this part to identify a new non-peptidic small molecule that represents the prototype for a new class of compounds which can selectively inhibit Hsp90's chaperone activity. Computational and theoretical results are in all cases benchmarked by experimental validations *in vitro* and *in vivo*.

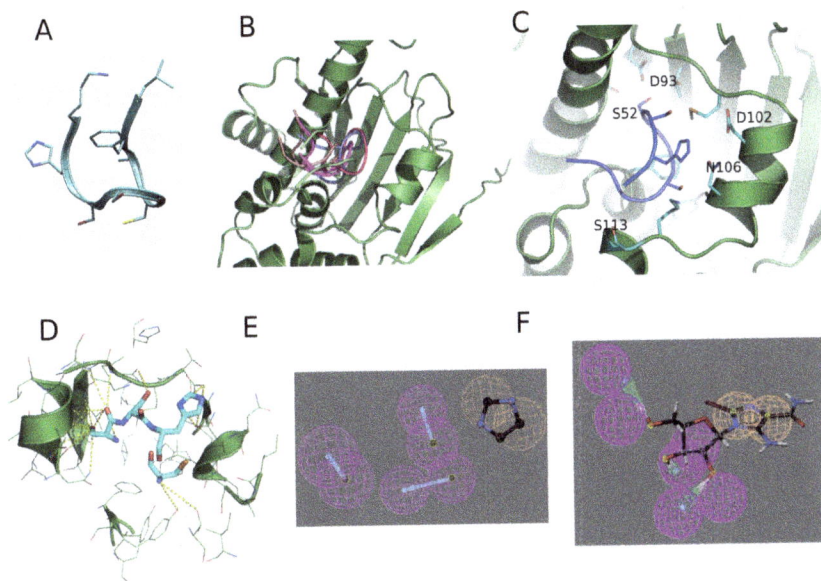

Fig. 13.1. Pictorial representation of the strategy for the design of Hsp90 inhibitors. (A) The representative structure of Shepherdin. (B) The five minimum free-energy structures of the Hsp90-Shepherdin complex from the Autodock runs, and (C) after the MD refinement. (D) The side chains responsible for most stabilising interactions in the complex and their translation into a pharmacophore model (E). (F) The structure of the lead satisfying the pharmacophoric constraints.

13.4.1.1 *Identifying Possible Binding Structures for Active Peptides*

1. *Shepherdin and Shepherdin RV:* The peptides of the Shepherdin series were identified by Altieri and co-workers based on a peptide scanning analysis of the Survivin sequence.

The modelling study of the peptide K79-L87 named Shepherdin and its retroinverso version L87-K79 (all D amminoacid) Shepherdin-RV start with a long MD simulation with the aim to identify the characteristic conformations of these peptides in solution. Analysis of the trajectories predicts that both Shepherdin and Shepherdin-RV have a dominant configuration characterised by a turn involving S82–G83 in Shepherdin and G83–S84 in Shepherdin-RV, and overall β-hairpin geometry (Fig. 13.1a). The most populated conformation of Shepherdin-RV shows a higher degree of compactness, with the aromatic ring of F80 packing on the turn region (Fig. 13.1a). The representative β-hairpin conformations of both peptides were subjected to multiple docking experiments on Hsp90 using the AutoDock programme package. In all cases, the peptides were predicted to dock into the ATP binding site of Hsp90 (Fig. 13.1b).

The geometry of the final complex is highly correlated with that of the complex between Hsp90 and GA,[122] with the turn region of the peptides closely tracing the ansa ring backbone of GA. Shepherdin and Shepherdin-RV make 13 and 18 predicted hydrogen bonds with the ATP pocket of Hsp90, respectively, involving the side chains of H80, S81, S82, the carbonyl group of G83, and the side chains of K87 and C82 (Shepherdin-RV). Except for D93, the complementary residues of Hsp90 predicted to make contact with Shepherdin and/or Shepherdin-RV largely overlap with amino acids implicated in GA binding,[122] including S113, which has been recently shown to contribute to stepwise accessibility of the ATP pocket of Hsp90 to GA. Shepherdin and Shepherdin-RV are predicted to assume more extended conformations than GA in the Hsp90 pocket, and bury a solvent accessible surface of 498 and 546 Å^2, respectively, as opposed to 402 Å^2 buried by GA.

To check these structural predictions, and validate experimentally that Shepherdin engaged Hsp90 differently from GA, we introduced experimentally targeted mutations in the ATP pocket of Hsp90, and tested their effect on Shepherdin binding. Individual substitution of N51, S52, D102, or S113 in the N domain of Hsp90 reduced binding to Shepherdin by 20%–60%, whereas mutagenesis of 'GA-specific' D93 had no effect, and a scrambled peptide did not bind wild-type or mutant Hsp90 (Fig. 13.1c).

2. *Simulations of Shepherdin-RV Mutants:* To investigate the impact of single point mutations on the structure-activity relationship properties of Shepherdin-RV, and to shed more light on the determinants of the interaction between the peptide and Hsp90, two mutant peptides (H80A and C84A) were simulated with long time scale all-atom MD simulations. A total of four simulation (two runs for each mutant) are calculated for the two peptide mutants. Two different initial conformations were used: one completely extended and the second one from the dominant Shepherdin-RV conformation found in the previous runs, and subjected to mutation. The purpose of the first simulation is to identify the characteristic conformations of these peptides in solution and the stability of the Shepherdin-RV β-hairpin conformation after the mutation.

In the case of the C84A mutant, simulations suggest that the mutation dramatically decreases the tendency of the peptide to form a stable hairpin-like structure. In the 100ns time span from the completely extended conformation neither the analysis of the time evolution of secondary structure, nor the structural cluster analysis are able to identify a hairpin conformation similar to that observed for the original sequence. The second simulation, from a preformed hairpin structure, shows that the mutant peptide retain the hairpin conformation for about 10ns and after that the turn geometry changes for a long period, before complete loss of the initial conformation.

An analogous behaviour is observed for the H80A mutant peptide. Both the analysis of the time evolution of secondary structure and the structural clustering suggest that the hairpin is not the dominant conformation in solution, despite being present for a smaller percentage.

These results clearly suggest that the bent, hairpin like conformation of Shepherdin is a fundamental determinant for recognition with the active site of Hsp90. It is worth noting at this point, that this type of conformation for the same sequence is also present in the native structure of Shepherdin, suggesting a certain level of structural pre-organisation for this sequence, optimised for binding to Hsp90.

3. *Shepherdin[79-83]:* The combination of theoretical analysis and experimental verification described in the previous paragraphs suggests that the minimal motif necessary for recognition should contain the HSSG sequence. Based on these considerations, a new short peptide, with sequence KHSSG spanning residues 79–83 of Survivin was synthesized and named Shepherdin[79–83].

To understand possible structure-activity relationships of this peptide and of its interaction with Hsp90, Shepherdin[79–83] was simulated in isolation in explicit water. The peptide did not populate any preferred ordered secondary structure, and so the hydrophilic side chains and the backbone carboxyl and amino groups tended to maximise their interactions with the surrounding water solvent. Cluster analysis of the 200-nanosecond simulations determined that the main conformational family of Shepherdin[79–83] was characterised by a slight bend geometry involving residues His-80, Ser-81 and Ser-82. Docking experiments on Hsp90 with this geometry predicted that Shepherdin[79–83] bound to the ATP pocket of Hsp90. Two different orientations of Shepherdin[79–83] were observed: one that corresponded to the global free-energy minimum structure of the Shepherdin[79–83]–Hsp90 complex and one that represented the most frequently obtained structure after statistical clustering of all the structures studied during the AutoDock simulations. The sites of contact between Hsp90 and Shepherdin[79–83] in either configuration overlapped. In the global free-energy minimum configuration, the side chain of His-80 in Shepherdin[79–83] made hydrophobic contacts with Ile-96 and hydrogen bonded with Gly-97 in Hsp90, the side chain of Ser-81 in Shepherdin[79–83] hydrogen bonded with the side chains of Asp-102 and Asn-106 in Hsp90, and Ser-82 in Shepherdin[79–83] hydrogen bonded with Asn-51 and Phe-138 in Hsp90. In the most frequently obtained Shepherdin configuration, His-80 in Shepherdin[79–83] formed a hydrophobic interaction with Ile-96 in

Hsp90 but was also involved in a new hydrogen bonding interaction with the side chain of Asp-54 in Hsp90; Ser-81 interacted with Asp-93 and Asn-106 in Hsp90, and Ser-82 interacted with Asn-106 and Asp-102 in Hsp90. Consistent with these molecular dynamics predictions, in biochemical experiments, Shepherdin[79–83] efficiently displaced ATP binding from the N-domain of recombinant Hsp90, whereas the scrambled peptide was ineffective.

4. *Characterization of Hsp90/Shepherdin Binding Interface*: The dominant conformations of Shepherdin in solution were investigated through all-atom, explicit solvent MD simulations for a total time span of 400 ns. Statistical cluster analysis showed that Shepherdin displays one main conformation, characterised by the presence of a turn involving residues G83–S84 and an overall hairpin geometry (see Plescia *et al.* 2005 for details). The remaining clusters were mainly extended conformations, with the peptide backbone groups involved in hydrogen bonding with water.

The most populated conformation was subjected to multiple blind docking experiments on Hsp90 using the AutoDock programme. In all cases, the peptide was predicted to bind within the ATP binding site of Hsp90 (Figs 13.1b and 13.1c). Analysis of the blind docking results through the procedure described by Hetenyi *et al.*[123] showed that low energy poses are all closely correlated with one another, with an RMSD from the global minimum structure lower than 2.5 Å. Control docking experiments were conducted with the extended structures representative of other clusters, but in those cases it was not possible to univocally identify any particular binding site on Hsp90.

The free-energy minimum structure of the Hsp90/Shepherdin complex was then subjected to two long, 54 and 73 ns, all-atom MD simulations. Analysis of the statistical and time-dependent distribution of the interactions between functional groups of the ligand and of the chaperone was carried out to develop pharmacophore models, keeping into account the motional and flexibility properties of both the ligand and the receptor. Shepherdin partially reoriented to increase the total number of stabilising contacts with the ATP binding pocket of Hsp90 (Figs 13.1c and 13.1d). Attention was focused in particular on the analysis of hydrogen-bonding, hydrophobic/aromatic and charge–

G. Morra, A. Genoni, G. Colombo

charge interactions, as these represent the most common types of intermolecular forces determining host/guest recognition in drugs.

The functional groups of Shepherdin involved in direct or water-mediated hydrogen bonds with the binding pocket of Hsp90 included the gammaOH functions of Ser84, Ser85 and the imidazole ring of His86 (Fig. 13.1d). The latter, in particular, could satisfy hydrogen-bonding conditions being involved in interactions both as an acceptor ($N\varepsilon$ atom) and as a donor (Nd-H functional group). The remaining hydrogen-bonding group on Shepherdin, Cys82, was involved to a lesser extent in intermolecular H-bonding interactions; however, its presence was shown experimentally to be necessary to ensure binding. Moreover, it displayed hydrophobic interactions with the side chains of Hsp90 Leu108 and the alkyl part of Asn109. Cys82 is also important for preserving the hairpin structure: mutations to Ala on the isolated peptide lead to loss of ordered hairpin structure.

To define the presence of hydrophobic/aromatic interactions, the contacts involving the side chains of Phe80 and His86 were monitored during simulation. Shepherdin Phe80 was found to be in contact mainly with the charged/polar side chains of Lys59, Asn52 and Asn55 on Hsp90, while Shepherdin His86 was not significantly involved in hydrophobic/aromatic contacts with Hsp90 residues.

Finally, the role of the positively charged ammonium group on the side chain of Shepherdin Lys87 was found to be only marginally involved in interactions with the backbone carbonyl oxygens of Hsp90 residues Phe135 and Gly136, being mostly exposed to the water solvent during MD simulations.

5. *Pharmacophoric Hypotheses and Small Molecule Identification:* Three different pharmacophore models were built and labelled PHARM1, PHARM2 and PHARM3 based on the results of MD simulations. The conformation of Shepherdin and the orientations of its side chain functional groups in the most populated structural cluster from MD trajectories of the complex were used as structural template (Fig. 13.1d). The distributions of dihedral values and distances among critical functionalities were used to define upper and lower boundaries for geometric constraints.

PHARM1 (Fig. 13.1e) consisted of four pharmacophoric points: three H-bonding donor functionalities mapped over the side chain OH group of Ser84, Ser85 and the SH group of the Cys82, plus one imidazole ring moiety mapped on the position of the corresponding ring of His86 (Figure 13.1e). Each imidazole atom was allowed to bear any substituent or be a bridgehead in the presence of a fused ring. PHARM2 consisted of two H-bonding donor groups corresponding to the g-OH of Ser84 and Ser85, one aromatic function centred on the position of the Phe80 benzene ring, and one hydrophobic function centred on the S atom of Cys82. PHARM3 had the same properties as PHARM1, augmented by the presence of a positive charge mapped on the position of the ammonium group of Lys87.

The three models described above were used as queries for a search of the NCI_3D database of molecules (containing approximately 250,000 compounds). The search with PHARM1 yielded 73 compounds, the search with PHARM2 yielded 42 compounds, while PHARM3 gave no hits. In experimental tests, the molecules corresponding to hits of PHARM2 proved to be extremely insoluble due to the presence of aromatic/hydrophobic groups and had thus no tumour-cell–killing effect. The search with PHARM1 yielded, among others, 20 hits reminiscent of the class of known purine-based inhibitors of the ATP-binding pocket of Hsp90. Interestingly, one of the non–purine-based hits that was found to be effective in experiments, AICAR (Fig. 13.1e), was not previously known to interfere with Hsp90 chaperone functions and was characterised by a novel molecular structure among Hsp90 antagonists. All of the data generated with simulations were punctually verified experimentally.

13.4.1.2 *Structure-dynamics Based Design of New Hsp90 Inhibitors*

In this study, we used structure-based rational studies to identify and characterise Shepherdin, a novel anti-cancer peptidomimetic modelled on the Survivin-Hsp90 binding interface.[121] All theoretical predictions were subjected to experimental verification and the activities of the peptides were evaluated both *in vitro* and *in vivo*, in a large multidisciplinary effort. Shepherdin engages the ATP pocket of Hsp90 with unique binding characteristics, destabilises Survivin plus several additional client proteins, and causes massive killing of tumour cells by apoptotic and nonapoptotic mechanisms.[95] Shepherdin is selective in its anti-tumour activity, and does not affect the viability of normal cells or tissues, including human hematopoietic progenitors. When administered *in vivo*, Shepherdin is safe and well tolerated, and inhibits growth of different tumour cell types without systemic or organ toxicity. Taken together, these features may make Shepherdin an attractive lead prodrug for 'targeted' cancer therapy.[95]

Although initially designed as a high-affinity (K_D ~80 nM) inhibitor of the Survivin–Hsp90 interaction, the data presented here suggest that Shepherdin may function as a more global antagonist of Hsp90 chaperone activity. This conclusion is based on the structure-function analysis of Shepherdin, and in particular its ability to expansively engage the chaperone ATP pocket, compete for the Hsp90-ATP complex and destabilise multiple Hsp90 client proteins in addition to Survivin, *in vivo*. Because of these features, Shepherdin appears ideally suited to interfere with the periodicity of Hsp90 ATPase cycles, by directly preventing ATP binding,[122] and/or by competing with cochaperone recruitment, especially that of p50[cdc],[37] which is required for ATPase activity and shares overlapping binding contacts with Shepherdin.[124] In this context, the simultaneous destabilization of Survivin levels,[120] combined with the acute collapse of the Hsp90 function,[120] would be expected to cause a general breakdown of multiple cell proliferation and cell survival pathways in tumour cells, suitable for therapeutic exploitation.[119]

When tested as an anti-cancer agent in tumour models, Shepherdin was selective and well-tolerated, sparing normal cells, preserving colony-forming ability of purified human hematopoietic progenitors and causing no organ or systemic toxicity after prolonged administration *in vivo*.

Experimental tests with the minimal sequence of five residues also confirm theoretical hypotheses. The results of MD simulations on the structure of Shepherdin[79–83] and of its complex with Hsp90 were challenged with competition experiments by use of enzyme-linked immunosorbent assay (ELISA). Apoptosis, Hsp90 client protein expression and mitochondrial dysfunction were evaluated in Acute Myeloid Leukemia (AML) types (myeloblastic, monocytic and chronic myelogenous leukemia in blast crisis), patient-derived blasts and normal mononuclear cells. Effects of Shepherdin[79–83] on tumour growth were evaluated in AML xenograft tumours in mice (n = 6). Organ tissues were examined histologically. Taken together, these results showed that Shepherdin[79–83] bound to Hsp90, inhibited formation of the Survivin-Hsp90 complex, and competed with ATP binding to Hsp90.

Based on this knowledge, we used a structure- and dynamics-based rational design to identify the non-peptidic small molecule AICAR as a structurally novel inhibitor of Hsp90 (Fig. 13.1f). The compound was selected to engage the ATP-binding pocket of the N-terminal domain of Hsp90, with binding and functional properties that mimic those of the peptidic antagonist of the Survivin-Hsp90 complex, Shepherdin. Accordingly, AICAR bound the Hsp90 N-domain, destabilised multiple Hsp90 client proteins *in vivo*, including Survivin, and exhibited broad antiproliferative activity in multiple tumour cell lines, although at higher concentrations than those required to obtain the same growth inhibitory effect with 17-AAG, with induction of apoptosis and inhibition of cell proliferation. Reminiscent of the selective anti-cancer activity of Shepherdin, AICAR did not affect proliferation of normal human fibroblasts.

In summary, Shepherdin has the molecular features of both an inhibitor of a critical protein–protein interaction in tumour cells, e.g., Survivin-Hsp90, and an enzymatic antagonist of Hsp90 ATPase cycles. Because of these combined features, plus its considerably higher potency compared to other Hsp90 inhibitors, e.g., 17-AAG, Shepherdin may provide a potent and selective new anti-cancer agent in humans, consistent with the use of peptidomimetics in targeted cancer therapy. In addition, we narrowed the Shepherdin binding interface to a short stretch of amino acids between H80 and C84 in the Survivin sequence. Previously, mutagenesis of H80 or C84 resulted in dominant negative phenotypes with mitotic defects and induction of apoptosis in tumour cells, thus further underscoring their critical roles in Survivin function. This small cluster of residues may thus provide a manageable platform for further derivatization of Shepherdin, as well as for chemical screenings to identify Shepherdin-like small molecules with enhanced, 'targeted' anti-cancer activity in humans. The strategy presented here can be used in a general way to generate small molecules from peptide leads, keeping flexibility, solvation and dynamics into account.

13.4.2 *Targeting Peptide Aggregation: MD as a Tool to Identify Potential Inhibitory Determinants*

In this section, we will discuss the possibility to use MD simulations to generate pharmacophoric models for the identification of potential anti-amyloidogenic drug-like molecules.[97] Protein and peptide self-aggregation and fibril formation are the central events in the pathogenesis of more than 20 human disorders known as amyloid diseases.[6] Therefore, the development of molecules able to inhibit amyloid aggregation represents a very active area of research. However, several challenges face rational design efforts of anti-amyloid drugs: the target structures may not be well-defined, they are generally highly plastic oligomers, and the contact area that a small molecule should cover is extended in most cases.

Recent studies have identified soluble beta-sheet oligomeric species as the main cause of cellular toxicity,[7–12] making them targets for the design of antiamyloidogenic drugs. However, the transient nature of these early oligomers has hampered the characterization of their structural-dynamical properties at the atomic level of resolution. The lack of detailed structural models for the different species on- and off-the amyloid pathway that might represent potential drug targets has seriously limited the potential of rational drug-design and hindered efforts to rationally improve the efficacy of lead molecules.

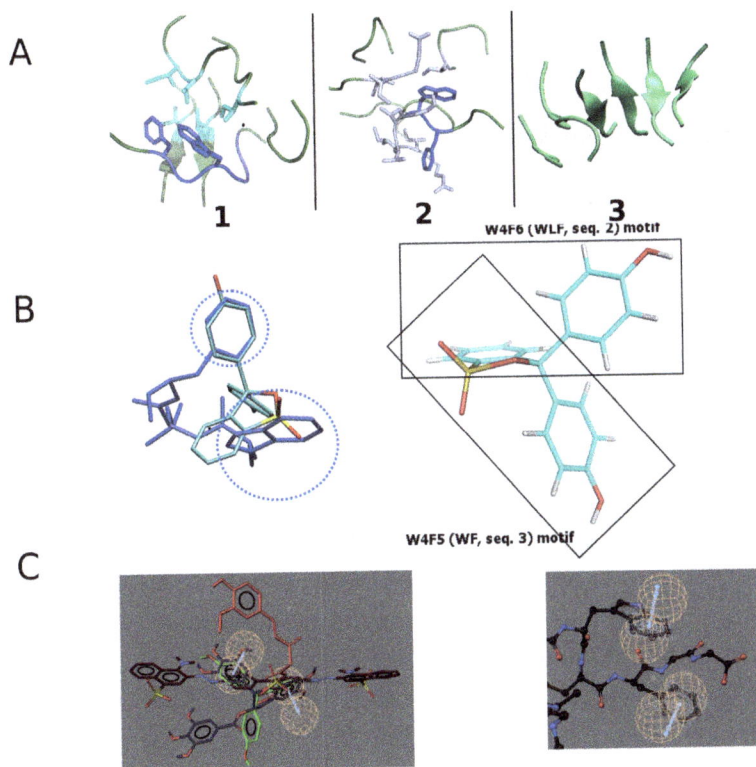

Fig. 13.2. MD based development of a pharmacophore model for the inhibition of peptide amyloid aggregation, starting from peptide leads. (A) Representative MD derived structure of the complex between the amyloidogenic peptide STVIIE and the active inhibitor sequences 1 and 2, and the inactive sequence 3. (B) Chemical functional similarities between the inhibitory cores of 1 and 2 and the small molecule Phenol-Red. (C) Similarities between the pharmacophore model derived from the analysis of 32 small molecule inhibitors of amyloid aggregation (left) and the model derived from the WXF motif of sequences 1 and 2 (right).

13.4.2.1 *Short Peptide Models of Aggregation and Pharmacophore Design*

Recent studies have shown that it is possible to design small peptides recapitulating the main supramolecular and toxicity features of naturally occurring amyloid proteins.[125–128] The small sizes of these systems make them also ideally suited for theoretical investigations. Indeed, molecular dynamics simulations of different point mutants of the *de novo* designed STVIIE helped explain sequence effects on the amyloidogenic behaviour in terms of molecular interactions.[129] Moreover, the toxic amyloid oligomeric state that should be ideally targeted is particularly suited for computer simulations. For example, the aggregation of small oligomers, ranging from dimers to octamers, has been examined using a number of different computational methods and protein energy models.[130–136]

All-atom MD simulations of short peptide systems in the presence of peptidic or peptidomimetic inhibitors may thus provide valuable information to illuminate the molecular requirements for inhibiting aggregation and for the consequent development of small-molecule inhibitors. Based on these considerations we combined experimental analysis of the amyloid inhibiting properties of short D-peptides with the results of several MD simulations of the inhibitor peptides in the presence of model oligomers of aggregating sequences. Specifically, we carried out extensive MD simulations of the interaction of selected D-peptide sequences that exhibit different inhibitory activity with small beta-sheet oligomers of the amyloidogenic sequence STVIIE, representing the initial oligomerization states of the target sequence.[129] Analysis of the MD trajectories actually suggests that oligomers are highly dynamic entities, with different possible arrangements and 3D structures accessible on the free energy landscape (Fig. 13.2a). Interestingly, however, structural analysis of the simulations suggests a possible common inhibitory core for active D-peptides whose conformational, stereochemical and physicochemical properties are actually shared by known small-molecule inhibitors of amyloid formation and cytotoxicity of IAPP and Aβ[137–139] (Figs 13.2a and 13.2b). These observations were then experimentally tested by performing inhibition assays of one of these molecules, Phenol Red (PR), against the amyloid peptide STVIIE used in this study (Fig. 13.2b). The results supported the hypothesis that Phenol Red is indeed an effective inhibitor of fibril formation for this sequence as well.

Based on these observations we used the D-peptide inhibitory core to generate a pharmacophore model that may be used to screen small-molecule libraries for new amyloid inhibitors. In parallel, we also generated pharmacophore models based on the common chemical features of a diverse training set of 32 non-peptidic small amyloid inhibitors. Interestingly, the main chemical features of the inhibitory core identified in the active D-peptides and from the small-molecule training set match optimally. Hence, we could propose that the stereospecific arrangement of chemical groups displayed by the inhibitory core can in turn be used as a template for the identification and/or design of new small-molecules with amyloid inhibition capabilities. Moreover, the specific pharmacophore models, and/or combinations of the different chemical features obtained from them, could then be used to optimise existing small molecule inhibitors of amyloidogenesis.

Importantly, the use of MD simulations of the aggregating peptides in the presence of the D-peptide inhibitors allowed us to investigate possible mechanisms of oligomer and fibril growth inhibition. The hydrophobic-aromatic interactions determined by the D-peptides induce a destabilising perturbation of the stereospecific interactions in the hydrophobic core that were observed as essential for the stability of the fibril nucleus in the absence of the D-peptide inhibitors[129] (Fig. 13.2c). They also induce a perturbation in the extended structure of STVIIE in the strands, breaking the register of hydrogen bonds necessary for nucleus stability. The combination of these factors determine a transition from an amyloid competent, ordered, extended beta-sheet structure into an alternative, non-amyloid competent, disordered oligomeric complex that lacks the conformational properties to evolve into a higher peptide assembly. These observations are exemplified by structures of the representatives of the most populate clusters from the ensembles sampled during the simulation (Fig. 13.2a) and have been confirmed by a test simulation with a larger fibril model. The inhibitory mechanism may thus be pictured as follows: the inhibitors bind directly to non-amyloidogenic oligomeric forms of the peptides preventing the conformational re-arrangements required to nucleate assembly to larger beta-sheet rich oligomers and fibrils. The assemblies are then redirected into off-pathway, innocuous oligomers.[140]

13.5 Conclusions and Outlook

The two examples we have reported herein show that the use of all-atom molecular dynamics can provide important new insights into different realms of molecular recognition problems, ranging from protein–peptide complex formation with a well defined receptor site, to the more plastic problem of peptide–peptide self organisation. MD simulations represent in fact a versatile, general-purpose tool to investigate recognition determinants and define (flexible) constraints for the identification of possible leads.

Moreover, the MD-based incorporation of protein flexibility in design or discovery projects has the potentiality to greatly expand the chemical and conformational space of the predicted ligands. The use of pharmacophore models taking flexibility and dynamics into account, rather than specific, predetermined chemical scaffolds, will actually increase the available chemical space for the hit list. Benchmarking simulation results with experimental data still remains an important priority in all these projects.

References

1. Agarwal P.K., Billeter S.R., Rajagopalan P.T.R., Benkovic S.J., Hammes-Schiffer S. (2002). *Proc Natl Acad Sci USA* 99: 2,794–2,799.
2. Boehr D.D., McElheny D., Dyson H.J., Wright P.E. (2006). *Science* 313: 1,638–1,642.
3. Osborne M.J., Schnell J., Benkovic S.J., Dyson H.J., Wright P.E. (2001). *Biochemistry* 40: 9,846–9,859.
4. Fischer E. (1894). *Ber Dtsch Chem Ges* 27: 2,985–2,993.
5. Tsai C.J., del Sol A., Nussinov R. (2008). *J Mol Biol* 378: 1–11.
6. Chiti F., Dobson C.M. (2006). *Annu Rev Biochem* 75: 333–366.
7. Barghorn S., Nimmrich V., Striebinger A., Krantz C., Keller P., Janson B., Bahr M., Schmidt M., Bitner R.S., Harlan J., Barlow E., Ebert U., Hillen H. (2005). *Journal of Neurochem* 95: 834–847
8. Walsh D.M., Hartley D.M., Kusumoto Y., Fezoui Y., Condron M.M., Lomakin A., Benedek G.B., Selkoe D.J., Teplow D.B. (1999). *J Biol Chem* 274: 25,945–25,952.
9. Hashimoto M., Rockenstein E., Crews L., Masliah E. (2003). *Neuromol Med* 4: 21–35.
10. Lorenzo A., Razzaboni B., Weir G.C., Yankner B.A. (1994). *Nature* 368: 756–760.
11. Hardy J., Selkoe D.J. (2002). *Science* 297: 353–356.

12. Bucciantini M., Giannoni E., Chiti F., Baroni F., Formigli L., Zurdo J.S., Taddei N., Ramponi G., Dobson C.M., Stefani M. (2002). *Nature* 416: 507–511.
13. Monod J., Jacob F. (1961). *Cold Spring Harbor Symp Quant Biol* 26: 389–401.
14. Monod J., Wyman J., Changeux J.P. (1965). *J Mol Biol* 12: 88.
15. Koshland D.E., Nemethy G., Filmer D. (1966). *Biochemistry* 5: 365.
16. Gunasekaran K., Ma B.Y., Nussinov R. (2004). *Proteins: Struct Funct Genet* 57: 433–443.
17. Hardy J.A., Wells J.A. (2004). *Curr Opin Struct Biol* 14: 706–715.
18. Swain J.F., Gierasch L.M. (2006). *Curr Opin Struct Biol* 16: 102–108.
19. Lindsley J.E., Rutter J. (2006). *Proc Natl Acad Sci USA* 103: 10,533–10,535
20. Lange O.F., Lakomek N.A., Fares C., Schroder G.F., Walter K.F.A., Becker S., Meiler J., Grubmuller H., Griesinger C., de Groot B.L. (2008). *Science* 320: 1,471–1,475.
21. Caflisch A. (2003). *Trends in Biotechnology* 21: 423–425.
22. Bahar I., Chennubhotla C., Tobi D. (2007). *Current Opinion in Structural Biology* 17: 633–640
23. Weber G. (1972). *Biochemistry* 11: 864–878.
24. Wang J., Verkhivker G.M. (2003). *Phys Rev Lett* 90: 188,181.
25. Frauenfelder H., Parak F., Young R.D. (1988). *Annu Rev Biophys Biophys Chem* 17: 451–479.
26. Eisenmesser E.Z., Millet O., Labeikovsky W., Korzhnev D.M., Wolf-Watz M., Bosco D.A., Skalicky J.J., Kay L.E., Kern D. (2005). *Nature* 438: 117–121.
27. Colombo G., Morra G., Meli M., Verkhivker G.M. (2008). *Proc Natl Acad Sci USA* 105: 7,676–7,681.
28. Kern D., Zuiderweg E.R.P. (2003). *Curr Opin Struct Biol* 13: 748–757.
29. Cooper A., Dryden D.T. (1984). *Eur Biophys J* 11: 103–109.
30. Suel G.M., Lockless S.W., Wall M.A., Ranganathan R. (2003). *Nat Struct Biol* 10: 59–69.
31. Shulman A.I., Larson C., Magelsdorf D.J., Ranganathan R. (2004). *Cell* 116: 417–429.
32. Fuentes E.J., Der C.J., Lee A.L. (2004). *J Mol Biol* 335: 1,105–1,115.
33. Ota N., Agard D.A. (2005). *J Mol Biol* 351: 345–354.
34. Chen J., Dima R.I., Thirumalai D. (2007). *J Mol Biol* 374: 250–266.
35. Chennubhotla C., Yang Z., Bahar I. (2008). *Mol BioSyst* 4: 287–292.
36. Morra G., Verkhivker G.M., Colombo G. (2009). *PLOS Comp Biol* (in press).
37. Ghosh A., Vishveshwara S. (2007). *Proc Natl Acad Sci USA* 104: 15,711–15,716.
38. Ghosh A., Vishveshwara S. (2008). *Biochemistry* 2008: 11,398–11,407.
39. Liu J., Nussinov R. (2008). *Proc Natl Acad Sci USA* 105: 901–906.
40. Verkhivker G.M., Bouzida D., Gehlhaar D.K., Rejto P.A., Freer S.T., Rose P.W. (2002). *Curr Opin Struct Biol* 12: 197–203.
41. Koshland D.E. (1958). *Proc Natl Acad Sci USA* 44: 98–104.
42. Ma B., Kumar S., Tsai C.J., Nussinov R. (1999). *Protein Eng* 12: 713–720.
43. Teodoro M.L., Kavraki L.E. (2003). *Current Pharmaceutical Design* 4: 1,635–1,648.

G. Morra, A. Genoni, G. Colombo

44. Teodoro M.L., Phillips G.N., Kavraki L.E. (2003). *Journal of Computational Biology* 10: 617–634.
45. Apostolakis J., Pluckthun A., Caflisch A. (1998). *Journal of Computational Chemistry* 19: 21–37.
46. Anderson A.C., O'Neil R.H., Surti T.S., Stroud R.M. (2001). *Chemistry & Biology* 8: 445–457.
47. Leach A.R. (1994). *Journal of Molecular Biology* 235: 345–356.
48. Desmet J., Demaeyer M., Hazes B., Lasters I. (1992). *Nature* 356: 539–542.
49. Hart P.E., Nilson N.J., Raphael B. (1968). *IEET T Syst Sci Cyb* 4: 100.
50. Schaffer L., Verkhivker G.M., (1998). *Proteins-Structure Function and Genetics* 33: 295–310.
51. Morris G.M., Goodsell D.S., Halliday R.S., Huey R., Hart W.E., Belew R.K., Olson A.J. (1998). *J Comp Chem* 19: 1,639–1,662.
52. Osterberg F., Morris G.M., Sanner M.F., Olson A.J., Goodsell D.S. (2002). *Proteins-Structure Function and Bioinformatics* 46: 34–40.
53. Jones G., Willett P., Glen R.C., Leach A.R., Taylor R. (1997). *Journal of Molecular Biology* 267: 727–748.
54. Verdonk M.L., Cole J.C., Hartshorn M.J., Murray C.W., Taylor R.D. (2003). *Proteins-Structure Function and Genetics* 52: 609–623.
55. Zavodszky M.I., Kuhn L.A. (2005). *Protein Science* 14: 1,104–1,114.
56. Schnecke V., Swanson C.A., Getzoff E.D., Tainer J.A., Kuhn L.A. (1998). *Proteins-Structure Function and Genetics* 33: 74–87.
57. Bottegoni G., Kufareva I., Totrov M., Abagyan R. (2008). *Journal of Computer-Aided Molecular Design* 22: 311–325.
58. Claussen H., Buning C., Rarey M., Lengauer T. (2001). *Journal of Molecular Biology* 308: 377–395.
59. Polgar T., Keseru G.M. (2006). *Journal of Chemical Information and Modeling* 46: 1,795–1,805.
60. Corbeil C.R., Englebienne P., Moitessier N. (2007). *Journal of Chemical Information and Modeling* 47: 435–449.
61. Zavodszky M.I., Ming L., Thorpe M.F., Day A.R., Kuhn L.A. (2004). *Proteins-Structure Function and Bioinformatics* 57: 243–261.
62. Zhao Y., Sanner M.F. (2007). *Proteins-Structure Function and Bioinformatics* 68: 726–737.
63. Wei B.Q., Weaver L.H., Ferrari A.M., Matthews B.W., Shoichet B.K. (2004). *Journal of Molecular Biology* 337: 1,161–1,182.
64. Lorber D.M., Udo M.K., Shoichet B.K. (2002). *Protein Science* 11: 1,393–1,408.
65. Su A.I., Lorber D.M., Weston G.S., Baase W.A., Matthews B.W., Shoichet B.K. (2001). *Proteins-Structure Function and Genetics* 42: 279–293.
66. Lorber D.M., Shoichet B.K. (1998). *Protein Science* 7: 938–950.
67. Meng E.C., Shoichet B.K., Kuntz I.D. (1992). *Journal of Computational Chemistry* 13: 505–524.
68. Sherman W., Day T., Jacobson M.P., Friesner R.A., Farid R. (2006). *Journal of Medicinal Chemistry* 49: 534–553.

69. Friesner R.A., Banks J.L., Murphy R.B., Halgren T.A., Klicic J.J., Mainz D.T., Repasky M.P., Knoll E.H., Shelley M., Perry J.K., Shaw D.E., Francis P., Shenkin P.S. (2004). *Journal of Medicinal Chemistry* 47: 1,739–1,749.
70. Jacobson M.P., Kaminski G.A., Friesner R.A., Rapp C.S., *Journal of Physical Chemistry B* 106: 11,673–11,680.
71. Jacobson M.P., Friesner R.A., Xiang Z.X., Honig B. (2002). *Journal of Molecular Biology* 320: 597–608.
72. Zentgraf M., Fokkens J., Sotriffer C.A. (2006). *Chemmedchem* 1: 1,355–1,365.
73. Sotriffer C.A., Dramburg I. (2005). *Journal of Medicinal Chemistry* 48: 3,122–3,125.
74. Knegtel R.M.A., Kuntz I.D., Oshiro C.M. (1997). *Journal of Molecular Biology* 266: 424–440.
75. Cavasotto C.N., Kovacs J.A., Abagyan R.A. (2005). *Journal of the American Chemical Society* 127: 9,632–9,640.
76. Pang Y.P., Kozikowski A.P. (1994). *Journal of Computer-Aided Molecular Design* 8: 669–681.
77. Lin J.H., Perryman A.L., Schames J.R., McCammon J.A. (2003). *Biopolymers* 68: 47–62.
78. Lin J.H., Perryman A.L., Schames J.R., McCammon J.A. (2002). *Journal of the American Chemical Society* 124: 5,632–5,633.
79. Kua J., Zhang Y.K., McCammon J.A. (2002). *Journal of the American Chemical Society* 124: 8,260–8,267.
80. Broughton H.B. (2000). *Journal of Molecular Graphics & Modelling* 18: 247–259.
81. Lerner M.G., Bowman A.L., Carlson H.A. (2007). *Journal of Chemical Information and Modeling* 47: 2,358–2,365.
82. Lerner M.G., Meagher K.L., Carlson H.A. (2008). *J Comp Aided Mol Des* 22: 727–736.
83. Bowman A.L., Nikolovska-Coleska Z., Zhong H.Z., Wang S.M., Carlson H.A. (2007). *J Am Chem Soc* 129: 12,809–12,814.
84. Carlson H.A., Masukawa K.M., Rubins K., Bushman F.D., Jorgensen W.L., Lins R.D., Briggs J.M., McCammon J.A. (2000). *J Med Chem* 43: 2,100–2,114.
85. Carlson H.A., Smith R.D., Khazanov N.A., Kirchhoff P.D., Dunbar J.B., Benson M.L. (2008). *Journal of Medicinal Chemistry* 51: 6,432–6,441.
86. Bowman A.L., Nikolovska-Coleska Z., Zhong H.Z., Wang S.M., Carlson H.A. (2007). *Journal of the American Chemical Society* 129: 12,809–12,814.
87. Damm K.L., Carlson H.A. (2007). *Journal of the American Chemical Society* 129: 8,225–8,235.
88. Bowman A.L., Lerner M.G., Carlson H.A. (2007). *Journal of the American Chemical Society* 129: 3,634–3,640.
89. Meagher K.L., Lerner M.G., Carlson H.A. (2006). *Journal of Medicinal Chemistry* 49: 3,478–3,484.
90. Meagher K.L., Carlson H.A. (2004). *Journal of the American Chemical Society* 126: 13,276–13,281.
91. Carlson H.A. (2002). *Current Opinion in Chemical Biology* 6: 447–452.

92. Carlson H.A. (2002). *Current Pharmaceutical Design* 8: 1,571–1,578.
93. Carlson H.A. (2000). Masukawa K.M., Rubins K., Bushman F.D., Jorgensen W.L., Lins R.D., Briggs J.M., McCammon J.A. (2000). *Journal of Medicinal Chemistry* 43: 2,100–2,114.
94. Carlson H.A., McCammon J.A. (2000). *Molecular Pharmacology* 57: 213–218.
95. Plescia J., Salz W., Xia F., Pennati M., Zaffaroni N., Daidone M.G., Meli M., Dohi T., Fortugno P., Nefedova Y., Gabrilovich D.I., Colombo G., Altieri D.C. (2005). *Cancer Cell* 7: 457–467.
96. Meli M., Pennati M., Curto M., Daidone M.G., Plescia J., Toba S., Altieri D.C., Zaffaroni N., Colombo G. (2006). *J Med Chem* 49: 7,721–7,730.
97. Esteras-Chopo A., Morra G., Moroni E., Serrano L., de la Paz M.L., Colombo G. (2008). *Journal of Molecular Biology* 383: 266–280.
98. Mangoni R., Roccatano D., Di Nola A. (1999). *Proteins-Structure Function and Bioinformatics* 35: 153–162.
99. Pak Y.S., Wang S.M. (2000). *Journal of Physical Chemistry B* 104: 354–359.
100. Berg B.A., Neuhaus T. (1992). *Physical Review Letters* 68: 9–12.
101. Nakajima N., Higo J., Kidera A., Nakamura H. (1997). *Chemical Physics Letters* 278: 297–301.
102. Nakajima N., Nakamura H., Kidera A. (1997). *Journal of Physical Chemistry B* 101: 817–824.
103. Trosset J.Y., Scheraga H.A. (1999). *Journal of Computational Chemistry* 20: 412–427.
104. *Ibid.*, 244–252.
105. Zacharias M., Sklenar H. (1999). *Journal of Computational Chemistry* 20: 287–300.
106. Keseru G.M., Kolossvary I. (2001). *Journal of the American Chemical Society* 123: 12,708–12,709.
107. Keseru G.M. (2001). *Journal of Computer-Aided Molecular Design* 15: 649–657.
108. Tatsumi R., Fukunishi Y., Nakamura H. (2004). *Journal of Computational Chemistry* 25: 1,995–2,005.
109. Machicado C., Lopez-Llano J., Cuesta-Lopez S., Bueno M., Sancho J. (2005), *Journal of Computer-Aided Molecular Design* 19: 421–443.
110. Friesner R.A., Murphy R.B., Repasky M.P., Frye L.L., Greenwood J.R., Halgren T.A., Sanschagrin P.C., Mainz D.T. (2006). *Journal of Medicinal Chemistry* 49: 6,177–6,196.
111. Lee M.R., Sun Y.X. (2007). *Journal of Chemical Theory and Computation* 3: 1,106–1,119.
112. Ewing T.J.A., Makino S., Skillman A.G., Kuntz I.D. (2001). *Journal of Computer-Aided Molecular Design* 15: 411–428.
113. Wang J.M., Kang X.S., Kuntz I.D., Kollman P.A. (2005). *Journal of Medicinal Chemistry* 48: 2,432–2,444.
114. Graves A.P., Shivakumar D.M., Boyce S.E., Jacobson M.P., Case D.A., Shoichet B.K. (2008). *Journal of Molecular Biology* 377: 914–934.

115. Gyurkocza B., Plescia J., Raskett G.M., Garlick D.S., Lowry P.A., Carter B.Z., Andreeff M., Meli M., Colombo G., Altieri D.C. (2006). *J Natl Cancer Inst* 98: 1,068–1,077.
116. Vogelstein B., Kinzler K.W. (2004). *Nat Med* 46: 789–799.
117. Paez J.G., Janne P.A., Lee J.C., Tracy S., Greulich H., Gabriel S., Herman P., Kaye F.J., Lindeman N., Boggon T.J., Naoki K., Sasaki H., Fujii Y., Eck M.J., Sellers W.R., Johnson B.E., Meyerson M. (2004). *Science* 304: 1,497–1,500.
118. Whitesell L., Lindquist S.L. (2005). *Nat Rev Cancer* 5: 761–772.
119. Neckers L., Ivy S.P. (2003). *Curr Opin Oncol* 15: 419–424.
120. Altieri D.C. (2003). *Nat Rev Cancer* 3: 46–54.
121. Fortugno P., Beltrami E., Plescia J., Fontana J., Pradhan D., Marchisio P.C., Sessa W.C., Altieri D.C. (2003). *Proc Natl Acad Sci USA* 100: 13,791–13,796.
122. Stebbins C.E., Russo A.A., Schneider C., Rosen N., Hartl F.U., Pavletich N.P. (1997). *Cell* 89: 239–250.
123. Hetenyi C., Spoel D.v.d. (2002). *Prot Sci* 11: 1,729–1,737.
124. Roe S.M., Prodromou C., O'Brien R., Ladbury J.E., Piper P.W., Pearl L.H. (1999). *J Med Chem* 42: 260–266.
125. Lopez de la Paz M., Goldie K., Zurdo J., Lacroix E., Dobson C.M., Hoenger A., Serrano L. (2002). *Proc Natl Acad Sci USA* 99: 16,052–16,057.
126. Pastor M.T., Kummerer N., Schubert V., Esteras-Chopo A., Dotti C.G., Lopez de la Paz M., Serrano L. (2008). *J Mol Biol* 375: 695–707.
127. Sawaya M.R., Sambashivan S., Nelson R., Ivanova M.I., Sievers S.A., Apostol M.I., Thompson M.J., Balbirnie M., Wiltzius J.J., McFarlane H.T., Madsen A.O., Riekel C., Eisenberg D. (2007). *Nature* 447: 453–457.
128. Zanuy D., Porat Y., Gazit E., Nussinov R. (2004). *Structure* 12: 439–455.
129. Lopez de la Paz M., de Mori G.M.S., Serrano L., Colombo G. (2005). *J Mol Biol* 349: 583–596.
130. Mousseau N., Derreumaux P. (2005). *Acc Chem Res* 38: 885–891.
131. Melquiond A., Gelly J.C., Mousseau N., Derreumaux P. (2007). *J Chem Phys* 126: 065101.
132. Soto P., Cladera J., Mark A.E., Daura X. (2005). *Angew Chem Int Ed* 44: 1,065–1,067.
133. Baumketner A., Shea J.E. (2006). *J Mol Biol* 362: 567–579.
134. Wu C., Lei H.X., Duan Y. (2005). *J Am Chem Soc* 127: 13,530–13,537.
135. Nguyen H.D., Hall C.K. (2004). *Proc Natl Acad Sci USA* 101: 16,180–16,185.
136. Colombo G., Soto P., Gazit E. (2007). *Trends Biotechnol* 25: 211–218.
137. Gazit E. (2005). *Febs J* 272: 5,971–5,978.
138. Porat Y., Mazor Y., Efrat S., Gazit E. (2004). *Biochemistry* 43: 14,454–14,462
139. Necula M., Breydo L., Milton S., Kayed R., van der Veer W.E., Tone P., Glabe C.G. (2007). *Biochemistry* 46: 8,850–8,860.
140. Ehrnhoefer D.E., Bieschke J., Boeddrich A., Herbst M., Masino L., Lurz R., Engemann S., Pastore A., Wanker E.E. (2008). *Nat Struct Mol Biol* 15: 558–566.

INDEX